AF361842

Artificial Neural Networks in Agriculture

Artificial Neural Networks in Agriculture

Editors

Sebastian Kujawa
Gniewko Niedbała

MDPI • Basel • Beijing • Wuhan • Barcelona • Belgrade • Manchester • Tokyo • Cluj • Tianjin

Editors
Sebastian Kujawa
Department of Biosystems
Engineering
Poznań University of Life
Sciences
Poznań
Poland

Gniewko Niedbała
Department of Biosystems
Engineering
Poznań University of Life
Sciences
Poznań
Poland

Editorial Office
MDPI
St. Alban-Anlage 66
4052 Basel, Switzerland

This is a reprint of articles from the Special Issue published online in the open access journal *Agriculture* (ISSN 2077-0472) (available at: https://www.mdpi.com/journal/agriculture/special_issues/Artificial_Neural_Networks_Agriculture).

For citation purposes, cite each article independently as indicated on the article page online and as indicated below:

LastName, A.A.; LastName, B.B.; LastName, C.C. Article Title. *Journal Name* **Year**, *Volume Number*, Page Range.

ISBN 978-3-0365-1580-9 (Hbk)
ISBN 978-3-0365-1579-3 (PDF)

Contents

About the Editors

Sebastian Kujawa

Sebastian Kujawa is employed as an Assistant Professor in the Department of Biosystems Engineering, Poznań University of Life Sciences, Poland. In 2009, he received a PhD degree from the Poznań University of Life Sciences. His scientific activity concerns the applications of computer image analysis and neural modeling in developing methods for condition assessments of dynamic biosystems. Dr. Kujawa is an author of over 50 publications in scientific journals and peer-reviewed conference proceedings. He is a member of the Board of Polish Society for Information Technology in Agriculture (POLSITA). He has served as a Guest Editor of a Special Issue in *Agriculture*: Neural Networks in Agriculture.

Gniewko Niedbała

Gniewko Niedbała is an Associate Professor at the Department of Biosystems Engineering, Poznań University of Life Sciences, Poland. He defended his doctorate in 2006 at the August Cieszkowski Agricultural University of Poznań, and performed his habilitation in 2019 at the Poznań University of Life Sciences. Presently, he is working on using artificial neural networks and machine learning in many aspects of agriculture and forestry. Between 2012 and 2016, he was a Board Member of the National Centre for Research and Development, Poland. He was a Guest Editor for a Special Issue of *Agriculture:* Neural Networks in Agriculture. He presently serves as a Guest Editor in *Agriculture* and *Energies*, is a Member of the Editorial Board in *Agronomy*, and a Member of the Reviewer Board in *Agriculture, Water,* and *Land*. In addition, he has authored and coauthored over 90 journal and conference papers and book chapters related to artificial intelligence in agriculture.

Editorial

Artificial Neural Networks in Agriculture

Sebastian Kujawa *,† and Gniewko Niedbała *,†

Department of Biosystems Engineering, Faculty of Environmental and Mechanical Engineering, Poznań University of Life Sciences, Wojska Polskiego 50, 60-627 Poznań, Poland
* Correspondence: sebastian.kujawa@up.poznan.pl (S.K.); gniewko.niedbala@up.poznan.pl (G.N.)
† These authors contributed equally to this work.

Abstract: Artificial neural networks are one of the most important elements of machine learning and artificial intelligence. They are inspired by the human brain structure and function as if they are based on interconnected nodes in which simple processing operations take place. The spectrum of neural networks application is very wide, and it also includes agriculture. Artificial neural networks are increasingly used by food producers at every stage of agricultural production and in efficient farm management. Examples of their applications include: forecasting of production effects in agriculture on the basis of a wide range of independent variables, verification of diseases and pests, intelligent weed control, and classification of the quality of harvested crops. Artificial intelligence methods support decision-making systems in agriculture, help optimize storage and transport processes, and make it possible to predict the costs incurred depending on the chosen direction of management. The inclusion of machine learning methods in the "life cycle of a farm" requires handling large amounts of data collected during the entire growing season and having the appropriate software. Currently, the visible development of precision farming and digital agriculture is causing more and more farms to turn to tools based on artificial intelligence. The purpose of this Special Issue was to publish high-quality research and review papers that cover the application of various types of artificial neural networks in solving relevant tasks and problems of widely defined agriculture.

Keywords: yield prediction; crop models; soil and plant nutrition; automated harvesting; model application for sustainable agriculture; precision agriculture; remote sensing for agriculture; decision supporting systems; neural image analysis; convolutional neural networks

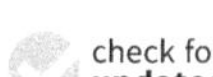
check for
updates

Citation: Kujawa, S.; Niedbała, G. Artificial Neural Networks in Agriculture. *Agriculture* **2021**, *11*, 497. https://doi.org/10.3390/agriculture11060497

Received: 20 May 2021
Accepted: 25 May 2021
Published: 27 May 2021

1. Introduction

Modern agriculture needs to have a high production efficiency combined with a high quality of obtained products. This applies to both crop and livestock production. To meet these requirements, advanced methods of data analysis are more and more frequently used, including those derived from artificial intelligence methods. Artificial neural networks (ANNs) are one of the most popular tools of this kind. They are widely used in solving various classification and prediction tasks. For some time they have also been used in the broadly defined field of agriculture. They can form part of the precision farming and decision support systems. Artificial neural networks can replace the classical methods of modeling, and are one of the main alternatives to classical mathematical models. The spectrum of the applications of artificial neural networks is very wide. For a long time now, researchers from all over the world have been using these tools to support agricultural production, making it more efficient and providing the highest-quality products possible. The purpose of this Special Issue was to publish high-quality research and review papers that cover the application of various types of artificial neural networks in solving relevant tasks and problems of what is widely defined as agriculture. The Special Issue covers 14 peer-reviewed research papers and 1 review paper.

2. Papers in the Special Issue

In the studies described in the first article [1], the following four features were adopted to develop artificial neural networks to identify grain weevil in wheat kernels: mass, equivalent diameter, humidity, and hardness. The authors developed a total of 100 such models. They concluded that the model based on radial basis function, and the structure of 4:10:1, was optimal in solving the problem of the weevil identification. This model was characterized by the RMSE for the test set at a level of 0.25.

In the studies presented in the second article [2], artificial neural networks were adopted to examine the impact of the variety and weather conditions on the concentration of ferulic acid, deoxynivalenol, and nivalenol in winter wheat grain. As a result of the studies, three independent neural models were developed, one for each of the parameters considered. These models were based on the multilayer perceptron topology with two hidden layers and 14 inputs. On the basis of sensitivity analyses conducted for the models, the authors pointed out that the experiment variant and wheat variety were two of the most important features determining the concentration of ferulic acid, deoxynivalenol, and nivalenol in winter wheat grain.

The authors of the studies discussed in the third article [3] proposed an ANN-based continual classification method that may be used in solving many potential classification tasks in the field of agriculture. The method is based on memory storage and retrieval, and it combines a convolutional neural network (CNN) and generative adversarial network (GAN). The evaluation of this method was carried out for two classification problems: identification of crop pests and plant leaves. The method was also compared with the regular CNN approach. The results show that the regular CNNs classify the categories clearly for single task problems, whereas in the continual tasks, they are prone to forgetting problems and cannot balance new and old tasks. The method proposed by the authors has the ability to accumulate knowledge and alleviate forgetting. Therefore it can better distinguish all categories from both old and new tasks.

In the studies described in the fourth article [4], the authors made an attempt to apply artificial neural networks to model the dynamic responses of plant growth as affected by a change in root zone temperature in hydroponic chili pepper plants. They used a non-linear autoregressive model with exogenous input (NARX) for this purpose. The performance of the developed model was evaluated by comparing the estimated dynamic response of plant growth values with the observed values. The RMSE and R^2 calculated during the evaluation were 0.49 g and 0.99, respectively. The authors assumed that the applied type of neural network was useful in identifying the complex process analyzed.

The authors of the studies presented in the fifth article [5] designed neural network models that were used to estimate the corn grain yield. The inputs of the models were: six vegetation indices (NDVI, NDRE, WDRVI, EXG, TGI, VARI), canopy cover, and plant density. The authors also analyzed the relative importance of the predictor variables. Crop information was acquired using an unmanned aerial vehicle (UAV). It was assumed that the spectral information acquired using remote sensors mounted on UAVs and its further processing in vegetation indices, canopy cover, and plant density allows the characterization and estimation of corn grain yield.

The studies discussed in the sixth article [6] were aimed to develop an accurate spectral model to evaluate the cultivated land quality (CLQ) based on the gross primary productivity (GPP) spectral indicator at different growth stages of late rice. The authors compared three types of models that were used for this purpose-linear model based on partial least squares regression (PLSR), and two non-linear models based on support vector regression (SVR), and genetic algorithm-based backpropagation model neural network (GA-BPNN). The CLQ measurements from 294 training samples and the corresponding GPP values were also used to develop the models. The results presented in the paper show that the GA-BPNN model was more accurate in evaluating CLQ than the other two models.

In the studies presented in the seventh article [7], a hybrid feature extraction procedure to address the feature extraction problem in machine learning models was proposed.

The procedure combines correlation-based feature selection (CFS) and random forest recursive feature elimination (RF-RFE). The proposed method was tested by implementing it in the development of paddy crop yield models that were based on the following machine learning algorithms: random forests, decision trees, and gradient boosting. The authors assumed that the validation of the described hybrid feature extraction method was profoundly satisfying. They also pointed out that the method may also be used in combination with artificial neural networks for agricultural applications.

In the eighth review article [8], all applications and potential applications of non-linear machine learning (ML) algorithms in classical and in vitro-based plant breeding methods have been discussed. Classification of genetic diversity, yield component analysis, yield stability and genotype $\times$ environment interaction (G $\times$ E) analysis, biotic and abiotic stress assessment, and predicting the outcome of mating designs and hybrid breeding programs through machine learning algorithms has been presented in the first part of the review. Applications of machine learning algorithms in in vitro regeneration studies, including optimizing basal culture media, finding the best combination(s) and interaction of medium additives and conditions, and predicting the outcome of in vitro-based plant biotechnology techniques such as artificial polyploidy induction and Agrobacterium-mediated gene transformation have been discussed in the second part of the review. In the third part of the review, the authors presented the application of ML in high-throughput phenotyping and precision agriculture.

The purpose of this study [9] was to develop a model using artificial neural networks to predict soybean harvest area, yield, and production compared with classical methods of time series analysis. The authors collected data from a time series (1961–2016) regarding soybean production in Brazil to perform the model and the classic methods. The results indicate that artificial neural networks are the best approach to predict soybean harvest area and production, while the classical linear function remains more effective to predict soybean yield. Furthermore, artificial neural networks present as a reliable model to predict time series and can help the stakeholders to anticipate the world soybean offer. Finally, the study shows that artificial neural networks can be effective in predicting a commodities index, even using a short time series, which in this case was 50 years. Research results are an important part of supporting grain and cereal production. Among them, soybean is sixth by production volume and fourth by both production area and economic value. The production is concentrated in Brazil and the USA, which are responsible for more than 70% of world production.

Study [10] proposed a weakly supervised paddy rice mapping approach based on long short-term memory (LSTM) network and dynamic time warping (DTW) distance. Standard temporal synthetic aperture radar (SAR) backscatter profiles for each land cover type were constructed. Weak samples were then labeled on the basis of their DTW distances to the standard temporal profiles. A time-series feature set was then created that combined multi-spectral Sentinel-2 bands and Sentinel-1 SAR vertical received (VV) band. Experiments showed that weakly supervised learning outperformed supervised learning in paddy rice identification when field samples were insufficient. With only 10% of field samples, weakly supervised learning achieved better results in producer's accuracy (0.981 to 0.904) and user's accuracy (0.961 to 0.917) for paddy rice. Training with 50% of field samples also presented minor improvement. Finally, a paddy rice map was generated with the weakly supervised approach trained on field samples and DTW-labeled samples. The proposed approach based on DTW distance can reduce field sampling cost since it requires fewer field samples. Validation results indicated that the proposed LSTM classifier is suitable for paddy rice mapping.

In the eleventh article [11], the relationship between soil electrical parameters (apparent soil electrical conductivity (ECa) and magnetic susceptibility (MS) measured at two depths, 0.5 and 1 m) and soil compaction was investigated with the use of artificial neural networks. A soil's compaction influences the yield; therefore, the delimitation of management zones connected with this parameter is very important in modern agriculture.

As the geospatial measurement of soil electrical parameters is reliable and economic, it can be a useful technique for the delimitation of management zones. The neural models of the highest quality were developed with the use of RBF neural networks for the prediction of soil compaction measured for soil layers at 0–0.5 and 0.4–0.5 m based on electrical parameters of the soil. The authors highlighted that both electrical parameters, ECa and MS, should be used in the process of the delimitation of management zones. This approach can be employed in precision agriculture, e.g., for the reduction in fuel consumption by variable-depth tillage technology based on soil compaction mapping.

Research article [12] describes the challenges date palm growers face in manual processes such as picking, grading, and packing. The authors highlighted that the main challenges are misclassification of dates, maturity level, size, and texture. This article assesses the CNN architectures: VGG-16, VGG-19, ResNet-50, ResNet-101, ResNet-152, AlexNet, Inception V3, and CNN from scratch, considering the learning transfer technique for their training and five hyperparameters to reduce time and improve accuracy in the selection process. The data set they used contains 1002 images corresponding to mature and immature dates in trays. The main results showed that the model obtained from the VGG-19 architecture with a batch of 128 and Adam optimizer with a learning rate of 0.01 presented the best performance with a precision of 99.32%. Therefore, the authors suggest the possibility of using the VGG-19 model to build computer vision systems that help producers improve their classification process.

The thirteenth paper [13] presented the results of a laboratory study about the average degree of coverage and the coverage unevenness coefficient. The research was conducted with a special spray track machine, which functioned as a self-propelled field sprayer. Three artificial plants as three replicates were placed on the machine's route. Water-sensitive papers were attached to artificial plants to form specific surfaces. Four types of nozzles were selected for testing: standard, air-induction single flat fan, anti-drift, and air-induction dual flat fan. The following parameters of nozzles' work were used for the research: pressure, height of boom, spray angle, perpendicular to the ground, and driving speed. The research was conducted with a constant dose of liquid. Based on the research results, it was found that the highest average coverage was obtained for single standard flat fan nozzles and dual anti-drift flat fan nozzles. At the same time, the highest values of unevenness were observed for these nozzles. Inverse relationships were obtained for air-induction nozzles.

In study [14], an alternative method for oil palm tree management is proposed by applying high-resolution imagery, combined with faster-RCNN, for the automatic detection and health classification of oil palm trees. This study used a total of 4172 bounding boxes of healthy and unhealthy palm trees, constructed from 2000 × 2000 pixel images. Of the total dataset, 90% was used for training, and 10% was prepared for testing using ResNet-50 and VGG-16. Three techniques were used to assess the models' performance: model training evaluation, evaluation using visual interpretation, and ground sampling inspections. The study identified three characteristics needed for detection and health classification: crown size, color, and density. The optimal altitude to capture images for detection and classification was determined to be 100 m. For oil palm tree detection, healthy tree identification, and unhealthy tree identification, ResNet-50 obtained F1 scores of 95.09%, 92.07%, and 86.96%, respectively, with respect to visual interpretation ground truth and 97.67%, 95.30%, and 57.14%, respectively, with respect to ground sampling inspection ground truth. ResNet-50 yielded better F1 scores than VGG-16 in both evaluations.

The research [15] presents an alternative to the use of pesticides in agriculture and horticulture for weed control. Weeds are a major yield-limiting factor in food production, and the use of pesticides levies heavy costs on human and environmental health. The proposed autonomous weeding robot uses artificial neural networks to detect weeds and plants, following it up with the desired action to restrict the growth of weeds. During the training phase, three types of plants were used to assess the efficiency of the robot in identification and detection tasks. The study reported the highest classification accuracy of 95% for plant specimens and 99% for weed specimens. The mean average precision (mAP) in object detection was found to be 31%. An urgent need for such alternative weed management solutions in view of the planned gradual phasing out of pesticides was expressed by farmers in northern Germany in conversation with the authors. It is pertinent to mention here the goal set by the EU's planned Organic Action Plan, which envisages at least 25% of the EU's agricultural land under organic (pesticide-free) farming by 2030.

3. Conclusions

The papers presented in this Special Issue cover a wide range of applications of various types of artificial neural networks in agriculture. We hope that these papers will stimulate further research in this domain.

Author Contributions: All editors equally contributed to organizing the Special Issue, to editorial work, and to writing this editorial. All authors have read and agreed to the published version of the manuscript.

Funding: This research received no external funding.

Institutional Review Board Statement: Not applicable.

Informed Consent Statement: Not applicable.

Data Availability Statement: No new data were created or analyzed in this study. Data sharing is not applicable to this article.

Acknowledgments: We thank the authors for submitting manuscripts of high quality and their willingness to further improve them after peer review, the reviewers for their careful evaluations aimed at eliminating weaknesses and their suggestions to optimize the manuscripts and the editorial staff of MDPI for the professional support and the rapid actions taken when necessary throughout the editorial process.

Conflicts of Interest: The authors declare no conflict of interest.

References

1. Boniecki, P.; Koszela, K.; Świerczyński, K.; Skwarcz, J.; Zaborowicz, M.; Przybył, J. Neural Visual Detection of Grain Weevil (*Sitophilus granarius* L.). *Agriculture* **2020**, *10*, 25. [CrossRef]
2. Niedbała, G.; Kurasiak-Popowska, D.; Stuper-Szablewska, K.; Nawracała, J. Application of Artificial Neural Networks to Analyze the Concentration of Ferulic Acid, Deoxynivalenol, and Nivalenol in Winter Wheat Grain. *Agriculture* **2020**, *10*, 127. [CrossRef]
3. Li, Y.; Chao, X. ANN-Based Continual Classification in Agriculture. *Agriculture* **2020**, *10*, 178. [CrossRef]
4. Aji, G.K.; Hatou, K.; Morimoto, T. Modeling the Dynamic Response of Plant Growth to Root Zone Temperature in Hydroponic Chili Pepper Plant Using Neural Networks. *Agriculture* **2020**, *10*, 234. [CrossRef]
5. García-Martínez, H.; Flores-Magdaleno, H.; Ascencio-Hernández, R.; Khalil-Gardezi, A.; Tijerina-Chávez, L.; Mancilla-Villa, O.R.; Vázquez-Peña, M.A. Corn Grain Yield Estimation from Vegetation Indices, Canopy Cover, Plant Density, and a Neural Network Using Multispectral and RGB Images Acquired with Unmanned Aerial Vehicles. *Agriculture* **2020**, *10*, 277. [CrossRef]
6. Zhu, M.; Liu, S.; Xia, Z.; Wang, G.; Hu, Y.; Liu, Z. Crop Growth Stage GPP-Driven Spectral Model for Evaluation of Cultivated Land Quality Using GA-BPNN. *Agriculture* **2020**, *10*, 318. [CrossRef]
7. Elavarasan, D.; Vincent, P.M.D.R.; Srinivasan, K.; Chang, C.-Y. A Hybrid CFS Filter and RF-RFE Wrapper-Based Feature Extraction for Enhanced Agricultural Crop Yield Prediction Modeling. *Agriculture* **2020**, *10*, 400. [CrossRef]
8. Niazian, M.; Niedbała, G. Machine Learning for Plant Breeding and Biotechnology. *Agriculture* **2020**, *10*, 436. [CrossRef]
9. Abraham, E.R.; Mendes dos Reis, J.G.; Vendrametto, O.; Oliveira Costa Neto, P.L.d.; Carlo Toloi, R.; Souza, A.E.d.; Oliveira Morais, M.d. Time Series Prediction with Artificial Neural Networks: An Analysis Using Brazilian Soybean Production. *Agriculture* **2020**, *10*, 475. [CrossRef]
10. Wang, M.; Wang, J.; Chen, L. Mapping Paddy Rice Using Weakly Supervised Long Short-Term Memory Network with Time Series Sentinel Optical and SAR Images. *Agriculture* **2020**, *10*, 483. [CrossRef]
11. Pentoś, K.; Pieczarka, K.; Serwata, K. The Relationship between Soil Electrical Parameters and Compaction of Sandy Clay Loam Soil. *Agriculture* **2021**, *11*, 114. [CrossRef]
12. Pérez-Pérez, B.D.; García Vázquez, J.P.; Salomón-Torres, R. Evaluation of Convolutional Neural Networks' Hyperparameters with Transfer Learning to Determine Sorting of Ripe Medjool Dates. *Agriculture* **2021**, *11*, 115. [CrossRef]
13. Cieniawska, B.; Pentos, K. Average Degree of Coverage and Coverage Unevenness Coefficient as Parameters for Spraying Quality Assessment. *Agriculture* **2021**, *11*, 151. [CrossRef]

14. Yarak, K.; Witayangkurn, A.; Kritiyutanont, K.; Arunplod, C.; Shibasaki, R. Oil Palm Tree Detection and Health Classification on High-Resolution Imagery Using Deep Learning. *Agriculture* **2021**, *11*, 183. [CrossRef]
15. Shah, T.M.; Nasika, D.P.B.; Otterpohl, R. Plant and Weed Identifier Robot as an Agroecological Tool Using Artificial Neural Networks for Image Identification. *Agriculture* **2021**, *11*, 222. [CrossRef]

Article

Plant and Weed Identifier Robot as an Agroecological Tool Using Artificial Neural Networks for Image Identification

Tavseef Mairaj Shah *,†, Durga Prasad Babu Nasika † and Ralf Otterpohl

Rural Revival and Restoration Egineering (RUVIVAL), Institute of Wastewater Management and Water Protection, Hamburg University of Technology, Eissendorfer Strasse 42, 21073 Hamburg, Germany; durga.nasika@tuhh.de (D.P.B.N.); ro@tuhh.de (R.O.)
* Correspondence: tavseef.mairaj.shah@tuhh.de
† These authors contributed equally to this work.

Abstract: Farming systems form the backbone of the world food system. The food system, in turn, is a critical component in sustainable development, with direct linkages to the social, economic, and ecological systems. Weeds are one of the major factors responsible for the crop yield gap in the different regions of the world. In this work, a plant and weed identifier tool was conceptualized, developed, and trained based on artificial deep neural networks to be used for the purpose of weeding the inter-row space in crop fields. A high-level design of the weeding robot is conceptualized and proposed as a solution to the problem of weed infestation in farming systems. The implementation process includes data collection, data pre-processing, training and optimizing a neural network model. A selective pre-trained neural network model was considered for implementing the task of plant and weed identification. The faster R-CNN (Region based Convolution Neural Network) method achieved an overall mean Average Precision (mAP) of around 31% while considering the learning rate hyperparameter of 0.0002. In the plant and weed prediction tests, prediction values in the range of 88–98% were observed in comparison to the ground truth. While as on a completely unknown dataset of plants and weeds, predictions were observed in the range of 67–95% for plants, and 84% to 99% in the case of weeds. In addition to that, a simple yet unique stem estimation technique for the identified weeds based on bounding box localization of the object inside the image frame is proposed.

Keywords: deep learning; artificial neural networks; image identification; agroecology; weeds; yield gap; environment; health

Citation: Shah, T.M.; Nasika, D.P.B.; Otterpohl, R. Plant and Weed Identifier Robot as an Agroecological Tool Using Artificial Neural Networks for Image Identification. *Agriculture* **2021**, *11*, 222. https://doi.org/10.3390/agriculture11030222

Academic Editors: Sebastian Kujawa, Gniewko Niedbała and Maciej Zaborowicz

Received: 31 December 2020
Accepted: 4 March 2021
Published: 8 March 2021

Publisher's Note: MDPI stays neutral with regard to jurisdictional claims in published maps and institutional affiliations.

1. Introduction

Growing food through agriculture involves different labor-intensive practices. Most of these practices have traditionally been performed manually. Weeding is one such agricultural practice. However, generally, as farming has become more industrialized—or that the industrialized agriculture has become the leitmotif for all to emulate—different practices evolved over time with the aim of increasing the efficiency of labor and increasing the productivity of the land. This involved efforts to increase the efficacy of the manual practices by using mechanical and chemical aids or in some cases to present alternate pathways for these practices without any direct manual intervention [1].

The growth of weeds is one of the largest biotic factors contributing to the yield gap in food crops [2,3]. In South Asia, it is the single largest biotic yield gap factor in rice production systems [4,5]. It has been reported that in sugarcane cultivation, weeds reduced the crop growth at early stages and have resulted in a yield loss of 27–35% [6]. In traditional farming systems, weeds have been manually removed from the crop field with a help of hands or with a hoe. Growing intercrops in between the main crop rows is also a potential strategy to control the growth of weeds. However, the rise in the use of agrochemicals multiple times (up to 300 times) in the last 50 years, to control the growth of weeds among other things, has shown a lot of negative effects on human and planetary health [7]. The incidence of herbicide resistance among certain weed populations is also a cause of concern in this regard [8,9].

It is in this backdrop that a transition to agroecology-based farming systems is being recommended internationally with an urgency never expressed before. Agroecology is the study of the ecology of food systems and applying this knowledge for the design of sustainable farming systems. Agroecology-based alternatives include organic farming and sustainable intensification strategies like the System of Rice Intensification [10]. The problem of weeds, however, persists in some of the proposed methodologies too. For example, the proliferation of weeds is an oft-cited critique of an agroecological methodology of growing rice, the System of Rice Intensification, which involves growing rice under alternate wetting and drying conditions, with earlier transplantation and wider spacing between the rice plants [11] (Figure 1). While as in the case of agrochemical-based farming, the problem of weeds leads to environmental hazards due to the use of pesticides, in the case of agroecological methodologies, the practices that are suggested to counter weed proliferation are not harmful to the environment. Such practices are however often labor intensive [10,12].

Figure 1. Weeds growing in between rice crop rows (Source: Author).

The excessive use of agrochemicals like pesticides including herbicides has become a burning topic of discussion in the past few years although the dangers associated with it have been discussed in the literature for a long time [13–16]. The presence of fertilizer residues in surface and groundwater and that of pesticide residues in food items has been well documented [15–20]. Their effects on human and planetary health have been detailed in different studies; with the use of fertilizers and pesticides has increased manifold over the past four decades particularly in developing countries [9,21,22]. On the other hand, lack of nutrients in the soil and pest proliferation continues to challenge farmers leading to a decline in productivity [23,24]. For example, increased weed proliferation due to excessive use of fertilizers has resulted in yield losses in farming systems in South Asia [2,9,18,25].

In agrarian societies, secondary practices in farming, associated with plant protection, have traditionally been done with the help of manual labor, much like the primary practices, those associated with sowing, planting and harvesting. In some parts of the world, farming practices like weeding are still done or were done until recently, manually. These practices have gradually phased out to a large extent and have been replaced by the use of chemical pesticides like herbicides and weedicides. As such, the use of agrochemical pesticides has become the norm [26].

So, one of the options to reverse the ecological damage of the pesticides would be to go back to manual weeding. However, agroecology does not simply advocate going back to earlier practices; it involves going back to roots armed with new knowledge and tools [27,28]. This is the motivation behind the AI-based weed identifier robot, the concept and design of which is detailed in the following sections. An AI-trained weeding robot could play a supporting role in agroecology in this regard, when designed keeping in view the needs of smallholder farms, in particular. As for conventional farming, by which the current dominant form of agriculture is referred to, such a robot could achieve the double goals of reducing pesticide use and controlling weed proliferation [11,26].

Different non-conventional yet non-chemical methods for weed identification and management have been proposed, thanks to the widening scope of technological advances [29]. In this regard, different technologies have been used for the task of precision weed management in agriculture, which includes the follows:

Aerial and Satellite Remote Sensing: Aerial remote sensing technologies operate from a certain height. Here the differential spectral reflectance of the plants and weeds and spectral resolution of the instrument (vision device) are the driving factors of identification [30]. In the case of a developing plant canopy or taller plants, such methods are hindered by their inability to differentiate through the lack of or improper visual access to the weeds growing on the ground. In the initial stages of the cropping season as well, random stubbles or crop residues might interfere with weed identification [31]. Inaccuracies due to spectral signal mixing have also been reported in aerial weed identification and hence hinders precision weed removal [32]. The major reported challenges in aircraft and satellite-enabled remote sensing for weed management in addition to the acquisition of high spatial and temporal imagery from higher altitudes is the acquisition of good imagery under cloudy conditions [31,32].

Unmanned Aerial Vehicles (UAVs): UAVs provide an edge over remote sensing methods as they operate from a height that is closer to the ground and provides high-resolution imagery in real-time. Images can be retrieved more frequently and largely independent of the weather conditions like clouds [29]. Although UAVs provide higher resolution imagery, they are beset with limitations such as high battery use during flight time and the high processing time of the imagery [33]. The operation of UAVs like drones is also often regulated by the government and hence their use and usability might get affected by local government regulations [34]. Huang et al. have proposed the serial application of a UAV-based weed identification system and a Variable Rate Spray (VRS) system for weed identification and management [33]. The integration of both the operative functions is limited by the payload carrying capacity of the UAV. However, the two operative functions could easily be integrated into the same machine, with a much higher carrying capacity, for example, in an on-ground robotic precision weed identification and removal system.

Robotics: The increasing scope of robotic technologies has made possible the deployment of robotics in weed identification and management [29]. With robotics, weed identification goes a step closer to the ground as compared to the previously discussed methods. Based on artificial intelligence, using artificial neural networks, weeds can be not just identified in real-time with higher spatial resolution but can also be tackled, physically, thermally, or biologically, in real-time with a robotic arm attached to the robot on the ground. In this regard, the application of machine learning using convolutional neural networks for the identification of plants/fruits at their different stages has also been reported [35].

In this study, a plant and weed identifier robot (precision weeding robot) has been conceptualized and its software designed, based on state-of-the-art deep learning techniques using artificial neural networks (convolution neural networks). Experiments were conducted on a dataset of over 200 images of three different plant species taken under different conditions and of different sizes at different growth stages. The neural network was trained to identify the plants and classify them as either weed or plant.

The robot is conceptualized for use in both small and big farms. However, the motivation behind rendering it low-cost and low-tech is to enable smallholders to be the primary beneficiaries. The importance of this approach stems from the fact that smallholder farmers are the primary producers of food for the majority of the world population [36]. A low-cost weeding robot that can identify and distinguish weeds from plants could be an addition to the agroecological interventions [28,37]. The robot can, based on the need, either remove the weeds or incorporate them into the soil. The option of fitting the robotic arm with other heads is also there, which can be used to spray trace elements or plant protection substances.

The construction of the autonomous farming robot mainly focussed on performing weeding operations is broadly divided into six phases for prototyping and carry out the initial tests:

Phase 1: Conceptualisation of the idea framework for the design of the robot.

Phase 2: Building and testing an artificially intelligent classifier that can distinguish a plant from a weed in real-time.

Phase 3: Design the method for estimation and extraction of the position of the identified weeds using computer vision techniques.

Phase 4: Building a mobile robotic platform prototype and install all necessary components and the robotic manipulator for developing and testing.

Phase 5: Design and develop control algorithms for moving the robot platform and the manipulator with the end effector towards the weed and perform different removal strategies.

Phase 6: Validation studies and iterative tests in the lab and in the field. Improving on the flaws and developing additional features and testing.

The ideas and results from the first three phases are described in the following sections.

2. Literature Review

2.1. Studies on Weed Killing Herbicides and Its Effects

Application of weedicides is the commonly used method for post-emergent control of weeds [38,39]. A study conducted in 2016 reported that, globally, the use of the single most commonly used herbicide Glyphosat increased 15-fold in a span of 20 years [40]. An increasing number of studies detail the concerns that arise with the usage of herbicides with respect to adverse effects on human health, soil nutrition, crop health, groundwater, and biodiversity [41]. Many governments are planning to ban the usage of such agrochemicals and are hence looking for alternative solutions in this regard [40]. The World Health Organisation (WHO) has reported sufficient evidence regarding the carcinogenicity of insecticides and herbicides, while its potential effect on human beings at the DNA (Deoxyribonucleic acid) and chromosome level has also been reported [42]. In a study, the US FDA (Food and Drug Administration) reported the presence of glyphosate residues in 63.1% of corn and 67% of soy samples, respectively [40]. A case study in 2017 reported that a detectable amount of glyphosate was found in the urine specimens of pregnant women leading them to have shorter pregnancy lengths [40]. Another study from Sri Lanka shows that drinking glyphosate contaminated water causes chronic kidney diseases [43]. In addition to being a health risk for humans, the use of pesticides has also been reported to cause a decrease in monarch butteries population [44], slow larvae growth in honey bees, and lead to their death when exposed to glyphosate [45]. The use of herbicides generally poses a slew of adverse non-target risks on the different components of the agroecosystems [46]. Hence exploring a non-chemical solution to the problem of weed proliferation is plausible.

2.2. Deep Machine Learning in Agriculture

Machine learning (ML) is a subset of the artificial intelligence domain that provides computers the ability to learn, analyze, and make their own decisions/predictions without being explicitly programmed. It is mainly categorized into predictive or supervised learning and unsupervised learning [47].

The goal of the supervised learning approach is to learn a mapping function from inputs x to outputs y, given a labeled N set of input-output pairs

$$D = \{(x_i, y_i)\}_{i=1}^{N} \tag{1}$$

Here, D is called the training set, and N is the number of training examples. In simple terms, we have few sample inputs and outputs and we use a mathematical algorithm to learn an underlying mapping function that maps input to the output. Hereby, the aim is to estimate the mapping function and predict the output when an entirely new set of input data is provided. Currently, supervised learning is widely used in many applications, such as classification, pattern recognition, and regression problems [47].

On the other hand, in unsupervised learning, we are only given inputs, and the goal is to find 'interesting patterns' in the data [11,47].

$$D = \{x_i\}_{i=1}^{N}$$

In simple terms, here, the algorithm is left to learn and analyze the underlying pattern without providing any input labeled data. The algorithm learns through structuring data patterns and predicts the output. Some of the examples of unsupervised learning are clustering and association problems [47]. Figure 2 shows a general block diagram of the machine learning approach.

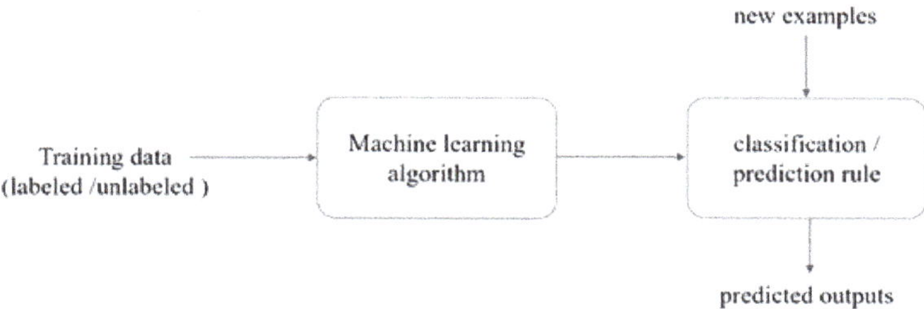

Figure 2. Machine learning approach.

Deep Learning is a subset of the Machine Learning approach in artificial intelligence. Artificial deep neural networks are one of the deep learning architectures, which provide a compelling supervised learning framework [48,49]. Machine learning and deep learning algorithms are applied in various agricultural operations, such as flower species recognition, disease prediction and detection in plants, crop yield forecasting, weed classification and detection, and plant species recognition and classification [50]. These are briefly described below.

2.2.1. Disease Identification

Crop diseases are a significant threat to the crop yield and the quality of the food produced, with adverse consequences on the livelihood of small-scale farmers and food security [51]. Globally, 80% of the food is grown majorly by the small-scale farmers, and among them, there is a reported yield loss of 50% due to crop diseases and pests [51]. Various types of microbial plant pathogens are the typical causative agents of plant diseases [20]. Different bio-control agents have been assessed and used against those pathogens to curb plant diseases [52]. However, a few decades back, research efforts were initiated for the early identification of plant and crop diseases at different agricultural institutes to help farmers in the prevention of crop diseases [51]. To carry out the prevention measures, early detection of the pathogens, and the diagnosis of crop diseases is essential. With technological advancements, today, these disease detection steps are carried out much more efficiently [53].

Artificial Intelligence technology, along with computer vision, image processing, object detection, and machine learning algorithms are widely used and analyzed and have proven to be effective in plant disease diagnosis and detection [53]. By utilizing popular architectures like AlexNet [23] and GoogleNet [24], Mohanty et al. reported a disease prediction accuracy of 99.35% upon the analysis of 26 diseases in 14 crop varieties [51]. In addition to that, a real-time disease detector proposed in the experimental study by Alvaro et al. in tomato plants helped to diagnose diseases at an early stage in tomato crops in comparison to various lab analyses [54]. Hence, deep machine learning-based interventions are making significant contributions to agricultural research.

2.2.2. Crop Yield Forecasting

For the purpose of planning and designing food supply chains, it is helpful to have an idea about the crop yield that can be expected for a particular cropping system. Accurate yield estimation also helps farmers to choose better crop management methodologies among the different available ones [55]. Conventionally, crop yield estimation is based on previous experience and seasonal weather conditions [55,56]. Such yield estimation approaches, however, are constrained by factors including climate variability and the changing soil and water dynamics and are hence often not well adapted to changing conditions [56]. In modern farming systems, the availability of time-series yield data, combined with many other sources of spatial agricultural farm data, can be utilized in designing machine learning algorithms that can contribute to better yield prediction models [56]. Support Vector Machines (SVM), Artificial Neural Networks (ANN's), Bayesian Networks (BN), Backpropagation Networks (BPN), Least Squared Support Vector Machines (LS-SVM), Convolutional Neural Networks (CNN) are some of the models that are used for yield prediction [50].

In a study, Support Vector Machine (SVM) algorithms used on coffee plantations to determine whether the seeds are harvestable or not helped farmers to optimize their economic plans and work schedules [50]. In another study, Unmanned Aircraft Systems (UAS) were used to collect the spatial and temporal remote sensing data, using an artificial neural network model to predict tomato crop yield which had a predictive accuracy of (R^2~0.78–0.89) [57]. R^2 is the coefficient of determination which is an evaluation metric that is commonly used in regression tasks. In another study, three factors, such as soil conditions, weather conditions, and management practices data (sowing dates) from the year 1980 to 2015, were collected and considered as inputs [58]. With that data, a CNN-RNN (Convolutional Neural Network-Recurrent Neural Networks) model was used to predict the yield in soybean and corn fields across 13 states in the United States. The model showed that soil and weather conditions are vital components in yield forecasting in addition to crop management practices [58]. In other recent research, it is reported that a deep learning-based 3D CNN model applied for soybean crop yield prediction outperformed the state-of-the-art machine learning techniques [59].

2.2.3. Plant Leaf Classification and Identification

Easy recognition of different plant species can be of great help to ecologists, biologists, taxonomists, and researchers in plant-related studies and for medical purposes [60,61]. Machine learning and computer vision algorithms are making considerable contributions in this field [50]. They help reduce the dependency on expert availability and save time in classification tasks [50]. Deep learning models that specifically deal with images are used in plant leaf identification and have outperformed conventional image processing techniques and machine learning algorithms [62].

In one research study, a proposed deep learning model that uses ResNet26 architecture could achieve recognition levels of 91.78% on the BJFU100 dataset that consists of 10,000 images of 100 classes [60,62]. In comparison to that, the same proposed model could achieve

99.65% in classifying 32 kinds of leaf structures of plants utilizing the publicly available Flavia leaf dataset [60,62,63]. Studies report that it is not just the colors and shape of the leaves that are used to classify the plants, rather plant leaf veins can also be used as input features in determining leaf identity and properties [62]. The increased usage of mobile technology has brought the above techniques to the stage of practical implementation, being integrated into the form of mobile applications. Few mobile applications like Flora-incognita, Pla@ntNet are able to recognize plants, fruits, flowers, and barks of the trees by just snapping a picture of it [64,65]. Currently, Pl@ntNet is able to recognize 27,909 varieties of plants and maintains a database of 1,794,096 images of different plants [64].

2.2.4. Weed Classification and Detection

Weed management in crops is a challenging task for farmers and poses a significant threat to crop yields if not done properly [50,66]. Weeds compete with crops for nutrients and usually grow faster, hence early identification and classification are crucial for a better crop yield [50,67,68]. Machine learning algorithms like SVM, ANN, have already been used for classifying and achieved high accuracy levels in different crops [50].

Utilizing the openly available dataset of plant seedlings provided by the Aarhus University of Denmark, Ashqar et al. developed a deep learning model that was able to classify 12 species of weeds over 5000 images with a precision of 99.48% [69]. In another study, Smith et al. used CNNs and transfer learning techniques to classify grass, dock, and clover and achieved a 94.9% accuracy in classifying weeds [70]. The transfer learning technique is a powerful tool that can be used over small datasets and can achieve a reasonable level of accuracies [70]. In another study, a fuzzy real-time classifier was developed for weed identification in sugarcane crops, with an accuracy level of 92.9% [6]. However, the latest deep learning architectures can improve the performance of the tools and can leverage the possibilities in exploring new ideas in weed control and management strategies [68]. Real-time identification of weeds can be a potent tool for robots in precise weeding. It can be a valuable addition to sustainable weed management systems [50,68]. Consequently, this could contribute towards offsetting the heavy usage of pesticides [67].

2.3. Artificial Neural Networks

As the name suggests, an artificial neural network (ANN) is a system that is inspired by the connections of neurons in human brains [71]. An artificial neuron is a single block mathematical entity that processes information and is essential in the functioning of a neural network [71]. Haykin stated that a typical neuron has three essential elements: a set of connection links that have their weights, a summation point, and an activation function. The neuron k can be mathematically described by the following equations [71].

$$u_k = \sum_{j=1}^{m} w_{kj} x_j$$

$$y_k = \Phi(u_k + b_k)$$

where u_k is linear combiner output; $w_{k1}, w_{k2}, w_{k3}, \ldots w_{km}$ are synaptic weights; $x_1, x_2, x_3, \ldots x_m$ are inputs; b_k is the bias that has the effect of lowering the input activation function; $\Phi(.)$ is the activation function; y_k is the output of the neuron. A typical mathematical model of the neuron is shown in Figure 3 [71].

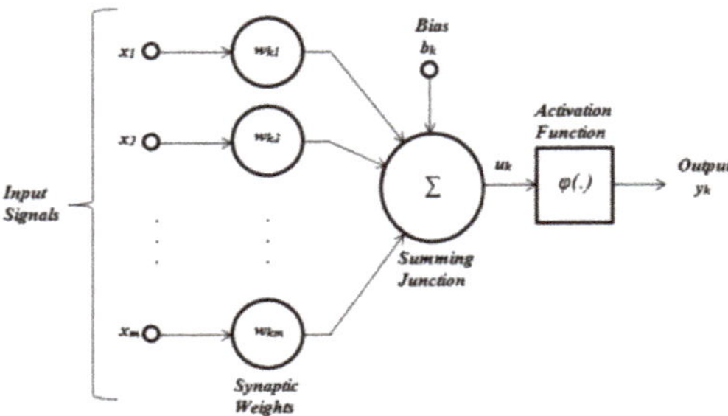

Figure 3. A non-linear mathematical model of an artificial neuron [71].

An artificial neural network is simply a collection of artificial neurons. Typically they are connected and organized in layers. A layer is made up of interconnected neurons that contain an activation function. A neural network consists of an input layer, an output layer, and one or more hidden layers. The input layer takes the inputs from the outside world and passes those inputs with a weighted connection to the hidden layers. The hidden layers then perform the computations and feature extractions and are activated by standard nonlinear activation functions such as tanh, ReLU (Rectified Linear Unit), sigmoid, softmax, and pass the values to the output layer. These types of networks are typically called feed-forward neural networks or multilayer perceptrons. Figure 4 shows a feed-forward neural network [72].

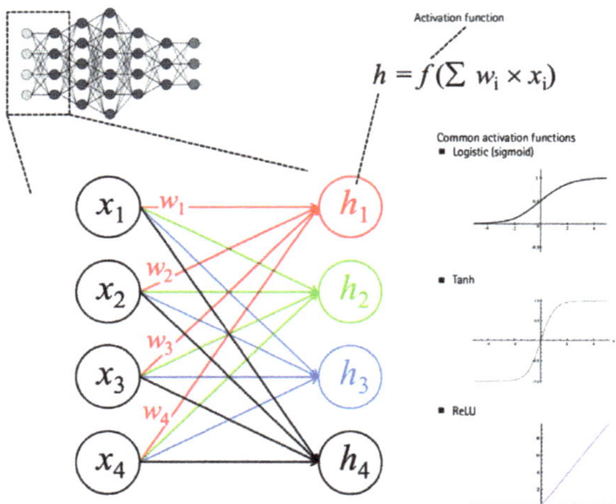

Figure 4. A feed-forward neural network [72].

When it comes to training a neural network, the focus is mainly put on minimizing the output prediction error by adjusting the weights on each connection in a backward manner. This process is called back-propagation [73]. The back-propagation algorithm then searches for the minimum value in the weight space using a stochastic gradient descent method. The obtained weights, which can minimize the loss/cost function, are then considered as a solution for the training problem and the training process culminates [73].

2.4. Convolution Neural Networks

The term convolutional neural network (CNN) denotes one of the deep neural network algorithms that mainly deal with computer vision-related tasks [48]. They are often used in applications like image classification, object detection, and instance segmentation problems.

The special feature of CNNs is that they are able to learn and understand the spatial or temporal correlation of the data. These are highly successful in practical applications Convolutional neural networks use a special kind of mathematical operation in one of its layers called convolution operation instead of a generic matrix multiplication [48].

A convolution neural network (ConvNet) typically consists of three layers, a convolutional layer, a pooling layer, and a fully connected or dense layer. By aligning all those layers in a sequence or stacking them up, CNN architectures can be built. Figure 5 illustrates a convolutional neural network. The convolution layer is the central building unit of CNNs. It consists of kernels that convolve independently on the input image resulting in a set of feature maps. Strides, depth, and zero paddings are the three parameters that control the size or volume of the activation map [74]. Here, stride represents the number of pixels it has to move over the input image at a time; depth represents the number of kernels that are used for convolution over the input image [74]. Convolving kernel over the input image results in a reduction of the size of the activation map and loss of information in the corners. The zero-padding concept adds zero values at the corners and helps to control the output volume of the activation map. Besides, to provide the network with the ability to understand complex data, every neuron is linked with a nonlinear activation function. ReLU is one of the frequently used activation functions because it provides the network with the ability to make accurate predictions [74].

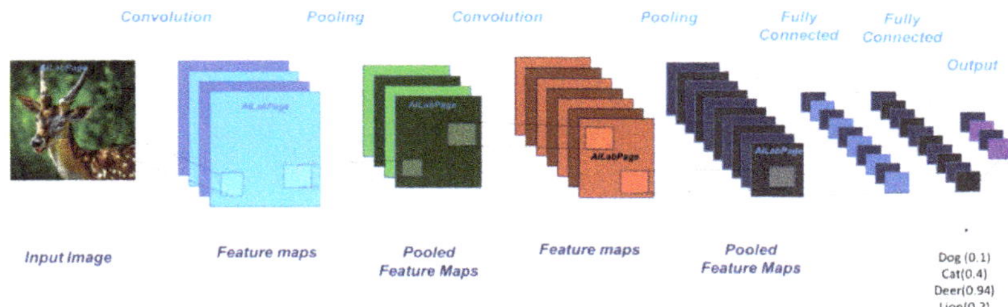

Figure 5. A convolutional neural network (CNN) [74].

The pooling layer mainly serves the purpose of reducing the spatial size representation to reduce training parameters and computing costs in the network and retains essential information when the images are larger. Pooling is also referred to as downsampling or subsampling. Pooling is done independently on each depth dimension of the image. However, the pooling layer also helps to reduce over-fitting during training. Among other types of pooling, max pooling with a 2×2 filter, and stride = 2 is commonly used in practice for better results [74].

2.5. State-of-the-Art Object Detection Methods

In case of image classification problems, the object recognition (detection, recognition or identification) part is the challenging part. It involves the classification of various objects in an image and localization of the detected objects by drawing some bounding boxes and assigning class label names for every bounding box [75]. The instance or semantic segmentation is another problem in computer vision, where instead of drawing a bounding box around the objects, they are indicated with specific pixels or masks [75].

Compared to machine learning methods of detecting objects, deep learning methods are highly successful and do not require manual feature extraction. Region-Based Convolutional Neural Network (R-CNN), You Only Look Once (YOLO), Single shot Multi Detector (SSD) are some of the techniques that are proposed for object identification and localization tasks, that can perform end to end training and detection [76–81].

R-CNN was proposed in 2014, and comprises three steps. Initially, a selective search algorithm is used to find the regions that may contain objects (approximately 2000 proposals) in an image [76,77]. Later on, a CNN is used for feature extraction and finally, the features are classified. However, the constraint here is that the whole ROI (Region of Interest) with objects is warped to a fixed size and provided as an input to the CNN [77]. This process is computationally heavy and has a slow object detection speed. To mitigate some of the flaws and make it work fast, the Fast R-CNN method was introduced [77]. Here, in the first stage, it uses a CNN to extract all the features and then an ROI pooling layer is used to extract features for a specific input region and feed the output to a fully connected layer that divides and passes it to two classifiers which perform classification and bounding box regression [77].

However, another method Faster R-CNN was proposed by Shaoqing Ren and colleagues and it outperformed both the previous models in terms of speed and detection [76]. They introduced the Regional Proposal Network (RPN) method and combined it as a single-mode [76]. It uses RPN to propose the regions and Fast R-CNN detector that uses proposed regions. Mask R-CNN is another method that is an extension to the Faster R-CNN for pixel-level semantic segmentation [78]. It was introduced as a third branch, based on the Faster R-CNN architecture, along with classification and localization. It is a fully connected network that predicts a segmentation mask in a pixel-to-pixel manner. Although it is fast, it is not optimized for speed and accuracy [78]. Figure 6 represents the summary of the R-CNN family of methods [82].

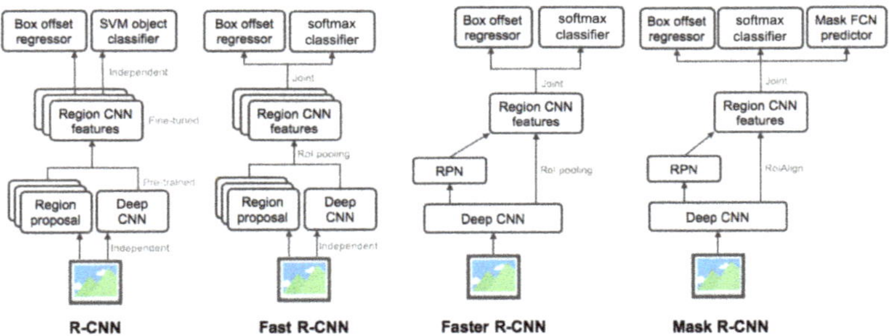

Figure 6. The Region-Based Convolutional Neural Network (R-CNN) family.

YOLO is another popular object detection method proposed by Redmon et al. that uses a different approach compared to the above R-CNN family of approaches [80]. A single neural network is used to predict class probabilities and bounding boxes from the images. Their base model and Fast YOLO model can process images in real-time at 45 fps and 155 fps with double mAP (mean Average Precision) [80]. Although it was reported to be fast and outperformed the state-of-the-art R-CNN's family techniques in terms of speed, it tends to make more localization errors [80].

SSD is another approach proposed by Wei Liu et al. to detect objects in images by using a single neural network [79]. It performs the generation of region proposals and also identifies the objects in the proposed region in a single shot. Whereas, RPN-based approaches use two shots, and are hence slower than SSD, have achieved an mAP higher than Faster R-CNN or YOLO [79].

2.6. Transfer Learning Technique

Transfer learning is a technique that is used in many machine learning and deep learning tasks. It has been defined in different ways. Goodfellow et al. define it as an approach of transferring the knowledge of a previously trained neural network model to the new model [48]. It has also been defined as an optimization that allows rapid progress when the model is learning for another task [83]. Mathematically, this can be defined as follows.

Definition: For a learning task L_s in the source domain D_s and a learning task L_t in the target domain D_t, transfer learning helps improving the performance of the predictive function $f_t(.)$ in target domain D_t by utilizing the knowledge acquired from D_s and T_s; where $D_s \neq D_t$ and $L_s \neq L_t$. Figure 7 represents transfer learning technique.

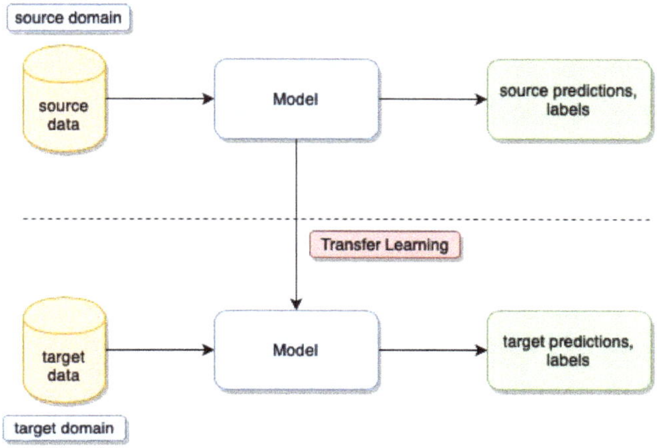

Figure 7. The transfer learning technique.

For instance, a neural network model that is trained to learn and recognize the images of animals or birds can be used to train and identify automotive cars or medical x-ray diagnostic images or any other set of images. Usually, this process comes in handy when there is less amount of data that is available to train for the second task. However, it also helps in accelerating the training process on the second task, compared to training from scratch, which may take weeks to achieve optimal performance. When the first task is trained to recognize some images, the low-level layers of the neural network model try to learn the basic features of the images. For example, contours, edges, circles are extracted by the low-level layers, which are called feature extractors. These feature extractors are a standard in the first stages of the neural network training and are the standard building blocks for most image recognition-related tasks. We utilize these feature extractors for the second task, and in the end, we use an image classifier to train and classify for our specific job. In our scenario, since the task is to recognize two classes i.e., plants and weeds, the transfer learning technique was utilized to perform experiments that are described in the next section.

The transfer learning technique, as described above, is proposed as the method to be utilized for the tasks of weed identification and classification as it has been reported to be suitable for tasks of autonomous identification and classification tasks [84]. Despite its widespread application in diverse fields like training self-driving cars to audio transcription, the transfer learning technique faces two major limitations. The phenomena of negative transfer and over-fitting are considered two major limitations of the transfer learning technique [85]. Negative transfer occurs when the model source domain data is dissimilar from target domain data. In other words, negative transfer can occur when the two tasks are too dissimilar [86]. As a result, the model does not perform well, leading to poor results. On the other hand, while doing transfer learning, the models are prone to overfitting, in absence of careful evaluation and tuning. Overfitting is however a general limitation for all prediction technologies [87]. These limitations can be overcome by carefully tuning the hyperparameters and choosing the right size (number of layers) of the neural network model.

3. Materials and Methods

This field of studies regarding the problem of weed infestation were carried out on rice farming systems in the Kashmir region in India. The robot development research is being undertaken at the Hamburg University of Technology (TU Hamburg), Hamburg, Germany under the research group Rural Revival and Restoration Engineering (RUVIVAL) at the Institute of Wastewater Management and Water Protection with the support of the Institute of Reliability Engineering.

3.1. Conceptualisation and High-Level Design of the Robot

The conceptualized mobile robot platform's intuitive design is shown in Figure 8 as a demonstration of how a robot platform might look once it is built in real-time. The design is developed using Onshape design software [88]. The robot was conceptualized initially to operate between rows of rice plants with a spacing of 25 cm, however, subsequently, it is planned that the robot shall be a modular one, as such operation can be adjusted to the row width and the height of the plants at different stages. The robot is intended to recognize weeds at an early BBCH stage, ideally at the leaf development stage. In this regard, the images of plants taken for training purposes also included plants at the sprouting stage. The conceptualized robot, as shown in the figure, has an electronics storage box where it has batteries, sensors, and a single-board computer. On top of the electronic box, there is a solar panel mount to provide a renewable source of energy for the robot's movement. Once the robot has successfully identified the weeds, an algorithm provides the position of the weeds in terms of real-world coordinates of the robotic platform relative to the image frame. After the transformations have taken place, a robotic manipulator picks up the real-world coordinates and performs inverse kinematics operations and drives the end effector to the desired position and performs weed control mechanisms like mechanical or thermal weed control, optionally mulching.

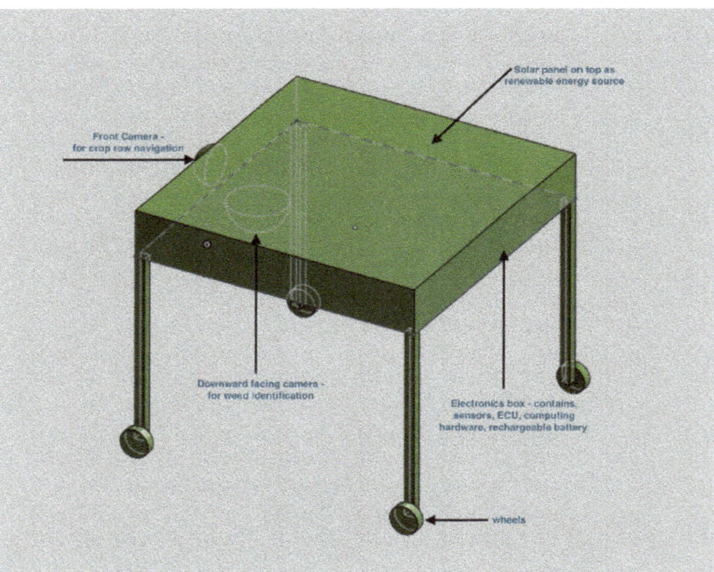

Figure 8. A rough representation of the idea of the plant and weed classifier robot.

The choice of robotic manipulators to perform mechanical weeding can vary depending on various factors such as kinematic structure, degrees of freedom, workspace, motion control, accuracy, and repeatability [89,90]. There is a possibility to mount three types of manipulators underneath the robotic platform.

1. Articulated arm
2. Cartesian robot

3. Parallel manipulator

Parallel manipulators have high rigidity, high payload/weight ratio, high speed and acceleration, high dynamic characteristics, and it is easier to solve inverse kinematics problems with them compared to serial manipulators [89,90]. On the downside, they have a limited and complex workspace. A parallel manipulator may still be one of the better choices for performing weeding action. Serial manipulators or articulated arms have a larger workspace, high inertia, low stiffness, low speeds, and accelerations and experience more difficulty in solving the inverse kinematics problem compared to parallel manipulators [89]. Cartesian robots are not considered an ideal choice because of their lesser number of applications on mobile platforms. At this point, we propose a parallel manipulator as the ideal choice based on its advantages and characteristics. However, it can still be an open question to agree on the perfect manipulator that can be mounted onto the robot to perform weeding acts. The following Figure 9 presents three degrees of freedom parallel delta manipulator (excluding the fourth degree of the end actuator) [90].

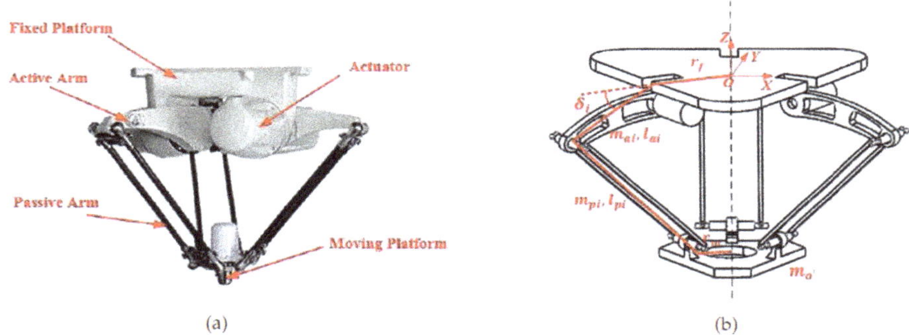

Figure 9. A schematic of a delta robot manipulator with three degrees of freedom: (**a**) A Delta robot with three degrees of freedom; (**b**) A three-dimensional model of a Delta robot with the different pa-rameters [90].

This robot is intended to be used as an agricultural tool together with other sustainable agricultural practices, which decrease the dependence of farmers on external inputs like mineral fertilizers and pesticides. Therefore, from a purely monetary perspective, the robot can decrease the input costs by decreasing labor requirements and eliminating the cost associated with pesticides, while increasing yield by bridging the yield gap resulting from weed infestation. An important aspect of the use of an autonomous weeding robot, from an agroecological perspective, is to reduce the ecological footprint of food production through the phasing out of chemical pesticides. This will also lead to better quality food and less contamination of soil and water due to agrochemical residues, as already discussed in the introduction. The environmental and societal damages of pesticide use have been estimated to be around $10 billion [91]. The costs and benefits of this intervention, hence, go much beyond the cost of procurement of the equipment and the benefit of labor savings due to robot deployment for weeding. The proposed weeding robot is conceptualized as a low-cost robotic machine, as compared to the robots that are available in the market, which are available in the range of $20,000 to $125,000 [92–94]. The prototype is being built with a cost estimation of $15,000 and the final robot upon industrial production is expected to be available to the farmers for a price under $10,000. In comparison, the monetary costs of pesticides for a smallholder with 10 hectare land under cultivation, is around $1750 per year at $70 per acre (2018, 2019) [95]. Pesticide costs are expected to further increase in the coming years with increased incidence of pesticide resistance. This means, if the robot is acquired by a farmer cooperative of five farmers who use it on sharing basis, the monetary cost of procuring the robot will be the same as the cost they would otherwise incur by using pesticides in one year, with environmental and human health benefits a strong motivation.

3.2. Hardware Design Approach of the Weeding Robot

Designing robot hardware that is operating under dynamic surroundings is often a challenging task. We can notice, a high-level, modular hardware design is presented and introduced in Figure 10. The robot ideally consists of a single-board computer along with all the required modules, peripherals, sensors, and actuators. Single boards computers have everything built on a single circuit board like RAM, processor, and peripherals. It has general-purpose input-output pins that are good at controlling sensors and actuators. There are many open-source single-board computer varieties available today. Depending on the choice of application, it is essential to choose one. Open source boards like Raspberry Pi have processors and have to ability to run Linux and distributed systems like Robot Operating System (ROS) [96,97].

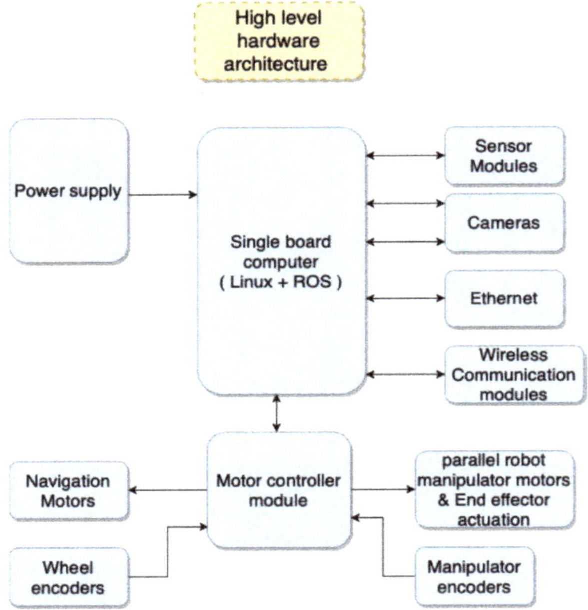

Figure 10. High-level hardware design block diagram for the weeding robot.

ROS is a lightweight middleware that is specifically designed for robotic applications. Its publish-subscribe design pattern is one of the featured patterns that enables asynchronous parallel processing from node to node communication. It has built-in packages that can solve inverse kinematics, forward kinematics, path planning, navigation, PID (proportional-integral-derivative) control, vision-related tasks. It also has graphical tools like Gazebo, a Rviz that helps to visualize the robot model for simulations [96].

Boards like Jetson Nano, Jetson TX2-Serie, Jetson Xavier NX, Jetson AGX Xavier-Series from NVIDIA [86], Coral dev board from Google has TPU(Tensor Processing Unit) and NPU(Neural Processing Unit) [98], that enables and accelerates them to use in AI-specific applications like object detection, image classification, instance segmentation for training and inferencing purposes [99]. These boards are cheaper and costs in the range of approximately 100$ to 800$. These boards will be analyzed and utilized for our robot building purpose in future work by keeping a low-cost reliable design in scope.

3.3. Software Design Approach of the Weeding Robot

Software for the weeding robot can be entirely developed in the ROS framework using high-level languages like C++ or python. A sensor interface provides all the inputs from the cameras and sensors on the robot. The perception interface deals with the identification of weeds, stem positions, and position estimation of the detected weeds. OpenCV libraries can be used in the perception interface for real-time weed identification. The navigation interface has closed-loop feedback control algorithms that help with path-planning between the crop rows. The robot interface takes the outputs from the feedback controllers and drives the robot in the crop field autonomously and manages the weeds in real-time using the delta manipulator. A high-level software block diagram for the weeding robot is presented in Figure 11.

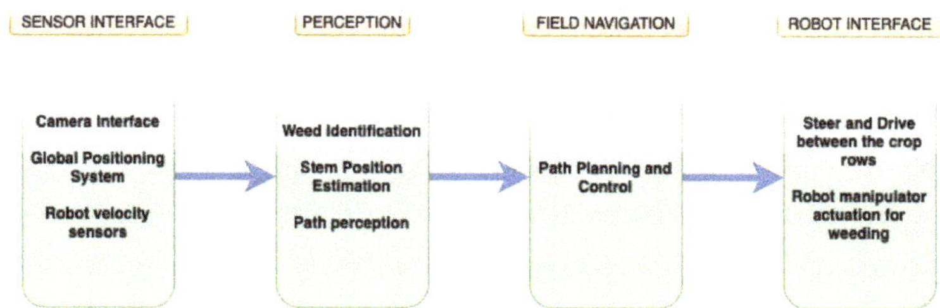

Figure 11. High-level software block diagram for the weeding robot.

Python is widely popular and is used for AI, Computer Vision, and Machine Learning applications. It has gained popularity over the last few years because of its simple syntax structure and versatile features. The open-source community developers are actively contributing to many libraries, which makes it easy for application or product developers to build a product without reinventing the wheel.

OpenCV is an open-source software library for computer vision applications. This library can be modified and used for commercial purposes under BSD-license. It comes with many built-in algorithms, for example, face recognition, object identification, tracking humans, and objects. This library is used broadly in all domains, including medicine, research labs, and defense.

3.4. Training and Implementation

3.4.1. Plant and Weed Identification Pipeline

The plant and weed identification pipeline process comprises three stages. Figure 12 represents the three stages. In the first stage, data was collected and preprocessed according to the input requirements of the neural network model. In the second stage, two neural network models were trained, evaluated, analyzed, and optimized. Finally, in the third stage, the best performing optimized model was exported for real-time identification of plants and weeds.

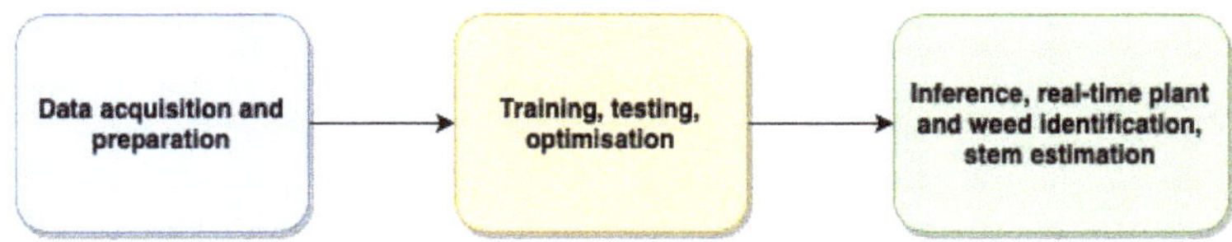

Figure 12. Proposed plant and weed identification pipeline.

3.4.2. Experimental Setup

Deep learning tasks are majorly dependent on data is essential for conducting experiments. The input data was based on three plant species: red radish (*Raphanus raphanistrum subsp. sativus or Raphanus sativus*), garden cress (*Lepidium sativum*), and common dandelion (*Taraxacum oficinale*) were considered for our experiments. The abundant availability of the common dandelion on lawns, and the fast growth of red radish and garden cress made us opt for them. The problem of plant and weed classification can be divided into two categories: binary and multi-class classification. By grouping the species separately into two categories, we considered this as a binary classification problem. Considering them individually, it becomes a multi-class classification problem. The end goal was to precisely locate any type of weeds in the soil. We treated the classification as a binary classification task. We merged edible radish and garden cress into one category (plants) and common dandelion (weed) into another category and carried out our classification tests.

It is a common phenomenon that weeds grow faster compared to edible plants and compete for more soil nutrients. As a result, during this crucial time at the beginning of the growth cycle, distinguishing between the plant and weed is essential. This can then be followed by weed management techniques. Based on that fact, a dataset of edible plant seedlings and weeds of different sizes under different surrounding conditions and backgrounds were prepared. Python programming language, Google's open-source TensorFlow object detection API were utilized to build, train, and analyze neural network models. The system overview used for training, testing, and inference is presented in the table below (Table 1).

Table 1. System overview.

CPU	AMD Ryzen 7 2700X 8x 3.70 GHz
Memory	16 GB DDR4 RAM 3000 MHz
GPU	NVIDIA 8 GB RAM
OS	Ubuntu 18.04 LTS 64-bit

3.4.3. Data Acquisition and Pre-Processing

Deep learning tasks require a considerable amount of input data as the main source for training the neural network models. For our problem, we made our dataset based on three plant species, for experimental purposes. A greenhouse was maintained in the laboratory, and we planted red radish and garden cress in mini-plots. We took photographs and compiled the dataset by taking RGB pictures of the growing plants using a mobile camera. The raw pictures collected were of pixel dimensions 4032 × 3024. Since they were high-resolution images, providing them directly as input to train the network would have been computationally expensive and hence the learning process would have been time-consuming. Therefore the raw images were converted to 800 × 600 dimensions and then used for pre-processing.

A complete set of 200 images consisting of photos taken from different perspectives and angles of plants and weeds was used for training and evaluation purposes. Figures 13 and 14 show some of the input image samples that were used for training the network.

Figure 13. Test weeds: two photographs of common Dandelion that were used in the training.

Figure 14. Test plants: A few photographs of Radish seedlings (**Left**) and Cress (**Right**) that were used in the training.

In order to train the network, the whole dataset was split into two, one for training and one for evaluation. The train/test split ratio was considered as 160/40 using the Pareto rule. When using the TensorFlow object detection API, we maintained a structure such as a workspace for all the configuration files and datasets. The whole process was divided into five steps based on the TensorFlow custom object detection process. Those five steps included preparing the workspace, annotating images, generate TFRecord file format input files, configure/train/optimize the model, and export the inference graph for testing.

For annotating images, an open-source labeling tool LabelImg was used to draw the bounding boxes. The annotations were saved in the PASCAL Visual Object Classes (VOC) format as XML files. A representation of the bounding boxes that were drawn around the edible plants and weeds is shown in Figure 15.

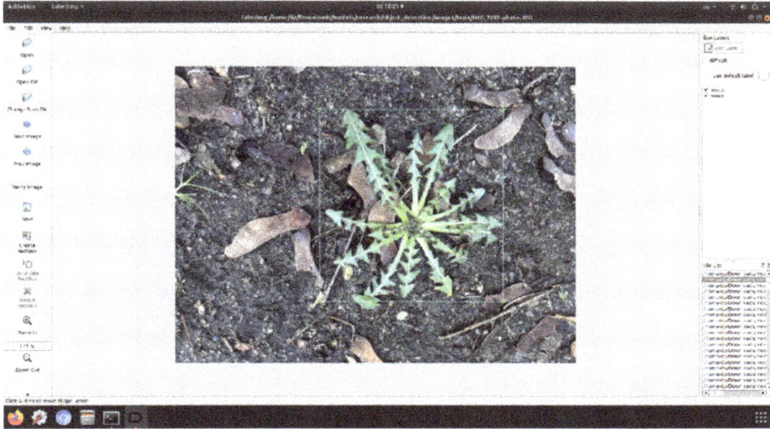

Figure 15. Annotating weed images using labeling software.

3.4.4. Training and Analysis of the Neural Network Model

By utilizing the transfer learning technique, two pre-trained models Faster R-CNN inceptionv2 and SSD inceptionv2 were chosen from the TensorFlow model zoo that were trained for the Common Objects in Context (COCO) dataset. For the weeding robot, a latency is preferred between the detection and interacting with the weed. Hence there was no primary requirement for higher detection speeds in our scenario. A reasonable detection speed with higher mean Average Precision (mAP) accuracies and higher confidence scores were preferred. The reported mAP accuracies and speed of the above two mentioned models on the COCO dataset were reasonably well and suitable for our plant and weed detection problem. Hence these models were adapted for training, optimization, or better generalization.

Generally, to come up with a model architecture, neural networks are stacked up in layers sequentially, which however can make the large network computationally expensive. A large neural network also comes with the downside of not being able to provide remarkable accuracies. Some of the available backbone architectures include AlexNet, VGG16/19, GoogLeNet, MobileNet, Inceptionv2, Inceptionv3, Inceptionv4, NASNet, ResNet, Xception, Inception-Resnet. The models that were trained in our experiments and analysis use Inceptionv2 architecture as feature extractors. Christian Szegedy and his colleagues had proposed GoogLeNet. It consists of a 22-layer deep convolutional neural network architecture that was considerably computationally efficient. Instead of stacking up layers sequentially and selecting filters, they proposed a block in between the layers named as "Inception" module. The inception module performed different kinds of filter operations in parallel. From these filter operations, we get different outputs that were concatenated 'depth' wise all together. This makes the network go wider rather than deeper. The single output obtained from the previous operation is then passed on to the next layer. The result of doing this operation was observed to be computationally less expensive. The inception block is represented in Figure 16.

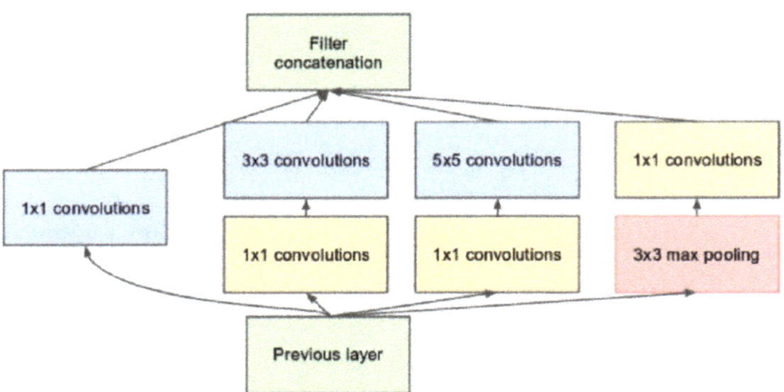

Figure 16. Inception module with dimension reduction.

Before the training process had started, the pre-trained Faster R-CNN inceptionv2 model configuration file that was trained on the COCO dataset was modified. In the custom configuration (Configuration 2), we set the total number of classes to 2, as it indicates the classification of plant and weed. The maximum detections per class and maximum total detections variables were set to 10. The network was then allowed to start the training process from the fine-tune checkpoint that comes with the unmodified model. The learning rate is considered as one of the essential hyperparameters that help to optimize the model to achieve better performance. Considering the unmodified learning rate and the number of steps that come with the pre-trained model, the model was over-fitting with a large deviation with increasing evaluation loss. By using the heuristics method and reducing the step size and keeping the learning rate constant, the model performed with a better generalization ability. For further evaluation and fine-tuning purposes, we also considered another higher learning rate value for the same model using the heuristics method. This configuration (Configuration 2) was tried to find if the model converges faster to 0 with better generalization capability.

Learning rate configuration 1:

```
learning_rate: {
manual_step_learning_rate {
        initial_learning_rate: 0.0002
        schedule {
          step: 20
          learning_rate: .00002
        }
        schedule {
          step: 40
          learning_rate: .000002
        }
      }
    }
```

Learning rate configuration 2:

```
learning_rate: {
manual_step_learning_rate {
        initial_learning_rate: 0.0003
        schedule {
          step: 20
          learning_rate: .00003
        }
        schedule {
          step: 40
          learning_rate: .000003
        }
      }
    }
```

Evaluation Metrics

Intersection Over Union (IOU): It is an evaluation metric based on the overlap between two bounding boxes. It requires a ground truth bounding box BG and a predicted bounding box BP. With this metric, we can determine if the detection is valid or invalid. IOU ranges from 0 to 1. The higher the number, the closer the boxes together. IOU is defined mathematically as the intersection of the overlapping bounding boxes area divided by the union of the overlapping bounding boxes area Figure 17.

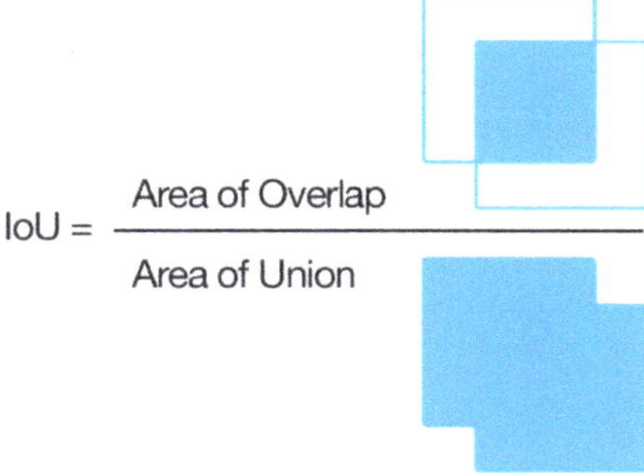

Figure 17. Graphical representation of Intersection Over Union (IOU) (Source: Adrian Rosebrock/Creative Commons).

When IOU scores were available, a threshold (example 0.5) was set for transforming the score into classifications. The IOU values that were above the threshold were considered positive predictions, and if it was below the threshold, they were considered as negative predictions.

Average Precision (AP): Average precision is another way to evaluate object detectors. It is a numerical metric that is the precision averaged across all the recall values between 0 and 1. It uses an 11 point interpolation technique to calculate the AP. It can be interpreted as the area under the precision x recall curve.

Mean Average Precision (mAP): The mAP is another and widely accepted metric to evaluate object detectors. It is merely the average of AP, i.e., it computes the AP for each class and averages them. Tensorboard app is used to visualize the mAP and AP values at different thresholds. The results are briefly discussed in the next sections.

3.4.5. Stem Position Extraction

Extracting the position of the stem is essential for the robotic manipulator for the precise weed management process. It can be done using semantic segmentation techniques, as described by Lottes et al. [100]. The approach reported though is computationally expensive and at best a predictive approach. In this work, a simple stem position extraction technique was formulated and proposed based on the bounding box localization, based on the fact that plants usually exhibit radial or bilateral symmetry. However, plants that are anchored to a single location exhibit an overall roughly radial symmetry. Based on that fact, we say that the center point of the detected bounding box around the weed should be the estimated stem position in the image frame. The accuracy of the stem position was directly proportional to how well the bounding box regressor localizes the complete weed or plant structure.

4. Results and Discussions

4.1. Training

Tensorboard is a powerful visualization tool for evaluating model performances. It was utilized in this work for obtaining the graphs and analyzing purposes. We consider COCO mAP at [0.5:0.95] IOU and mAP at a 0.5 IOU threshold to evaluate the model's performance.

4.1.1. Case 1: Configuration 1

In this case, we considered learning rate configuration 1. With that configuration, the Faster R-CNN inceptionv2 COCO model was trained and fine-tuned up to 200 k steps. The model performed considerably well and reached a maximum overall mAP [0.5:0.95]IOU of 30.94% at 149.6 k step (Figure 18). The maximum mAP at the 0.5 IOU threshold was 61.5% at 149.6 k (Figure 19). At the 200 k step, the maximum overall mAP[0.5:0.95]IOU was reached, at 30.57%. The maximum mAP at the 0.5 IOU threshold was 61.29%. These values were considered suitable given the comparatively less amount of data that the model was trained with. The graphs corresponding to the model performance are shown in the following figures. Graphs were generated at a smoothing value of 0.6 to show the overall trend of the training and evaluation process.

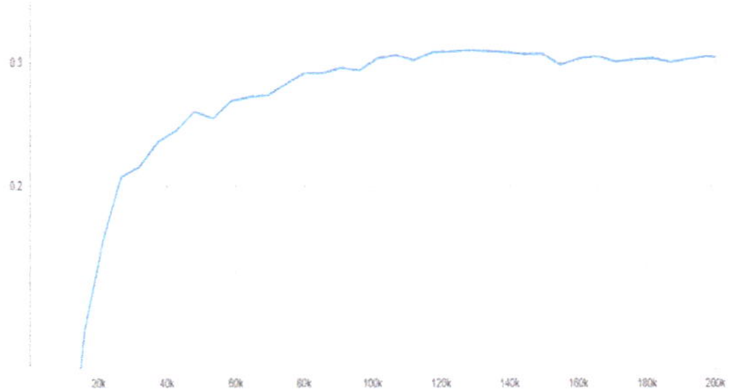

Figure 18. Overall mean Average Precision (mAP) at (0.5:0.95) IOU, X-axis: steps, Y-axis: mean average precision.

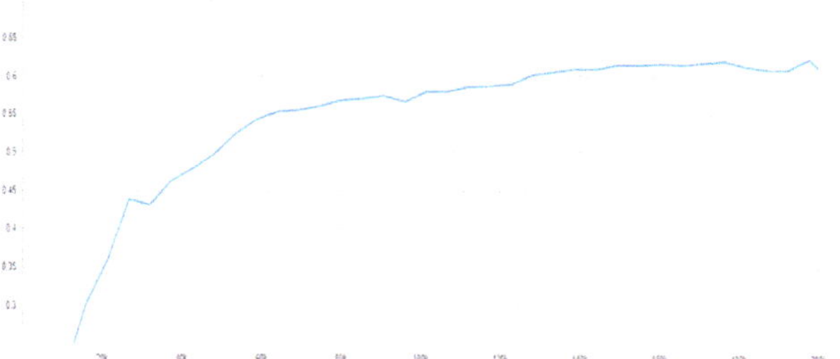

Figure 19. mAP at 0.5 IOU, X-axis: steps, Y-axis: mean average precision.

Steps on *X*-axis: One gradient update is considered as a training or evaluation step (iteration). It represents the number of batch-size images that are processed during a single iteration. For instance, we considered 200 images, and our batch size is set to 1 image in training configuration. That means one image was processed during one step, and gradients were updated once. Now the model takes 200 steps to complete the processing of the entire dataset. As the model processed the entire dataset, we say the model completed one epoch.

Y-axis: *Y*-axis in the following graphs corresponds to their respective losses and mAP of the model.

By observing Figures 20 and 21 for training and evaluation loss, we say the model is performing better as the two loss learning curves show a decreasing trend without huge variations. The X-axis represents the number of training and evaluation steps the model was trained with, while the Y-axis represents the training and evaluation loss recorded at each step respectively. Approximately at 150 k step, the model's total evaluation loss had reached a minimum of 0.61, and from after that, we observe a very slight increase in the loss values, this indicates the model was trying to overfit slowly and indicates it may not be feasible to train further. The training was stopped at 200 k, and the nearest checkpoint recorded at 200 k step was exported and inferencing was done. This performance was cross-verified with the pre-trained configuration, as it stated at 200 k steps were enough for the model to perform better. Although the model performed quite well on the new unknown images, there was scope in the optimization of the model by tuning the model's hyperparameters.

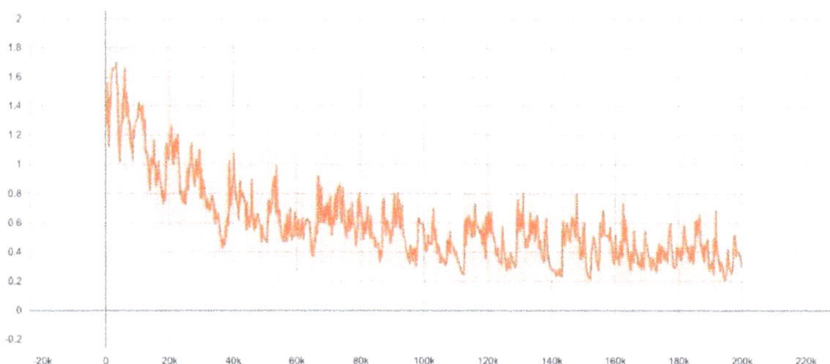

Figure 20. Training loss, X-axis: steps, Y-axis: training loss.

One of the observations from the analysis and experiments was: If the data considered was very low, data augmentation techniques such as flipping the images can help increase the mAP. The transfer learning technique was evaluated and justified that it can be quite helpful and quick when training on a new classification task instead of training the network from scratch or initializing with random weights. Hyperparameters such as the learning rate can be tuned further to increase mAP. Having high graphical processing units and performing a grid search or random search method can help us find optimal hyperparameters, but the process may be computationally expensive and time-consuming.

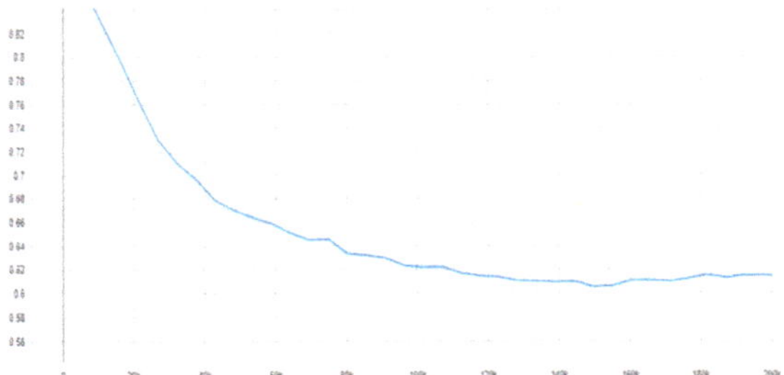

Figure 21. Total evaluation loss, X-axis: steps, Y-axis: evaluation loss.

In order to establish fully the notion that our model was finely well-tuned, the losses for the RPN network and the final classifier were also considered. By observing Figures 22 and 23, the decreasing trend of the box classifier classification and localization loss indicates that the final classifier is good at classifying and localizing the detected plant and weed objects. In Figures 22 and 23, X-axis represents the number of evaluation steps and Y-axis represents the classification loss and localization loss recorded at each step respectively.

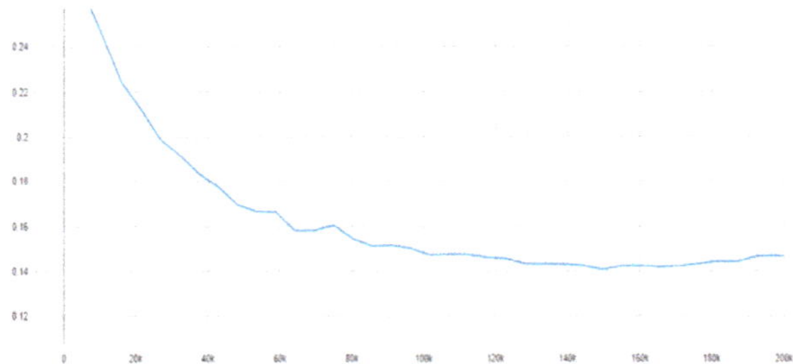

Figure 22. BoxClassifier: classification loss, X-axis: steps, Y-axis: classification loss.

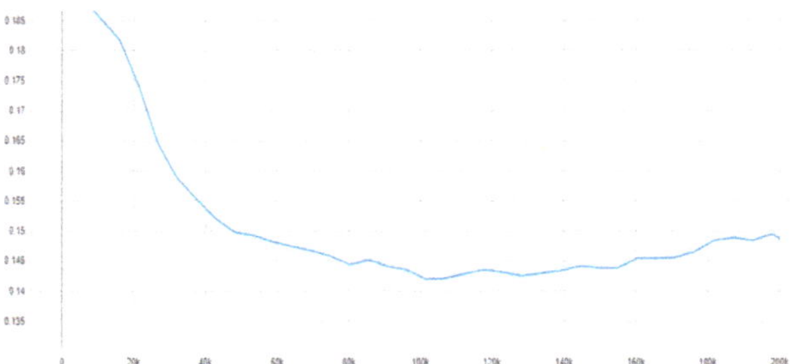

Figure 23. BoxClassifier: localisation loss, X-axis: steps, Y-axis: localisation loss.

The final ground truths and detections of various sizes of weeds and plants at the 200 k evaluation step are presented in Figures 24–26 corresponding to common dandelion (weed), garden cress and radish respectively. It is worth noticing the model gave predictions with good detection scores.

Figure 24. Left: detection (97%), **Right**: groundtruth—larger object.

Figure 25. Left: detection (98%), **Right**: groundtruth—smaller object.

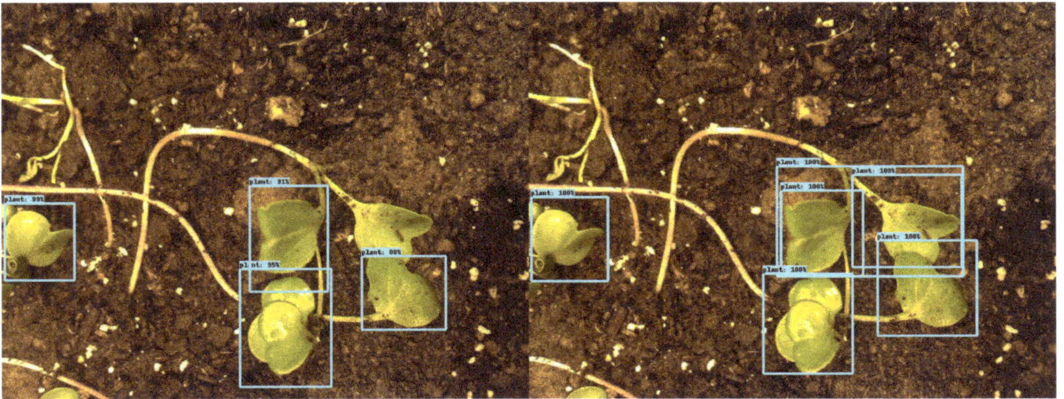

Figure 26. Left: detection (88–99%), **Right**: groundtruth—medium sized object.

4.1.2. Case 2: Configuration 2

In this case, we considered learning rate configuration 2. With this configuration, the training process was faster and achieved higher mAP values. With a lesser amount of steps in this configuration, the results obtained were not comparatively better than the results of the learning rate configuration 1. However, the model was overfitting and trying to memorize when trained for a longer time. It was one of the reasons the model achieved a higher overall mAP of 34.82% at (0.5:0.95) IOU (Figure 27) and mAP of 63% at 0.5 IOU threshold at 200 k step (Figure 28). The resultant graphs during the training and evaluation process are shown below. The graphs were generated at a smoothing value of 0.6 for showing the overall trend of the training and evaluation process.

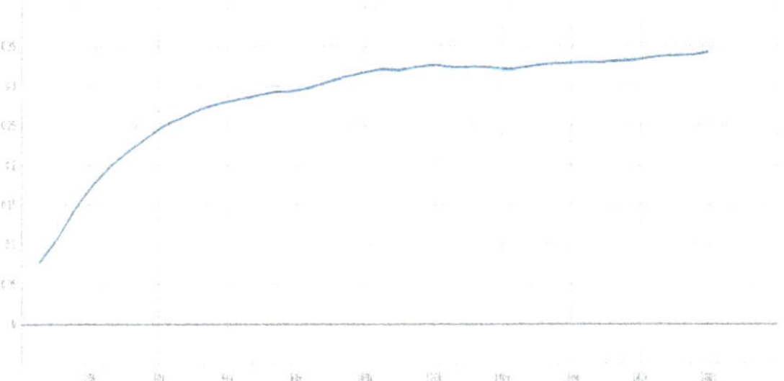

Figure 27. Overall mAP at (0.5:0.95) IOU, X-axis: steps, Y-axis: mean average precision.

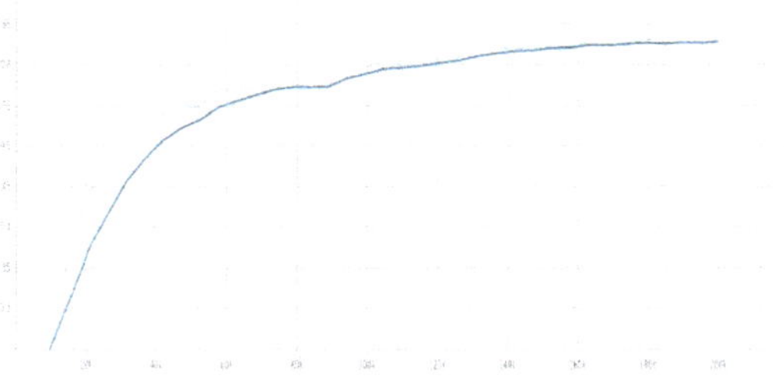

Figure 28. mAP at 0.5 IOU, X-axis: steps, Y-axis: mean average precision.

By observing the loss curves in Figures 29 and 30, the localization loss is increasing after the 60 k step. Ideally, all the loss curves should be in decreasing trend, and any large deviations of any loss are considered not suitable for generalization. Considering that, in this case, we should stop training at this point. Hence we can say the chosen learning rate hyperparameter may not be ideal for inferencing purposes compared to case 1 results. With that, case 1 results were considered for inferencing purposes, and the results are reported discussed in the following section.

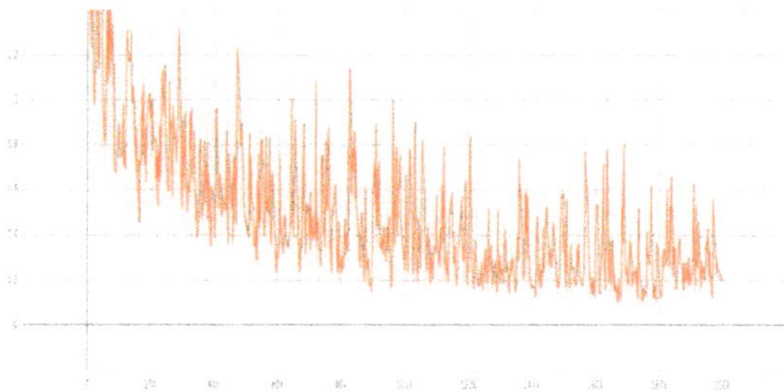

Figure 29. Training loss, X-axis: steps, Y-axis: training loss.

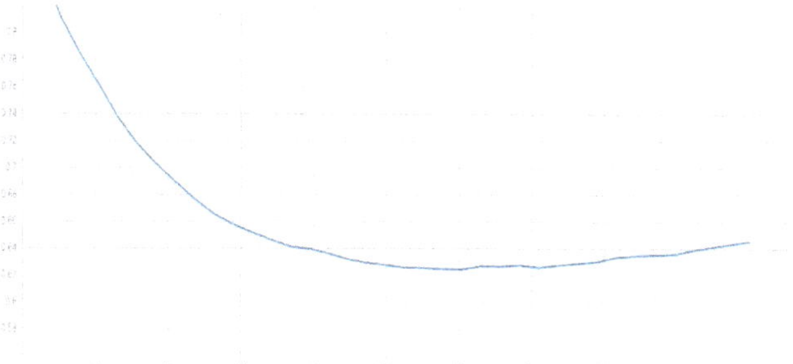

Figure 30. Total evaluation loss, X-axis: steps, Y-axis: evaluation loss.

4.2. Plant and Weed Identification

After the model was trained, it was used for inference on real-time data for plant and weed identification. For inferencing a new set of images, the model was saved and exported. For exporting the frozen graph, TensorFlow object detection API's inbuilt "export inference graph.py" script was used. The python script was modified accordingly to our task. The same training hardware setup and a Logitech stereo camera were used for real-time identification of plants and weeds. A completely new set of images was provided for predictions. The predicted output images are shown in Figures 31–33.

Figure 31. Plant and weed identification: detection of weed (92%).

Figure 32. Plant and weed identification in black soil.

Figure 33. Plant and weed identification in brown soil under artificial light.

4.3. Extracted Stem Positions

With the previously described stem estimation technique, we tested our method in real-time. We observed that the estimated stem positions were close enough (83–97%) to the original stem positions of the weed. The result of the extracted stem position in the image frame is presented in Figure 34.

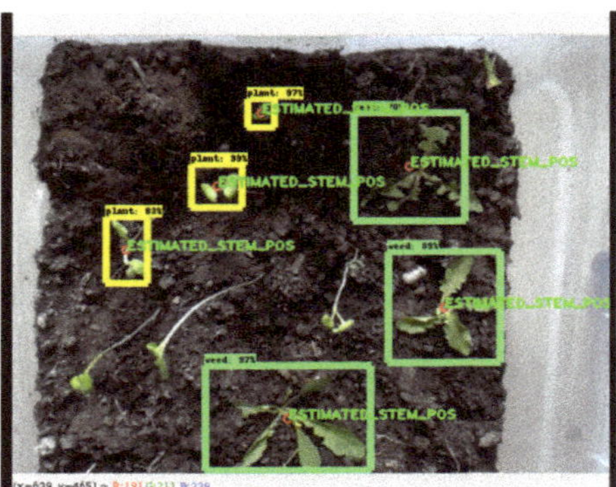

Figure 34. Estimated stem position of the weeds and plants using the trained robot.

4.4. Discussion of Results

The use of convolutional neural network-based models has been reported in different areas of agriculture, including disease identification, classification on the basis of ripeness of fruits, plant recognition using leaf images, and identification of weeds [35,101–104]. The application of convolutional neural networks (CNNs) using the transfer learning technique has also been reported in recent literature in the case of crop/fruit (age) classification. Perez-Perez et al. (2021) reported accuracy of 99.32% in the case of identification of different ripening stages of Medjoul dates [35]. This specific work points to the possibility of tuning the hyperparameters to achieve higher performance parameters with the proposed weeding robot as has been mentioned regarding the results of the current study. In recent years other studies have reported classification of plants through plant and leaf image recognition using convolution neural networks with accuracies up to 99% [103,104]. Sladojevic et al. (2016) reported the use of CNNs for disease recognition by leaf image classification with precision up to 98% [102].

With respect to the classification of different plant species, with the aim of site-specific weed management, Dyrmann et al. (2016) trained a CNN on a set of 10,413 images of 22 different plant species and were able to achieve a classification accuracy of up to 86% [101]. In the reported study, although the number of species classified was high, the images that were considered in the dataset were of plants at the same growth stage i.e., the seedling stage [101]. This makes the classification easier due to the same plant and leaf structure and hence higher accuracies are expected. However, in the case of weed removal applications, multiple weeding procedures might be needed at different times during a crop season, hence training a neural network with images of plants and weeds at different growth stages was done in the current study. The methodology is also reported in a recent study reported in literature where a crop field at two different growth stages was used to train the neural network, achieving an accuracy of 99.48% [105]. The classification accuracies achieved in the current study hence fall in the range of accuracies found in various studies reported in recent literature. The current study adds further value to the research by reporting the mean Average Precision (mAP) of the object detection tasks performed by the trained model. The mAP is an important metric to evaluate object detection models including both classification and localization tasks. Table 2 gives an overview of three other studies on CNNs for plant/weed/fruit classification that have reported comparable results together with the current study.

Table 2. Comparison of studies with reported training of CNNs for plant classification and identification tasks.

Reference	Number of Species	Growth Stages	Number of Images (Dataset)	Highest Classification Accuracy	Object Detection: Mean Average Precision (mAP)
Perez-Perez et al. (2021)	1	Ripe and Unripe	1002	99.32%	n.a.
Dyrmann et al. (2016)	22	Seedling	10,413	86.2%	n.a.
Asad and Bais (2020)	2	Two	906	99.48%	n.a.
Current study	3	Multiple	200	Plant: 95% Weed: 99%	31%

5. Conclusions

The weed identifier robot is proposed as a non-chemical solution to the rampant problem of weed infestation in food crop farming systems. Research and implementation of a plant and weed identification system using deep learning and state-of-the-art object detection methods was done. Transfer learning technique was explored and the deep learning model was further analysed, evaluated and justified for better generalization. It was seen that deep learning architectures are much better than conventional machine learning architectures in terms of image identification and predictive performance. A simple unique stem estimation technique was proposed which extracted their positions in the image frame. Consequently, the paper also offers a high-level hardware and software design architecture proposal of a cost-effective autonomous weeding robot.

The developed plant and weed identification system was presented and tested on the real-world data and good confidence scores on classification and identification were achieved. It can be concluded that higher values of mAP could be achieved with more steps with the right hyperparameters. Real-time identification was done using a Logitech web camera and it was observed that the model was good at identifying and distinguishing between plants and weeds. The stem position estimation approach was tested and it was found that accuracies were directly dependent on the bounding box localization during identification. Based on our observation, we conclude that this technique also reduces the amount of computation when compared with other methods. In addition to building the prototype and validation studies, future work in this direction could include investigations on choosing a method to find the right hyperparameters for optimization of the identification function of the robot. Further studies could explore 3D position estimation methods to determine the position from the center of the identified weed in the 2D image frame to the real-world robot frame.

Author Contributions: D.P.B.N., did the main work in the research, through programming, experiments and implementation part of the work presented in this paper. T.M.S., did the main work in the writing and putting together the contents of this manuscript, in addition to supervising the experiments. R.O., ideated and supervised the research work and gave feedback during the course the research. All authors have read and agreed to the published version of the manuscript.

Funding: This research received no external funding.

Institutional Review Board Statement: Not applicable.

Informed Consent Statement: Not applicable.

Data Availability Statement: The data generated in this study are presented in the article. For any clarifications, please contact the corresponding author.

Acknowledgments: We acknowledge support for the Open Access fees by Hamburg University of Technology (TUHH) in the funding programme Open Access Publishing. We acknowledge support of Hamburg Open Online University (HOOU) for the grant to develop the prototype of this robot. We acknowledge the support the Institute of Reliability Engineering, TUHH for their logistical support in this research. We acknowledge the suggestions made by the editors and reviewers that led to vast improvements in the quality of the submitted manuscript.

Conflicts of Interest: The authors declare no conflict of interest.

References

1. Varma, P. Adoption of System of Rice Intensification under Information Constraints: An Analysis for India. *J. Dev. Stud.* **2018**, *54*, 1838–1857. [CrossRef]
2. Delmotte, S.; Tittonell, P.; Mouret, J.-C.; Hammond, R.; Lopez-Ridaura, S. On farm assessment of rice yield variability and productivity gaps between organic and conventional cropping systems under Mediterranean climate. *Eur. J. Agron.* **2011**, *35*, 223–236. [CrossRef]
3. Shennan, C.; Krupnik, T.J.; Baird, G.; Cohen, H.; Forbush, K.; Lovell, R.J.; Olimpi, E.M. Organic and Conventional Agriculture: A Useful Framing? *Annu. Rev. Environ. Resour.* **2017**, *42*, 317–346. [CrossRef]
4. John, A.; Fielding, M. Rice production constraints and "new" challenges for South Asian smallholders: Insights into de facto research priorities. *Agric. Food Secur.* **2014**, *3*, 1–16. [CrossRef]
5. Hazra, K.K.; Swain, D.K.; Bohra, A.; Singh, S.S.; Kumar, N.; Nath, C.P. Organic rice: Potential production strategies, challenges and prospects. *Org. Agric.* **2018**, *8*, 39–56. [CrossRef]
6. Sujaritha, M.; Annadurai, S.; Satheeshkumar, J.; Kowshik Sharan, S.; Mahesh, L. Weed detecting robot in sugarcane fields using fuzzy real time classifier. *Comput. Electron. Agric.* **2017**, *134*, 160–171. [CrossRef]
7. Zahm, S.H.; Ward, M.H. Pesticides and childhood cancer. *Environ. Health Perspect.* **1998**, *106*, 893–908. [CrossRef]
8. Chitra, G.A.; Muraleedharan, V.R.; Swaminathan, T.; Veeraraghavan, D. Use of pesticides and its impact on health of farmers in south India. *Int. J. Occup. Environ. Health* **2006**, *12*, 228–233. [CrossRef]
9. Wilson, C. Environmental and human costs of commercial agricultural production in South Asia. *Int. J. Soc. Econ.* **2000**, *27*, 816–846. [CrossRef]
10. Uphoff, N. SRI: An agroecological strategy to meet multiple objectives with reduced reliance on inputs. *Agroecol. Sustain. Food Syst.* **2017**, *41*, 825–854. [CrossRef]
11. Wayayok, A.; Soom, M.A.M.; Abdan, K.; Mohammed, U. Impact of Mulch on Weed Infestation in System of Rice Intensification (SRI) Farming. *Agric. Agric. Sci. Procedia* **2014**, *2*, 353–360. [CrossRef]
12. Krupnik, T.J.; Rodenburg, J.; Haden, V.R.; Mbaye, D.; Shennan, C. Genotypic trade-offs between water productivity and weed competition under the System of Rice Intensification in the Sahel. *Agric. Water Manag.* **2012**, *115*, 156–166. [CrossRef]
13. RCEP. *Royal Commission for Environmental Pollution 1979 Seventh Report. Agriculture and Pollution*; RCEP: London, UK, 1979.
14. Moss, B. Water pollution by agriculture. *Philos. Trans. R. Soc. B Biol. Sci.* **2008**, *363*, 659–666. [CrossRef] [PubMed]
15. James, C.; Fisher, J.; Russell, V.; Collings, S.; Moss, B. Nitrate availability and hydrophyte species richness in shallow lakes. *Freshw. Biol.* **2005**, *50*, 1049–1063. [CrossRef]
16. Mehaffey, M.H.; Nash, M.S.; Wade, T.G.; Ebert, D.W.; Jones, K.B.; Rager, A. Linking land cover and water quality in New York City's water supply watersheds. *Environ. Monit. Assess.* **2005**, *107*, 29–44. [CrossRef]
17. Sala, O.E.; Chapin, F.S.; Armesto, J.J.; Berlow, E.; Bloomfield, J.; Dirzo, R.; Huber-Sanwald, E.; Huenneke, L.F.; Jackson, R.B.; Kinzig, A.; et al. Global biodiversity scenarios for the year 2100. *Science* **2000**, *287*, 1770–1774. [CrossRef]
18. Pimentel, D.; Pimentel, M. Comment: Adverse environmental consequences of the Green Revolution. In *Resources, Environment and Population-Present Knowledge, Future Options, Population and Development Review*; Davis, K., Bernstam, M., Eds.; Oxford University Press: Oxford, UK, 1991.

19. Pimentel, D.; Acquay, H.; Biltonen, M.; Rice, P.; Silva, M.; Nelson, J.; Lipner, V.; Horowitz, A.; Amore, M.D. Environmental and Economic Costs of Pesticide Use. *Am. Inst. Biol. Sci.* **1992**, *42*, 750–760. [CrossRef]

20. Orlando, F.; Alali, S.; Vaglia, V.; Pagliarino, E.; Bacenetti, J.; Bocchi, S.; Bocchi, S. Participatory approach for developing knowledge on organic rice farming: Management strategies and productive performance. *Agric. Syst.* **2020**, *178*, 102739. [CrossRef]

21. Barker, R.; Herdt, R.W.; Rose, H. *The Rice Economy of Asia*; The Johns Hopkins University Press: Baltimore, MD, USA, 1985.

22. FAO. *FAO Production Yearbooks (1961–1988)*; FAO Statistics Series; FAO: Rome, Italy, 1988.

23. Chen, Z.; Shah, T.M. *An Introduction to the Global Soil Status*; RUVIVAL Publication Series; Schaldach, R., Otterpohl, R., Eds.; Hamburg University of Technology: Hamburg, Germany, 2019; Volume 5, pp. 7–17.

24. Kopittke, P.M.; Menzies, N.W.; Wang, P.; McKenna, B.A.; Lombi, E. Soil and the intensification of agriculture for global food security. *Environ. Int.* **2019**, *132*, 105078. [CrossRef]

25. Nawaz, A.; Farooq, M. Weed management in resource conservation production systems in Pakistan. *Crop Prot.* **2016**, *85*, 89–103. [CrossRef]

26. Sreekanth, M.; Hakeem, A.H.; Peer, Q.J.A.; Rashid, I.; Farooq, F. Adoption of Recommended Package of Practices by Rice Growers in District Baramulla. *J. Appl. Nat. Sci.* **2019**, *11*, 188–192. [CrossRef]

27. Holt-Giménez, E.; Altieri, M.A. Agroecology, food sovereignty, and the new green revolution. *Agroecol. Sustain. Food Syst.* **2013**, *37*, 90–102. [CrossRef]

28. Wezel, A.; Bellon, S.; Doré, T.; Francis, C.; Vallod, D.; David, C. Agroecology as a science, a movement and a practice. A review. *Agron. Sustain. Dev.* **2009**, *29*, 503–515. [CrossRef]

29. Bajwa, A.A.; Mahajan, G.; Chauhan, B.S. Nonconventional weed management strategies for modern agriculture. *Weed Sci.* **2015**, *63*, 723–747. [CrossRef]

30. Zhang, C.; Kovacs, J.M. The application of small unmanned aerial systems for precision agriculture: A review. *Precis. Agric.* **2012**, *13*, 693–712. [CrossRef]

31. Lamb, D.W.; Weedon, M. Evaluating the accuracy of mapping weeds in fallow fields using airborne digital imaging: Panicum effusum in oilseed rape stubble. *Weed Res.* **1998**, *38*, 443–451. [CrossRef]

32. Medlin, C.R.; Shaw, D.R. Economic comparison of broadcast and site-specific herbicide applications in nontransgenic and glyphosate-tolerant Glycine max. *Weed Sci.* **2000**, *48*, 653–661. [CrossRef]

33. Huang, Y.; Reddy, K.N.; Fletcher, R.S.; Pennington, D. UAV low-altitude remote sensing for precision weed management. *Weed Technol.* **2018**, *32*, 2–6. [CrossRef]

34. Freeman, P.K.; Freeland, R.S. Agricultural UAVs in the US: Potential, policy, and hype. *Remote Sens. Appl. Soc. Environ.* **2015**, *2*, 35–43.

35. Pérez-Pérez, B.D.; García Vázquez, J.P.; Salomón-Torres, R. Evaluation of Convolutional Neural Networks' Hyperparameters with Transfer Learning to Determine Sorting of Ripe Medjool Dates. *Agriculture* **2021**, *11*, 115. [CrossRef]

36. Graeub, B.E.; Chappell, M.J.; Wittman, H.; Ledermann, S.; Kerr, R.B.; Gemmill-Herren, B. The State of Family Farms in the World. *World Dev.* **2016**, *87*, 1–15. [CrossRef]

37. Wezel, A.; Casagrande, M.; Celette, F.; Vian, J.F.; Ferrer, A.; Peigné, J. Agroecological practices for sustainable agriculture. A review. *Agron. Sustain. Dev.* **2014**, *34*, 1–20. [CrossRef]

38. Harker, K.N.; O'Donovan, J.T. Recent Weed Control, Weed Management, and Integrated Weed Management. *Weed Technol.* **2013**, *27*, 1–11. [CrossRef]

39. Melander, B.; Rasmussen, I.A.; Bàrberi, P. Integrating physical and cultural methods of weed control—Examples from European research. *Weed Sci.* **2005**, *53*, 369–381. [CrossRef]

40. Benbrook, C.M. Trends in glyphosate herbicide use in the United States and globally. *Environ. Sci. Eur.* **2016**, *28*, 3. [CrossRef] [PubMed]

41. Helander, M.; Saloniemi, I.; Saikkonen, K. Glyphosate in northern ecosystems. *Trends Plant Sci.* **2012**, *17*, 569–574. [CrossRef]

42. IARC. *IARC Monographs Volume 112: Evaluation of Five Organophosphate Insecticides and Herbicides*; IARC: Lyon, France, 2017.

43. Jayasumana, C.; Paranagama, P.; Agampodi, S.; Wijewardane, C.; Gunatilake, S.; Siribaddana, S. Drinking well water and occupational exposure to Herbicides is associated with chronic kidney disease, in Padavi-Sripura, Sri Lanka. *Environ. Health* **2015**, *14*, 6. [CrossRef]

44. Monarch Butterfles: The Problem with Herbicides. Available online: https://www.sciencedaily.com/releases/2017/05/170517143600.htm (accessed on 23 January 2021).

45. Motta, E.V.S.; Raymann, K.; Moran, N.A. Glyphosate perturbs the gut microbiota of honey bees. *Proc. Natl. Acad. Sci. USA* **2018**, *115*, 10305–10310. [CrossRef]

46. Kanissery, R.; Gairhe, B.; Kadyampakeni, D.; Batuman, O.; Alferez, F. Glyphosate: Its environmental persistence and impact on crop health and nutrition. *Plants* **2019**, *8*, 499. [CrossRef] [PubMed]

47. Murphy, K.P. *Machine Learning: A Probabilistic Perspective*; MIT Press: Cambridge, MA, USA, 2012; ISBN 978-0-262-01802-9.

48. Goodfellow, I.; Bengio, Y.; Courville, A. *Deep Learning*; MIT Press: Cambridge, MA, USA, 2016; ISBN 978-0262035613.

49. Brownlee, J. What Is Deep Learning? Available online: https://machinelearningmastery.com/what-is-deep-learning/ (accessed on 23 January 2021).

50. Liakos, K.; Busato, P.; Moshou, D.; Pearson, S.; Bochtis, D. Machine Learning in Agriculture: A Review. *Sensors* **2018**, *18*, 2674. [CrossRef]

51. Mohanty, S.P.; Hughes, D.P.; Salathé, M. Using deep learning for image-based plant disease detection. *Front. Plant Sci.* **2016**, *7*, 1–10. [CrossRef] [PubMed]
52. Narayanasamy, P. *Biological Management of Diseases of Crops*; Springer: Dordrecht, The Netherlands, 2013; ISBN 978-94-007-6379-1.
53. Saleem, M.H.; Potgieter, J.; Arif, K.M. Mahmood Arif Plant Disease Detection and Classification by Deep Learning. *Plants* **2019**, *8*, 468. [CrossRef] [PubMed]
54. Fuentes, A.; Yoon, S.; Kim, S.; Park, D. A Robust Deep-Learning-Based Detector for Real-Time Tomato Plant Diseases and Pests Recognition. *Sensors* **2017**, *17*, 2022. [CrossRef] [PubMed]
55. Raun, W.R.; Solie, J.B.; Johnson, G.V.; Stone, M.L.; Lukina, E.V.; Thomason, W.E.; Schepers, J.S. In-season prediction of potential grain yield in winter wheat using canopy reflectance. *Agron. J.* **2001**, *93*, 131–138. [CrossRef]
56. Filippi, P.; Jones, E.J.; Wimalathunge, N.S.; Somarathna, P.D.S.N.; Pozza, L.E.; Ugbaje, S.U.; Jephcott, T.G.; Paterson, S.E.; Whelan, B.M.; Bishop, T.F.A. An approach to forecast grain crop yield using multi-layered, multi-farm data sets and machine learning. *Precis. Agric.* **2019**, *20*, 1015–1029. [CrossRef]
57. Ashapure, A.; Oh, S.; Marconi, T.G.; Chang, A.; Jung, J.; Landivar, J.; Enciso, J. Unmanned aerial system based tomato yield estimation using machine learning. In *Proceedings Volume 11008, Autonomous Air and Ground Sensing Systems for Agricultural Optimization and Phenotyping IV*; SPIE: Baltimore, MD, USA, 2019; p. 22. [CrossRef]
58. Khaki, S.; Wang, L.; Archontoulis, S.V. A CNN-RNN Framework for Crop Yield Prediction. *Front. Plant Sci.* **2020**, *10*, 1–14. [CrossRef]
59. Russello, H. Convolutional Neural Networks for Crop Yield Prediction Using Satellite Images. Master's Thesis, University of Amsterdam, Amsterdam, The Netherland, 2018.
60. Sun, Y.; Liu, Y.; Wang, G.; Zhang, H. Deep Learning for Plant Identification in Natural Environment. *Comput. Intell. Neurosci.* **2017**, *2017*, 1–6. [CrossRef]
61. Du, J.-X.; Wang, X.-F.; Zhang, G.-J. Leaf shape based plant species recognition. *Appl. Math. Comput.* **2007**, *185*, 883–893. [CrossRef]
62. Grinblat, G.L.; Uzal, L.C.; Larese, M.G.; Granitto, P.M. Deep learning for plant identification using vein morphological patterns. *Comput. Electron. Agric.* **2016**, *127*, 418–424. [CrossRef]
63. Wu, S.G.; Bao, F.S.; Xu, E.Y.; Wang, Y.-X.; Chang, Y.-F.; Xiang, Q.-L. A Leaf Recognition Algorithm for Plant Classification Using Probabilistic Neural Network. In Proceedings of the 2007 IEEE International Symposium on Signal Processing and Information Technology, Giza, Egypt, 15–18 December 2007; pp. 11–16. [CrossRef]
64. Goëau, H.; Bonnet, P.; Baki, V.; Barbe, J.; Amap, U.M.R.; Carré, J.; Barthélémy, D. Pl@ntNet Mobile App. In Proceedings of the 21st ACM international conference on Multimedia, Barcelona, Spain, 21–25 October 2013; pp. 423–424.
65. Wäldchen, J.; Mäder, P. Machine learning for image based species identification. *Methods Ecol. Evol.* **2018**, *9*, 2216–2225. [CrossRef]
66. Liebman, M.; Baraibar, B.; Buckley, Y.; Childs, D.; Christensen, S.; Cousens, R.; Eizenberg, H.; Heijting, S.; Loddo, D.; Merotto, A.; et al. Ecologically sustainable weed management: How do we get from proof-of-concept to adoption? *Ecol. Appl.* **2016**, *26*, 1352–1369. [CrossRef]
67. Farooq, A.; Jia, X.; Hu, J.; Zhou, J. Knowledge Transfer via Convolution Neural Networks for Multi-Resolution Lawn Weed Classification. In Proceedings of the 2019 10th Workshop on Hyperspectral Imaging and Signal Processing: Evolution in Remote Sensing (WHISPERS), Amsterdam, The Netherlands, 24–26 September 2019; Volume 2019.
68. Dadashzadeh, M.; Abbaspour, G.Y.; Mesri, G.T.; Sabzi, S.; Hernandez-Hernandez, J.L.; Hernandez-Hernandez, M.; Arribas, J.I. Weed Classification for Site-Specific Weed. *Plants* **2020**, *9*, 559. [CrossRef]
69. Ashqar, B.A.; Abu-Nasser, B.S.; Abu-Naser, S.S. Plant Seedlings Classification Using Deep Learning. *Int. J. Acad. Inf. Syst. Res.* **2019**, *46*, 745–749.
70. Smith, L.N.; Byrne, A.; Hansen, M.F.; Zhang, W.; Smith, M.L. Weed classification in grasslands using convolutional neural networks. *Int. Soc. Opt. Photonics* **2019**, *11139*, 1113919. [CrossRef]
71. Simon, H. *Neural Networks: A Comprehensive Foundation*; McMaster University: Hamilton, ON, Canada, 2005; p. 823.
72. Park, S.H. Artificial Intelligence in Medicine: Beginner's Guide. *J. Korena Soc Radiol.* **2018**, *78*, 301–308. [CrossRef]
73. Chollet, F. *Deep Learning with Python*; Manning: New York, NY, USA, 2018; Volume 361.
74. Convolutional Neural Networks (CNNs/ConvNets). Available online: https://cs231n.github.io/convolutional-networks/ (accessed on 23 January 2021).
75. Brownlee, J. A Gentle Introduction to Object Recognition with Deep Learning. Available online: https://machinelearningmastery.com/object-recognition-with-deep-learning/ (accessed on 23 January 2021).
76. Ren, S.; He, K.; Girshick, R.; Sun, J. Faster R-CNN: Towards Real-Time Object Detection with Region Proposal Networks. *IEEE Trans. Pattern Anal. Mach. Intell.* **2017**, *39*, 1137–1149. [CrossRef] [PubMed]
77. Girshick, R. Fast r-cnn. In Proceedings of the IEEE International Conference on Computer Vision, Santiago, Chile, 11–18 December 2015; pp. 1440–1448.
78. He, K.; Gkioxari, G.; Dollár, P.; Girshick, R. Mask R-CNN. *IEEE Trans. Pattern Anal. Mach. Intell.* **2020**, *42*, 386–397. [CrossRef]
79. Liu, W.; Anguelov, D.; Erhan, D.; Szegedy, C.; Reed, S.; Fu, C.-Y.; Berg, A.C. SSD: Single Shot MultiBox Detector. In *ECCV 2016: Computer Vision–ECCV 2016*; Lecture Notes in Computer Science; Springer: Cham, Germany, 2016; Volume 9905, pp. 21–37. ISBN 9783319464473. [CrossRef]
80. Redmon, J.; Divvala, S.; Girshick, R.; Farhadi, A. You Only Look Once: Unified, Real-Time Object Detection. In Proceedings of the 2016 IEEE Conference on Computer Vision and Pattern Recognition (CVPR), Las Vegas, NV, USA, 27–30 June 2016; Volume 2016, pp. 779–788.

81. Girshick, R.; Donahue, J.; Darrell, T.; Malik, J. Rich feature hierarchies for accurate object detection and semantic segmentation. In Proceedings of the IEEE Conference on Computer Vision and Pattern Recognition, Columbus, OH, USA, 23–28 June 2014; pp. 580–587.
82. Weng, L. Object Detection for Dummies Part 3: R-CNN Family. Available online: https://lilianweng.github.io/lil-log/2017/12/31/object-recognition-for-dummies-part-3.html (accessed on 23 January 2021).
83. Olivas, E.S.; Guerrero, J.D.M.; Martinez-Sober, M.; Magdalena-Benedito, J.R.; Serrano, L. *Handbook of Research on machine Learning Applications and Trends: Algorithms, Methods, and Techniques*; IGI Global: Hershey, PA, USA, 2009; ISBN 1605667676.
84. Kaya, A.; Keceli, A.S.; Catal, C.; Yalic, H.Y.; Temucin, H.; Tekinerdogan, B. Analysis of transfer learning for deep neural network based plant classification models. *Comput. Electron. Agric.* **2019**, *158*, 20–29. [CrossRef]
85. Williams, J.; Tadesse, A.; Sam, T.; Sun, H.; Montanez, G.D. Limits of Transfer Learning. In Proceedings of the International Conference on Machine Learning, Optimization, and Data Science, Siena, Italy, 19–23 July 2020; Springer: Berlin/Heidelberg, Germany, 2020; pp. 382–393.
86. Pan, S.J.; Yang, Q. A survey on transfer learning. *IEEE Trans. Knowl. Data Eng.* **2009**, *22*, 1345–1359. [CrossRef]
87. Zhao, W. Research on the deep learning of the small sample data based on transfer learning. *AIP Conf. Proc.* **2017**, *1864*, 20018.
88. Onshape. Available online: https://www.onshape.com/en/ (accessed on 23 January 2021).
89. Pandilov, Z.; Dukovski, V. Comparison of the characteristics between serial and parallel robots. *Acta Tech. Corviniensis-Bull. Eng.* **2014**, *7*, 143.
90. Wu, L.; Zhao, R.; Li, Y.; Chen, Y.-H. Optimal Design of Adaptive Robust Control for the Delta Robot with Uncertainty: Fuzzy Set-Based Approach. *Appl. Sci.* **2020**, *10*, 3472. [CrossRef]
91. Pimentel, D. Environmental and Economic Costs of the Application of Pesticides Primarily in the United States. *Environ. Dev. Sustain.* **2005**, *7*, 229–252. [CrossRef]
92. Siemens, M.C. Robotic weed control. In Proceedings of the California Weed Science Society, Monterey, CA, USA, 23 June 2014; Volume 66, pp. 76–80.
93. Ecorobotix. Available online: https://www.ecorobotix.com/en/autonomous-robot-weeder/ (accessed on 9 February 2021).
94. Pilz, K.H.; Feichter, S. How robots will revolutionize agriculture. In Proceedings of the 2017 European Conference on Educational Robotics, Sofia, Bulgaria, 24–28 April 2017.
95. Schnitkey, G. Historic Fertilizer, Seed, and Chemical Costs with 2019 Projections. *Farmdoc Daily*, 5 June 2018; 102.
96. ROS Documentation. Available online: http://wiki.ros.org/Documentation (accessed on 23 January 2021).
97. Raspberry Pi. Available online: https://www.raspberrypi.org/products/ (accessed on 23 January 2021).
98. Coral Dev Board. Available online: https://coral.ai/products/dev-board/ (accessed on 23 January 2021).
99. Embedded Systems for Next-Generation Autonmous Machines. Available online: https://www.nvidia.com/de-de/autonomous-machines/embedded-systems/ (accessed on 23 January 2021).
100. Lottes, P.; Behley, J.; Chebrolu, N.; Milioto, A.; Stachniss, C. Robust joint stem detection and crop-weed classification using image sequences for plant-specific treatment in precision farming. *J. F. Robot.* **2020**, *37*, 20–34. [CrossRef]
101. Dyrmann, M.; Karstoft, H.; Midtiby, H.S. Plant species classification using deep convolutional neural network. *Biosyst. Eng.* **2016**, *151*, 72–80. [CrossRef]
102. Sladojevic, S.; Arsenovic, M.; Anderla, A.; Culibrk, D.; Stefanovic, D. Deep Neural Networks Based Recognition of Plant Diseases by Leaf Image Classification. *Comput. Intell. Neurosci.* **2016**, *2016*, 3289801. [CrossRef] [PubMed]
103. Jeon, W.-S.; Rhee, S.-Y. Plant Leaf Recognition Using a Convolution Neural Network. *Int. J. Fuzzy Log. Intell. Syst.* **2017**, *17*, 26–34. [CrossRef]
104. Abdullahi, H.S.; Sheriff, R.; Mahieddine, F. Convolution neural network in precision agriculture for plant image recognition and classification. In Proceedings of the 2017 Seventh International Conference on Innovative Computing Technology (INTECH), Luton, UK, 16–18 August 2017; Volume 10.
105. Asad, M.H.; Bais, A. Weed detection in canola fields using maximum likelihood classification and deep convolutional neural network. *Inf. Process. Agric.* **2020**, *7*, 535–545. [CrossRef]

 *agriculture*

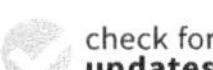

Article

Oil Palm Tree Detection and Health Classification on High-Resolution Imagery Using Deep Learning

Kanitta Yarak [1], Apichon Witayangkurn [2,*], Kunnaree Kritiyutanont [3], Chomchanok Arunplod [4] and Ryosuke Shibasaki [2]

[1] Department of Geoinformatics, Rambhai Barni Rajabhat University, Chanthaburi 22000, Thailand; kanitta.y@rbru.ac.th

[2] Center for Spatial Information Science, The University of Tokyo, Chiba 277-8568, Japan; shiba@csis.u-tokyo.ac.jp

[3] Department of Information and Communication Technologies, School of Engineering and Technology, Asian Institute of Technology, Pathumthani 12120, Thailand; kunnaree.ait@gmail.com

[4] Department of Geography, Faculty of Social Sciences, Srinakharinwirot University, Bangkok 10110, Thailand; chomchanok@g.swu.ac.th

* Correspondence: apichon@iis.u-tokyo.ac.jp

Abstract: Combining modern technology and agriculture is an important consideration for the effective management of oil palm trees. In this study, an alternative method for oil palm tree management is proposed by applying high-resolution imagery, combined with Faster-RCNN, for automatic detection and health classification of oil palm trees. This study used a total of 4172 bounding boxes of healthy and unhealthy palm trees, constructed from 2000 pixel × 2000 pixel images. Of the total dataset, 90% was used for training and 10% was prepared for testing using Resnet-50 and VGG-16. Three techniques were used to assess the models' performance: model training evaluation, evaluation using visual interpretation, and ground sampling inspections. The study identified three characteristics needed for detection and health classification: crown size, color, and density. The optimal altitude to capture images for detection and classification was determined to be 100 m, although the model showed satisfactory performance up to 140 m. For oil palm tree detection, healthy tree identification, and unhealthy tree identification, Resnet-50 obtained F1-scores of 95.09%, 92.07%, and 86.96%, respectively, with respect to visual interpretation ground truth and 97.67%, 95.30%, and 57.14%, respectively, with respect to ground sampling inspection ground truth. Resnet-50 yielded better F1-scores than VGG-16 in both evaluations. Therefore, the proposed method is well suited for the effective management of crops.

Keywords: high-resolution imagery; deep learning; oil palm tree; CNN; Faster-RCNN

Citation: Yarak, K.; Witayangkurn, A.; Kritiyutanont, K.; Arunplod, C.; Shibasaki, R. Oil Palm Tree Detection and Health Classification on High-Resolution Imagery Using Deep Learning. *Agriculture* **2021**, *11*, 183. https://doi.org/10.3390/agriculture11020183

Academic Editors: Sebastian Kujawa and Gniewko Niedbała

Received: 28 January 2021
Accepted: 20 February 2021
Published: 23 February 2021

Publisher's Note: MDPI stays neutral with regard to jurisdictional claims in published maps and institutional affiliations.

1. Introduction

Oil palm trees are one of Thailand's most essential economic crops considering it has the highest oil production when compared to other oil-producing plants such as soybean, peanut, sunflower, and rapeseed. Palm oil can be processed into various products such as cooking oil, soap, margarine, and sweetened condensed milk. In addition, it is also used as a raw material in the manufacturing of biodiesel and pulp. Oil palm trees grow well in tropical climates, which are often found in countries situated in equatorial regions. Thus, the oil palm is a crop that is widely cultivated by farmers in Southern Thailand.

Precision agriculture requires reliable data on the current situation at the right time. Therefore, the automated detection of oil palm trees and health disorder recognition is an alternative method for farmers to manage their resources using technology instead of a manual approach. The method also provides information on plant growth and health, which is especially useful to track the age and survival rate of plants that will contribute to the oil palm tree production in the future. Oil palm tree detection and enumeration are mostly performed using high-resolution imagery. For instance, many researchers have used high-resolution satellite images [1] and unmanned aerial vehicle (UAV) images [2]. UAVs have played a vital role in remote sensing in recent years as they can provide high-resolution images when there is no cloud cover. Users can set the altitude and time to fly. Aliero et al. used UAVs for automated counting of oil palm trees based on crown properties and the plants' response to radiation. Moreover, spatial analysis and morphological analysis were also used in their study [3]. Daliman et al. used Haar-based rectangular windows and support vector machines (SVMs) to detect oil palm trees on the WorldView-2 satellite imagery dataset [4]. Manandhar et al. presented a methodology for object detection with aerial imagery by applying shape feature characteristics for oil palm tree detection and counting. They used circular autocorrelation of the polar shape matrix to represent images as the shape feature and used a linear SVM to standardize and reduce the feature dimensions. Finally, they used local maximum detection on the spatial distribution of standardized features for oil palm tree detection [5].

Deep learning is one of the various machine learning procedures with a mechanism similar to that of the human brain. In addition, it is commonly applied to analyze visual imagery. Recently, much attention has been paid to this method and it has been applied in many fields, such as image recognition [6], handwriting recognition [7], medical, and healthcare [8]. Moreover, deep learning has also been applied to agricultural management to reduce production costs resulting in more effective agricultural production. For example, our method was used to detect and enumerate agricultural populations, including the classification of diseased plants. Cheang et al. proposed a system for the counting and positioning oil palm trees using a convolutional neural network (CNN) to classify the oil palm dataset on a high-resolution satellite image with a sliding window technique [9]. Li et al. proposed using deep learning to detect plants instead of manual detection methods. They used data from a manual count to train and improve the performance of the CNN system. Then, all samples were predicted on images using the sliding window technique [10]. Sladojevic et al. studied the development of plant disease patterns from leaf images using deep convolutional networks. The results of their study demonstrated the ability to distinguish diseased plants from healthy plants [11]. Mubin et al. used a geographic information system (GIS) and CNN named LeNet on WorldView-3 images for young and mature oil palm detection [12]. They used a training dataset with a mini-batch of size 20 and used GIS software to display and create maps of oil palm tree prediction.

There are various popular deep learning algorithms, such as recurrent neural networks (RNNs), long short-term memory networks (LSTMs), and CNNs. RNNs and LSTMs have similar capabilities and are widely used in time series problems/forecasting, but LSTM can be trained for tasks that require long-term memory. CNN is the deep neural network most commonly used in computer vision and object detection [13]. Many CNN models are available in the public repository. VGG-16 [14] and Resnet-50 [15] are examples of the CNN models commonly used for image classification. In recent years, a selective search based on regions was introduced to improve the speed and performance of a detector. The most common models include R-CNN, Fast R-CNN [16], and Faster R-CNN [17]. Faster R-CNN performed best among those three in terms of accuracy and detection speed.

The use of remote sensing in conjunction with these deep-learning techniques is common in the agricultural industry in many countries. Most research on oil palm tree detection uses satellite imagery and focuses solely on detection and counting. For example, Zheng et al. used Faster-RCNN, one of the most popular networks for object detection, to detect tree crowns from satellite images [18]. In contrast, in this study, high-resolution images from UAVs are used instead. Due to this approach, data surveys can be performed without time restrictions and under any cloud conditions. Our study focuses on oil palm health classification (healthy or unhealthy) rather than just detection and automatic counting as other research. Therefore, it is useful for modern precision agriculture, which focuses on reducing farmer work processes and unnecessary production costs.

This study proposes a technique to automatically detect and count oil palm trees and recognize oil palm health from high-resolution images using CNN with Faster-RCNN structure. The training data included over 4000 images of individual oil palm trees with healthy and unhealthy classes. We evaluated the model by conducting model training evaluation and comparing prediction results with visual and ground inspections. The model was also tested with images taken at different altitudes.

2. Materials and Methods

2.1. Data Used

Detection and health classification of oil palm trees by using the deep learning need to be performed using high-resolution images and require field survey data to study important physical characteristics of oil palm trees and evaluate the reliability of the results. Thailand is the 3rd largest producer of palm oil in the world [19]. Southern Thailand is the region with the most oil palm plantations. The study areas included three oil palm plantations, two of which were located in Surat Thani and the other one was located in Krabi, shown in Figure 1, with approximately 40 Rai (64,000 square meters) in each plantation area. Surat Thani and Krabi are the two provinces with the largest oil palm production, ranked first and second, respectively. The recommended practice for oil palm plantation is to allocate an area of 9 m × 9 m for each oil palm tree. This results in a regular pattern of oil palm positions within an area.

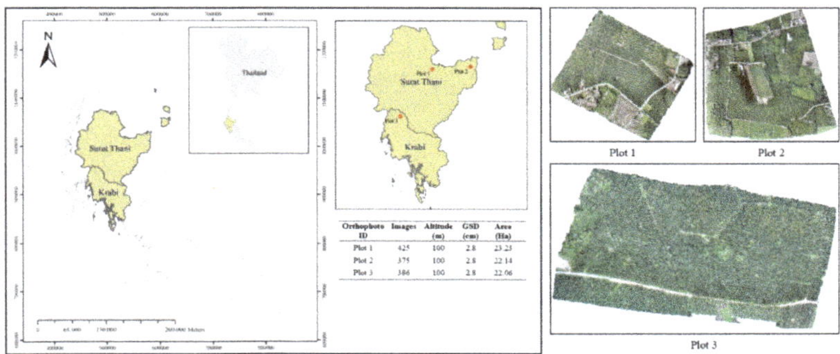

Orthophoto ID	Images	Altitude (m)	GSD (cm)	Area (Ha)
Plot 1	425	100	2.8	23.23
Plot 2	375	100	2.8	22.14
Plot 3	386	100	2.8	22.06

Figure 1. Location of study area.

The dataset used in this study consists of high-resolution RGB images taken from a UAV by a DJI Phantom 4 Pro camera with a 20 megapixel resolution using a D-Log color profile, as illustrated in Figure 2. All the images used were captured at an altitude of 100 m, and the ground sampling distance (GSD) was approximately 2.8 cm.

Moreover, field survey data were obtained by performing a health assessment using a paper survey and ground photography. Field observation data such as crown size, crown density, crown color, and examples of problematic oil palm trees were used to study the correlation between oil palm trees in plantations and UAV images.

2.2. Methodology

The methodology section of this paper presents the research process used during this study. First, data collection is discussed, together with the data preparation process used to prepare the high-resolution image dataset for use as training and testing data. Second, the method used to develop training data and a model for counting and health classification is addressed. Last, we covered model evaluation, which included three techniques: model training inspection, visual inspection, and ground inspection.

Figure 2. Unmanned aerial vehicle (UAV) image of the study area.

2.2.1. Data Collection and Data Preparation

The study area survey was critical given that the survey data would be used to check and monitor physical characteristic changes and oil palm tree health. The data collection process focused on three obvious external physical characteristics that could easily be observed: color (the level of green in the canopy), crown density, and crown size. Moreover, other general information such as age, height, diameter at breast height (DBH), nutritional deficiency symptoms, diseases at various levels, and the approximate amount of production were recorded for every sample tree. All the data above could be used as indicators of oil palm health and abnormalities. A tree under stress was also classified as unhealthy. Figure 3 showed the example of unhealthy trees. A laser range finder collected information on height and crown size (measured along the *x*- and *y*-axes). The DBH was measured using a tape measure or rope with scale. A sample of surveyed oil palm trees was chosen using a systematic sampling method, sampling approximately 25% of the total area. The sample trees were chosen by selecting every fourth tree in a row, starting with the first tree of the row.

Figure 3. Examples of unhealthy trees with nutrient deficiencies and *Ganoderma*.

The ground truth labels were acquired through sample observations using the following criteria: crown size was determined by measuring the crown's radius, and the crown color was assessed by classifying the amount of green in the canopy at different heights (lower fronds, middle fronds, and upper fronds) into three levels. We also separated crown density into three levels: low, moderate, and high, as shown in Figure 4. The measurements and labeling were assessed in collaboration with agricultural officers specializing in oil palms. To obtain additional information, a sample leaf from the 17th frond was collected for nutrient inspection in the laboratory. However, the aim of this study is to emphasize external indicators and focus on large-scale measurements before considering other methods such as the use of a multispectral camera or spectroradiometer for leaf inspection. Thus, obscure health conditions in oil palms may be less well detected, and this one of the limitations of this study.

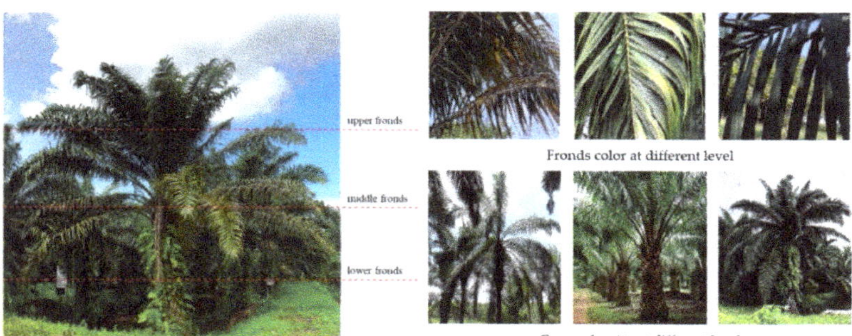

Figure 4. Examples of ground truth observations.

The preparation of the processed UAV images with red, green, and blue (RGB) bands began by converting the format of orthophotos from .TIF to .JPG. The image size used as data for training and testing had an extreme effect on memory and data processing. Thus, each image was divided into smaller images of similar sizes. Therefore, each image was split into equal-sized 2000 pixel × 2000 pixel images without taking image overlap into account, as shown in Figure 5.

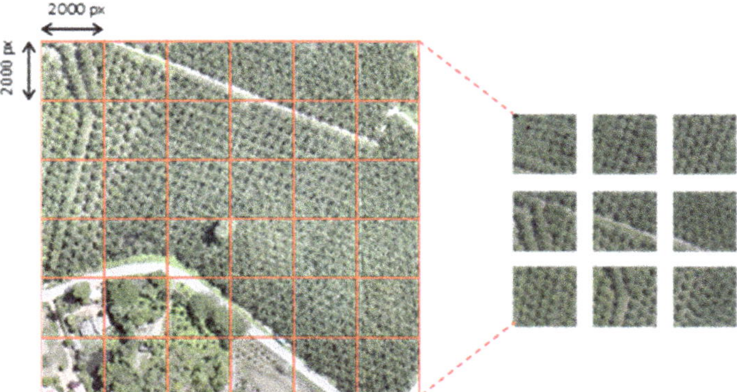

Figure 5. UAV images with 2000 pixel × 2000 pixel size.

2.2.2. Training Data Development and Oil Palm Tree Classification Model

In accordance with the development of training data, for automatic oil palm tree detection and counting based on deep learning techniques, this method required data that were already classified (labeled data). The data used for training were divided into two classes, i.e., healthy oil palm trees and unhealthy oil palm trees, as illustrated in Figure 6. In the train set, the class of healthy oil palm trees included healthy oil palm trees on images in various conditions, such as blurred images of healthy oil palm trees and healthy oil palm trees with incomplete canopies (however, more than 50% of the canopy is still present). It also included healthy oil palm trees that overlapped other objects, other plants, and other backgrounds. For the unhealthy oil palm tree class, the dataset included unhealthy oil palm trees on images in various conditions similar to that of the healthy dataset. To reduce the problem of detecting the same tree in multiple images, the bounding boxes were drawn around the regions that enclose more than 50% of the canopy.

(**a**) Healthy oil palm tree

(**b**) Unhealthy oil palm tree

Figure 6. Examples from the (**a**) healthy and (**b**) unhealthy datasets.

During the preparation of a training set, including healthy and unhealthy oil palm trees, the palm trees were separated from other objects that were not of interest by finding the Xmin, Ymin, Xmax, and Ymax values for each canopy boundary. These values were determined by using the program LabelImg and drawing a bounding box surrounding each canopy along with providing the class definition. The program generated the training dataset as the Pascal VOC dataset in .xml format. This format was then converted into a text file (.txt), with each line containing "filepath, Xmin, Ymin, Xmax, Ymax, class_name" before being used for processing. Additionally, the testing dataset was prepared the same as the training dataset, with the testing dataset accounting for testing was 10% of the total dataset.

According to Figure 7, the processing procedure for classifying and counting oil palm trees began with data collection and training data development. Then it followed with the selection of CNN architecture, including parameter optimization. The CNN architecture was then trained and tested using the dataset with Python3. Another criterion for architecture selection was the availability of the software, which enables people from multiple disciplines to benefit from the findings of this study.

As a result, Faster-RCNN was chosen as the underlying network architecture, and the two models selected for this research were VGG-16 and Resnet-50. These models are common options for Faster-RCNN and were mostly used as baseline models for further improvements. The models were implemented using the TensorFlow and Keras deep-learning libraries for Python3. Model training was performed on a PC running Windows 10 and equipped with a GeForce RTX 2080 Ti. The parameters were optimized using 1000 iterations per epoch with a batch size of 1, as supported by Faster RCNN. The training was halted at 39 epochs for the VGG-16 network and 40 epochs for the Resnet-50 network when no further increase in accuracy was noted.

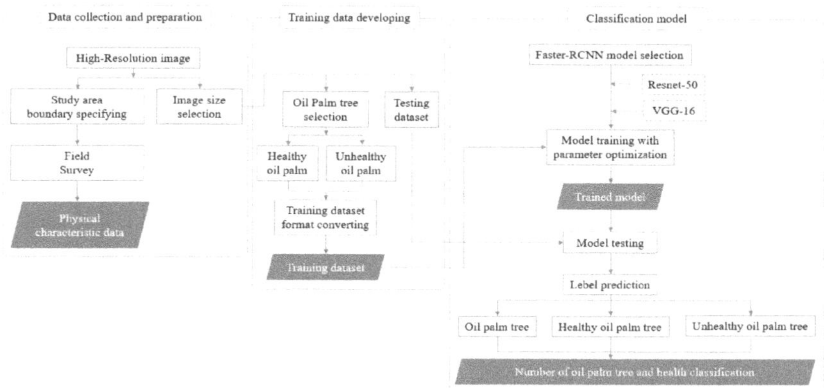

Figure 7. Methodology workflow.

2.2.3. Evaluating Model Performance

The evaluation of performance after processing was evaluated in three parts. A preliminary assessment of the effectiveness of the model was done by observing the accuracy and loss function values. After processing, classifier accuracy for bounding boxes from the region proposal network (RPN), four loss values, and elapsed times were compared for this research. Next, visual inspection, which evaluated the model's accuracy by comparing the number of oil palm trees predicted by the model to the number counted on UAV imagery. The performance evaluation of an object detection model commonly uses precision, recall, and F1-score without considering intersection over union (IoU), which is more suitable for image segmentation evaluation. Moreover, it takes a long time and is difficult to generate labeled training data for an IoU-based evaluation, especially for oil palm trees, because of the overlapping boundaries of the oil palms. Therefore, a confusion matrix was used to describe the achieved model classification showing true positive (TP), false negative (FN), false positive (FP), and true negative (TN). Then, the values from the confusion matrix were used to measure performance by calculating precision, recall, and F1-score defined by the following equations.

$$Precision = \frac{TP}{TP + FP} \tag{1}$$

$$Recall = \frac{TP}{TP + FN} \tag{2}$$

$$F1 - score = 2 \times \frac{Precision \times Recall}{recision + Recall} \tag{3}$$

The ground inspection was performed by comparing the predicted results from the CNN model with the physical characteristics of oil palm tree samples observed from surveying and ground photography. This method showed the consistency between the predicted results of the model and the oil palm trees in plantations. The result obtained from this evaluation was the ratio of all predicted oil palms to the total number of oil palms surveyed. Further, it compared the proportion of healthy and unhealthy oil palms predicted by the model to healthy and unhealthy oil palms from the survey.

3. Results and Discussion

3.1. Physical Characteristics and Data Preparation

The physical characteristics of oil palm trees are essential, as they indicate good or bad health. The health of oil palm trees on high-resolution imagery at vertical angles can be observed using three crucial characteristics: crown size, crown color, and crown density. The crown size relates to the age of the oil palm tree. Young oil palms (less than eight years) have a small crown size, becoming larger as they grow, and reaching full size at about eight years. Aside from the oil palm's age, the canopy's size can also be affected by exposure to *Ganoderma* disease. Concerning color, most healthy oil palm trees have a green canopy. In contrast, unhealthy oil palms tend to have different colored canopies ranging from yellowish green to brown, resulting from water or essential nutrient deficiency. Additionally, crown density is another important physical feature as the number of fronds can inform whether the oil palm tree is healthy or not. Specifically, healthy oil palm trees have very dense fronds and when viewed from a vertical angle it is almost impossible to see the ground below. However, problematic oil palm trees have fewer fronds; therefore, it is possible to see the ground below. Figure 8 presents a few examples of the important physical characteristics indicative of oil palm health.

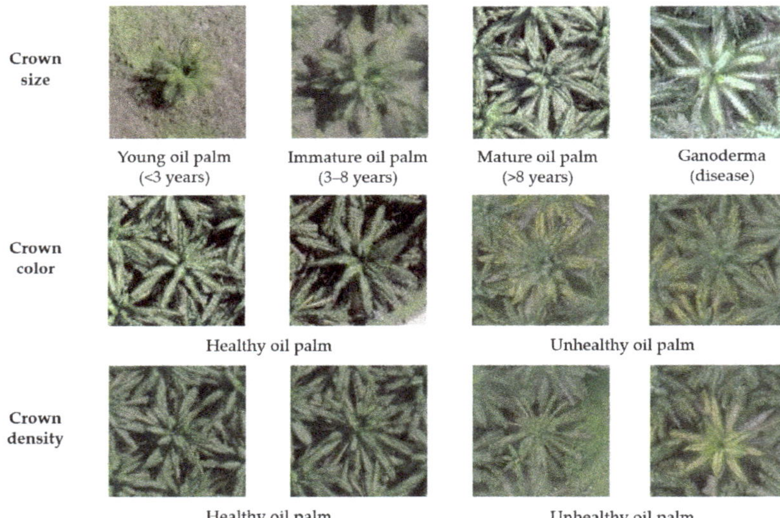

Figure 8. Examples of important physical characteristics of oil palm trees.

The general characteristics of healthy, mature oil palm trees are usually dark green leaves, about 18–25 fronds, and a diameter of approximately 7.5 m [20]. Nutritional deficiencies are most often the cause for the deterioration of oil palm health. Nutrients that play an essential role in changing the physical characteristics of oil palms are nitrogen, phosphorus, potassium, magnesium, and boron. The epidemic disease in the oil palms, called *Ganoderma*, is also a significant problem that stunts growth and reduces production.

In this study, observing the physical characteristics on the UAV images and surveying oil palms on plots suggested that most of the oil palm trees were healthy. The healthy palm trees in the images appeared with large crown sizes, had a dark green color, and a higher frond density than unhealthy oil palms. However, the classification of health according to significant external features was determined by all the characteristics mentioned above. For example, a young oil palm with a small crown size and low density could be a healthy oil palm. On the contrary, oil palms with large crown sizes and low density were classified as unhealthy.

Furthermore, the images oil palm trees that had differences in color, crown density, and crown size when compared to healthy oil palm trees were investigated further. The survey found that images showing a yellow canopy indicated an oil palm with nitrogen deficiency. Similarly, canopies with a greenish orange color indicated a palm with potassium deficiency. It was also found that most of the deficient palm trees were not only lacking in one nutrient but often lacked multiple nutrients. In the case of trees with severe nutritional deficiencies, reductions in crown density and size were noted. Additionally, some oil palm trees presented on the images with tiny crown sizes, low frond density, and colors of light green and yellow. When surveying and investigating these oil palm trees in the field, it was discovered that this resulted from the palm trees being affected by an advanced stage of *Ganoderma*. This disease causes leaves to dry out and to eventually drop down against the trunk, resulting in reduced crown size and density.

A dataset of high-resolution images from UAVs with visible bands (RGB) was used for training and testing on a deep learning model. The images used were images taken in May 2019 at an altitude of 100 m with a spatial resolution of approximately 2.8 cm. The images of each study area were converted into .JPG format before splitting them into equally sized images of 2000 pixels × 2000 pixels each. This was determined to be the optimal size to maximize memory usage and processing performance. The dataset comprised 133 images total, 116 were assigned to the training set and 17 to the test set.

3.2. Training Data Development and Model for Oil Palm Tree Classification

The training dataset contained the position of each oil palm tree, known as bounding boxes. From the 116 images in the training dataset, 3780 bounding boxes were created and used for training. The dataset comprised 3035 healthy oil palms and 745 unhealthy oil palms. The testing dataset contained 392 bounding boxes, with 325 healthy oil palms and 67 unhealthy oil palms sourced from 17 images. Table 1 shows a breakdown of the 4172 bounding boxes used in this study. Data augmentation was used to reduce the impact of the imbalanced data.

Table 1. The total amount of bounding boxes for the training and testing datasets used in this study.

Dataset	Healthy	Unhealthy	Total
Number of training bounding boxes	3035	745	3780
Number of testing bounding boxes	325	67	392

This study used the same dataset to train and test two different base models, namely Resnet-50 and VGG-16, both of which were based on the Faster-RCNN structure. The training and testing results were compared to assess the ability and efficiency of the model. Additionally, we conducted experiments to test the ability of automatic oil palm tree detection at different flying altitudes and settings. We flew the UAV at low altitudes of 50 m, 80 m, and 90 m. Image mosaicking was an issue because palm trees have a similar pattern, and lower attitude images mostly contained palm trees. This resulted in unmosaicking in many areas. The overlap and sidelap parameters were adjusted, but this did not solve the issue. In addition to the flights at these altitudes, we performed flights using different color profiles and at different times to determine the best lighting. The experimental results demonstrated that flying with a D-log color profile before 3:00 PM was a suitable scenario.

The experiment was performed using UAV images taken at altitudes of 100, 120, 140, 160, 180, and 200 m. The results are shown in Figure 9, and Table 2 summarizes the experimental results.

Table 2. Automatic detection of oil palm trees at various altitudes.

Altitude (m.)	Actual Oil Palm Tree	Detected Oil Palm Tree	Accuracy (%)
100	856	852	99.53
120	856	836	97.66
140	856	769	89.84
160	856	422	49.30
180	856	75	8.76
200	856	3	0.35

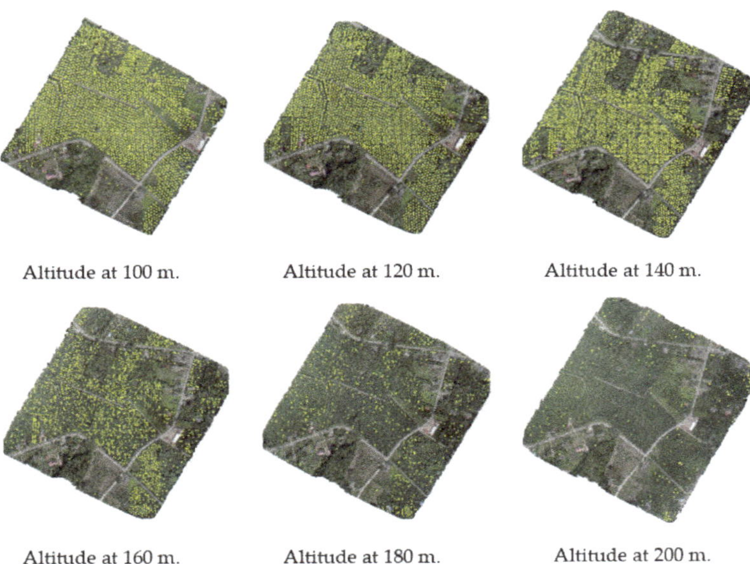

Altitude at 100 m.   Altitude at 120 m.   Altitude at 140 m.

Altitude at 160 m.   Altitude at 180 m.   Altitude at 200 m.

Figure 9. Oil palm tree detection at different flying altitudes.

Accuracy was evaluated by comparing the results with actual data derived from the visual interpretation of high-resolution images. Thus, this assessment emphasizes counting the number of oil palms. Table 2 indicates that oil palm trees are best detected on UAV images taken at a 100 m altitude, with accuracy measuring at 99.53%. Therefore, images taken at a height of 100 m were used in this study to detect and classify oil palm tree health. After the training, the results were tested using the test dataset that were prepared to be approximately 10% of the training set. The Resnet-50 network model predicted 331 healthy palms, and only 51 oil palm trees were predicted as unhealthy. Therefore, this model detected a total of 382 oil palm trees on 17 images. The VGG-16 network model predicted 268 healthy palms, and 103 oil palm trees were predicted as unhealthy. In summary, this model detected a total of 371 oil palm trees on 17 images. The prediction results of each image for the two models used are shown in Table 3. Example images of prediction results from Resnet-50 and VGG-16 are shown in Figure 10.

Table 4 shows the comparison of the actual counts and each model's predictions. The results show that both models were useful in terms of oil palm tree detection (healthy and unhealthy) as the number of predicted palm trees is very close to the actual count, with the Resnet-50 model performing slightly better. When comparing health classification, the number of healthy and unhealthy oil palms detected by the Resnet-50 model was more accurate than those of the VGG-16 network. Further, the accuracy and precision ratios were calculated and will be discussed in the section evaluating model performance.

The initial assessment results also indicated that errors occurred because the oil palm trees were obstructed by other tree canopies. Coconut and other trees with similar physical characteristics to those of oil palm trees were another cause of errors. The number of detectable and undetectable oil palm trees was affected by palms located on the edges of an image, where some parts of the crown area extended across two images. Moreover, the crown size, especially in young oil palms with small crown sizes, was another cause of detection error. This was due to the small number of young oil palm samples with small crowns, as the study focused on large number of mature oil palms that appear in the study area. However, the errors related to crown size could be addressed to improve performance by increasing the number of young oil palms in the training data. Figure 11 illustrates examples of misclassification cases.

Table 3. The predictable number from Resnet-50 and VGG-16.

Image	Predictable Number from Resnet-50 Network			Predictable Number from VGG-16 Network		
	Oil Palm	Healthy	Unhealthy	Oil Palm	Healthy	Unhealthy
1	42	42	0	40	40	0
2	41	41	0	41	41	0
3	39	39	0	36	36	0
4	13	13	0	9	7	2
5	5	5	0	4	3	1
6	19	19	0	22	17	5
7	11	11	0	18	12	6
8	9	8	1	9	7	2
9	45	21	24	41	8	33
10	27	17	10	26	14	12
11	1	1	0	8	3	5
12	40	27	13	33	12	21
13	23	20	3	21	13	8
14	3	3	0	3	3	0
15	12	12	0	11	3	8
16	12	12	0	12	12	0
17	40	40	0	37	37	0
Total	382	331	51	371	268	103

(a) Resnet-50

(b) VGG-16

Figure 10. Example of resulting images of testing with (**a**) Resnet-50 and (**b**) VGG-16.

Table 4. The comparison of results to the actual count.

Dataset	Oil Palm	Healthy Oil Palm	Unhealthy Oil Palm
Actual count	392	325	67
Resnet-50	382	331	51
VGG-16	371	268	103

a.   b.   c.   d.

Figure 11. Examples of (**a**) young oil palm, (**b**) oil palm under other trees, (**c**) coconut tree, and (**d**) other trees.

3.3. Oil Palm Tree Classification Model Performance Evaluation

The model's performance was evaluated in three main sections that evaluated the accuracy of data training, comparing the prediction results with visual interpretation and field surveys by using precision, recall, and F1-score as measures.

3.3.1. Model Training Inspection

Figure 12 shows the evaluation of the model performance focusing on accuracy and loss. The results indicate that the Resnet-50 network achieved higher performance than the VGG-16 network in all aspects. However, both models' values increase or decrease similarly. Therefore, evaluating model performance with multiple indicators is advised to assist in choosing an appropriate model.

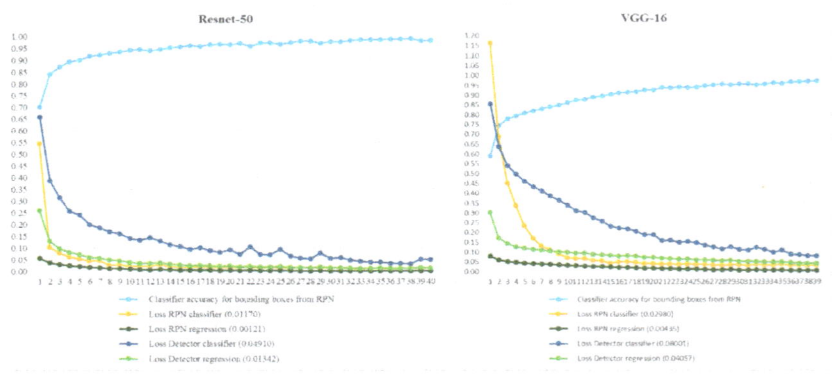

Figure 12. Comparison of model training performance.

Apart from accuracy and loss, the time used to train the model is another indicator that can assess a model's ability. Figure 13 illustrates the comparison of the time spent on each epoch for the models used. From the graph, it can be observed that Resnet-50 used approximately 10 min less time to process than VGG-16. Thus, the total processing time between the two-model differed with about 5 h, with Resnet-50 using approximately 40 h, while VGG-16 took approximately 45 h.

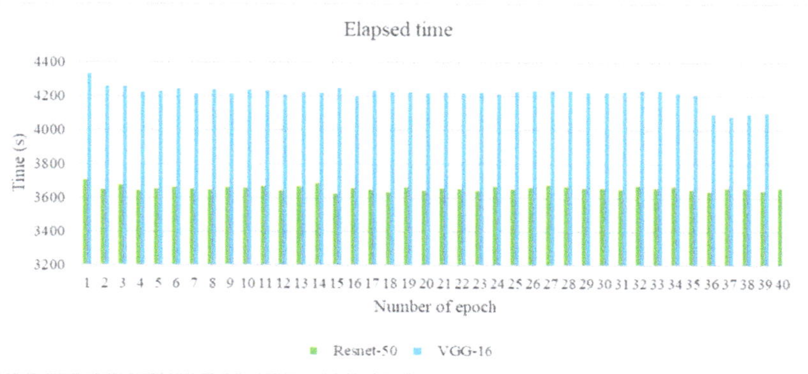

Figure 13. Processing time.

3.3.2. Visual Inspection

In this section we discussed the model performance when evaluated by comparing the results of both models with actual data derived from visual interpretation of high-resolution images in another area. As mentioned before, the effectiveness was measured using precision, recall, and F1-scores, as shown in Figure 14 below.

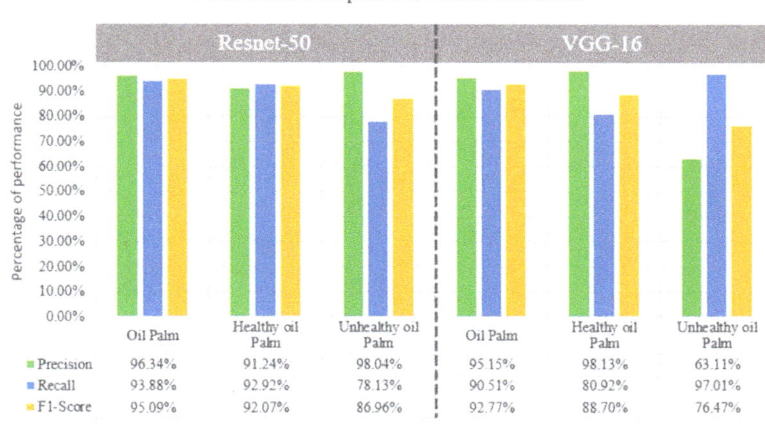

Figure 14. Comparative performance of oil palm tree classification between Resnet-50 and VGG-16 after compared with visual interpretation.

Each model's performance was evaluated by dividing the prediction results into three classes: oil palm tree detection, healthy, and unhealthy. Overall, the Resnet-50 network showed superior performance when comparing F1-scores. However, the Resnet-50 model had lower precision than VGG-16, when identifying healthy oil palms. Resnet-50 achieved 91.24%, while VGG-16 had a high value of 98.13%.

3.3.3. Ground Inspection

This evaluation was performed by using the results of the model showing the performance, i.e., Resnet-50. The results were compared with the field survey's sampling data to verify the predicted results from the model. In addition, this evaluation method determined the pattern of physical characteristics and their associated symptoms that the model can detect and classify. A total of 251 samples of oil palm trees were obtained from field survey sampling. Among these, 236 were healthy palm trees and 15 were unhealthy palm trees. The prediction results obtained from the Resnet-50 model comprised 223 healthy palm trees and 6 unhealthy palm trees. The performance evaluation was performed by calculating precision, recall, and F1-score.

Figure 15 illustrates the performance of Resnet-50. The precision of the predicted results was 100% for both the oil palm tree and unhealthy oil palm tree classes, while precision for the healthy oil palm tree class was 96.54%. For recall, the model achieved the highest percentage in the oil palm class, at 95.62%, with the healthy and unhealthy palm tree classes achieving 94.09% and 40.00%, respectively. For the F1-score, the class of oil palm tree had the highest performance, at 97.76%. Then followed the healthy and unhealthy oil palm tree classes with percentages of 95.30 and 57.14, respectively.

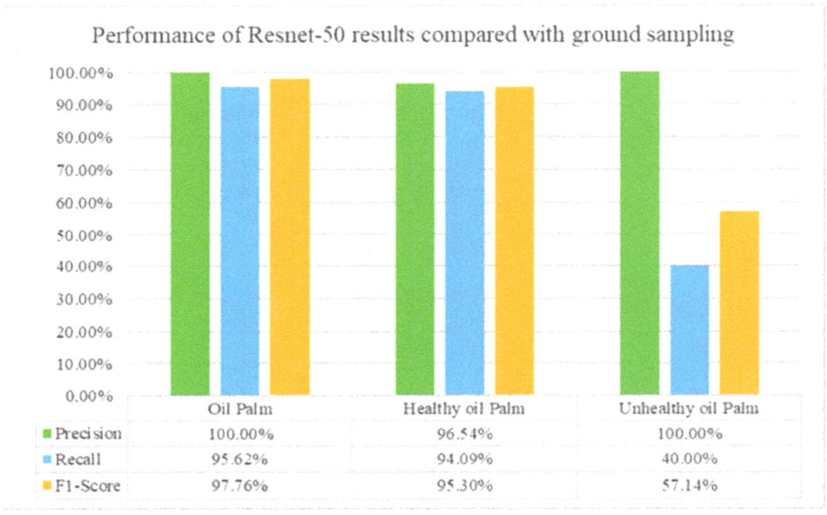

Figure 15. Comparative performance of Resnet-50 results and ground sampling.

Evaluating the model's predictive performance demonstrated that the model was best used for oil palm tree detection. The model showed predictive errors for health classification, especially in unhealthy oil palm. However, this model can be used for preliminary detection of health issues, since it is effective in identifying significant physical symptoms in palm trees.

According to the survey of 15 unhealthy oil palms on the plot, there were six oil palms with nitrogen deficiency, 12 oil palms with potassium deficiency, and seven oil palms with boron deficiency. Moreover, there were also two oil palms suffering from magnesium deficiency and two oil palms with *Ganoderma* disease. Of the six palms that were predicted to be unhealthy, two were lacking nitrogen, four were lacking potassium, two had boron deficiency, one had boron deficiency, and the last had *Ganoderma* disease. From the results, it can be concluded that the most common problem of unhealthy oil palm trees is a lack of nitrogen. However, the abnormality of the oil palm is usually not caused by the lack of one nutrient, but several. In addition, *Ganoderma* is another significant factor affecting the health classification since it always shows apparent physical symptoms. Based on the evaluation, the model can accurately predict up to 50% of all the oil palm trees that face this problem.

4. Conclusions

Since oil palms are one of the most important economic plants in Thailand, developing technology that can help to manage and maintain them is important. It can be a tool for farmers to become more efficient and increase their income. Currently, UAVs are widely used in agriculture as it can take many images with high spatial resolution in a short time, while traditional methods such as field surveying take longer. This research studied the detection and health classification of oil palms by using high-resolution imagery in conjunction with deep learning. Our study used Faster RCNN for object detection and evaluated the Resnet-50 and VGG-16 models.

The research used three important physical characteristics for detection and health classification: crown size, crown color, and crown density. These characteristics could indicate the age of the oil palm, nutrient deficiencies, and the presence of an epidemic disease, named *Ganoderma*.

In evaluating model performance, the accuracy from model training indicated that the Resnet-50 model was more accurate than VGG-16, and had fewer errors. Moreover, training on Resnet-50 was approximately 5 h faster than on VGG-16. The evaluation of the test prediction results was done by comparing them with both visual interpretations and field survey results. Next, precision, recall, and F1-score were calculated and evaluated.

Based on the study results and performance assessments, it can be concluded that the Resnet-50 network performed better in detection and health classification than the VGG-16 network. Further, the analysis of results highlighted that primarily the unhealthy palm trees faced potassium deficiencies and infection with *Ganoderma*. Additionally, the results showed that the model was often unable to detect young palm trees due to their smaller crown sizes.

In conclusion, our study showed that our proposed method could be used in the effective management of oil palm trees in Thailand. By tracking the number and health of oil palm trees this method can reduce fieldwork and the number of laborers. In addition, it can help to reduce the cost of production as treatments and fertilizer can only be applied in areas where it is needed. This will be beneficial to both farmers and organizations.

For the recommendation, the short processing time is an advantage of the Faster-RCNN structure, but it requires a large amount of varied training datasets. Therefore, we recommend that future studies increase the amount and variety of datasets, including varied image sizes, as this will improve model performance. Our study found errors in oil palm tree detection often occurred at the edge of the image. Consequently, increasing the overlap between images will result in a reduction of prediction errors. Our study focused on only three significant physical characteristics for oil palm tree health classification, all of which can be detected in RGB images. Thus, only preliminary classification of the health of oil palm trees can be done. For future studies, a multispectral camera could be used to enhance health classification.

Author Contributions: Conceptualization, A.W. and K.Y.; methodology, A.W. and K.Y.; software, K.K.; investigation, C.A.; writing—original draft preparation, K.Y.; writing—review and editing, A.W.; supervision, R.S.; funding acquisition, C.A. All authors have read and agreed to the published version of the manuscript.

Funding: This research was funded by Thai Research Organization (TRON) of The National Research Council of Thailand (NRCT) and Agricultural Development Agency (ARDA).

Institutional Review Board Statement: Not applicable.

Informed Consent Statement: Not applicable.

Data Availability Statement: The data presented in this study are available within the article.

Conflicts of Interest: The authors declare no conflict of interest.

References

1. Santoso, H.; Tani, H.; Wang, X. A simple method for detection and counting of oil palm trees using high-resolution multispectral satellite imagery. *Int. J. Remote Sens.* **2016**, *37*, 5122–5134. [CrossRef]
2. Kalantar, B.; Idrees, M.O.; Mansor, S.; Halin, A.A. Smart Counting—Oil Palm tree inventory with UAV. *Coordinates* **2017**, *13*, 17–22.
3. Aliero, M.M.; Mansur, M.A.; Al-Doksi, J. The Usefulness of Unmanned Airborne Vehicle (UAV) Imagery for Automated Palm Oil Tree Counting. *J. For.* **2014**, *1*, 1–12.
4. Daliman, S.; Abu-Bakar, S.A.R.; Azam, S.H.M.N. Development of young oil palm tree recognition using Haar- based rectangular windows. *IOP Conf. Ser. Earth Environ. Sci.* **2016**, *37*, 12041. [CrossRef]
5. Manandhar, A.; Hoegner, L.; Stilla, U. PALM TREE DETECTION USING CIRCULAR AUTOCORRELATION OF POLAR SHAPE MATRIX. *ISPRS Ann. Photogramm. Remote. Sens. Spat. Inf. Sci.* **2016**, *III*, 465–472. [CrossRef]
6. Yang, W.; Liu, Q.; Wang, S.; Cui, Z.; Chen, X.; Chen, L.; Zhang, N. Down image recognition based on deep convolutional neural network. *Inf. Process. Agric.* **2018**, *5*, 246–252. [CrossRef]
7. Perwej, Y.; Chaturvedi, A. Neural Networks for Handwritten English Alphabet Recognition. *Int. J. Comput. Appl.* **2011**, *20*, 1–5. [CrossRef]
8. Tajbakhsh, N.; Shin, J.Y.; Gurudu, S.R.; Hurst, R.T.; Kendall, C.B.; Gotway, M.B.; Liang, J. Convolutional Neural Networks for Medical Image Analysis: Full Training or Fine Tuning? *IEEE Trans. Med. Imaging* **2016**, *35*, 1299–1312. [CrossRef]
9. Cheang, E.K.; Cheang, T.K.; Tay, Y.H. Using Convolutional Neural Networks to Count Palm Trees in Satellite Images. *arXiv* **2017**, arXiv:1701.06462.
10. Li, W.; Fu, H.; Yu, L.; Cracknell, A. Deep Learning Based Oil Palm Tree Detection and Counting for High-Resolution Remote Sensing Images. *Remote Sens.* **2016**, *9*, 22. [CrossRef]
11. Sladojevic, S.; Arsenovic, M.; Anderla, A.; Culibrk, D.; Stefanovic, D. Deep Neural Networks Based Recognition of Plant Diseases by Leaf Image Classification. *Comput. Intell. Neurosci.* **2016**, *2016*, 1–11. [CrossRef] [PubMed]
12. Mubin, N.A.; Nadarajoo, E.; Shafri, H.Z.M.; Hamedianfar, A. Young and mature oil palm tree detection and counting using convolutional neural network deep learning method. *Int. J. Remote Sens.* **2019**, *40*, 7500–7515. [CrossRef]
13. Shrestha, A.; Mahmood, A. Review of Deep Learning Algorithms and Architectures. *IEEE Access* **2019**, *7*, 53040–53065. [CrossRef]
14. Simonyan, K.; Zisserman, A. Very deep convolutional networks for large-scale image recognition. *arXiv* **2014**, arXiv:1409.1556.
15. He, K.; Zhang, X.; Ren, S.; Sun, J. Deep residual learning for image recognition. *arXiv* **2015**, arXiv:1512.03385.
16. Girshick, R. Fast R-CNN. In Proceedings of the 2015 IEEE International Conference on Computer Vision (ICCV), Santiago, Chile, 7–13 December 2015; pp. 1440–1448.
17. Ren, S.; He, K.; Girshick, R.; Sun, J. Faster R-CNN: Towards real-time object detection with region proposal networks. In Proceedings of the 28th International Conference on Neural Information Processing Systems, Montreal, QC, Canada, 8–13 December 2015; MIT Press: Cambridge, MA, USA, 2015; pp. 91–99.
18. Zheng, J.; Li, W.; Xia, M.; Dong, R.; Fu, H.; Yuan, S. Large-Scale Oil Palm Tree Detection from High-Resolution Remote Sensing Images Using Faster-RCNN. In Proceedings of the IGARSS 2019—2019 IEEE International Geoscience and Remote Sensing Symposium, Yokohama, Japan, 28 July–2 August 2019; pp. 1422–1425.
19. Chuasuwan, C. Palm Oil Industry. 2018; pp. 4–8. Available online: http://www.tft-earth.org/wp-content/uploads/2017/02/TFT-Palm-Oil-Industry-Transformation-Paper.pdf (accessed on 10 March 2020).
20. Mandang, T.; Sinambela, R.; Pandianuraga, N.R. Physical and mechanical characteristics of oil palm leaf and fruits bunch stalks for bio-mulching. *IOP Conf. Ser. Earth Environ. Sci.* **2018**, *196*, 012015. [CrossRef]

Article

Average Degree of Coverage and Coverage Unevenness Coefficient as Parameters for Spraying Quality Assessment

Beata Cieniawska * and Katarzyna Pentos

Institute of Agricultural Engineering, Wrocław University of Environmental and Life Sciences,
37b Chełmońskiego Street, 51-630 Wrocław, Poland; katarzyna.pentos@upwr.edu.pl
* Correspondence: beata.cieniawska@upwr.edu.pl; Tel.: +48-666-988-949

Abstract: The purpose of the research was to determine the influence of selected factors on the average degree of coverage and uniformity of liquid spray coverage using selected single and dual flat fan nozzles. The impact of nozzle type, spray pressure, driving speed, and spray angle on the average degree of coverage and coverage unevenness coefficient were studied. The research was conducted with special spray track machinery designed and constructed to control and change the boom height, spray angle, driving speed, and spray pressure. Based on the research results, it was found that the highest average coverage was obtained for single standard flat fan nozzles and dual anti-drift flat fan nozzles. At the same time, the highest values of unevenness were observed for these nozzles. Inverse relationships were obtained for air-induction nozzles. Maximization of coverage with simultaneous minimization of unevenness can be achieved by using a medium droplet size for single flat fan nozzles (volume median diameter (VMD) = 300 µm) and coarse droplet size for dual flat fan nozzles (VMD = 352 µm), with low driving speed (respectively 1.1 m·s^{-1} and 1.6 m·s^{-1}) and angling of the nozzle by 20° in the opposite direction to the direction of travel.

Keywords: average degree of coverage; coverage unevenness coefficient; optimization; neural network

check for
updates

Citation: Cieniawska, B.; Pentos, K. Average Degree of Coverage and Coverage Unevenness Coefficient as Parameters for Spraying Quality Assessment. *Agriculture* **2021**, *11*, 151. https://doi.org/10.3390/agriculture 11020151

Academic Editors: Sebastian Kujawa, Gniewko Niedbała and Massimo Cecchini

Received: 29 December 2020
Accepted: 9 February 2021
Published: 12 February 2021

Publisher's Note: MDPI stays neutral with regard to jurisdictional claims in published maps and institutional affiliations.

1. Introduction

The use of a chemical method of plant protection results in high yields with significant qualitative values. However, adverse effects may occur during the application of plant protection products. The use of chemicals may cause natural environment pollution, as well as operators of sprayers, are exposed to contact with spray liquid. Also, agricultural products can be contaminated by chemicals what is dangerous for consumers [1–5]. However, the spraying process is still one of the most difficult agrotechnical processes. Efforts have therefore been made to reduce the negative effects for human health and the environment. These activities include the use of air-induction nozzles, the use of decision support tools [1,6–8], and injection into trunk trees of plant protection products [9]. In addition, the use of the latest advanced sprayers equipped with various sensors reduces the risk of hazards [10,11]. Scientists have also presented results of research on the exposure of bystanders to the drift liquid. Based on the conducted research, it was shown that adjustment parameters of a sprayer to meteorological conditions result in a reduction of exposure of bystanders. The experiments were carried out in vineyards located in mountainous areas during windless weather [12]. Similar studies were conducted by Butler-Ellis et al. and Kennedy et al. in field and orchard conditions. Scientists developed models that describe pesticide exposure in the short and long terms [13–15]. Moreover, the necessity to conduct experiments in regard to epidemiological research was emphasized [16]. The nature of problems in the field of plant protection and the need to take action to reduce them was emphasized by the Food and Agriculture Organization FAO, which established the year 2020 as the International Year of Plant Health.

Research into the modeling of spray drift and the optimization of the spraying process has progressed considerably in recent years. On the basis of the conducted research, scientists have emphasized the importance of knowledge about spray drift. The researchers pointed out, above all, the need for further work on drift curves under field conditions. Emphasis was also placed on the establishment of a system for EU Member States to prevent diffuse contamination.

Baldoin, Friso, and Pezzi, in their studies, used two adjuvants and performed the experiments at wind speed 1, 3, 5 m/s^{-1}, relative humidity 30%, 50%, and 70% at a temperature of 27 °C. On the basis of the conducted experiments, it was found that by the use of adjuvants, the droplet spectrum was modified by eliminating fine and very fine droplets. Thus, the predictions of mathematical modeling were confirmed [17–19], while Griesang et al. conducted a study to assess the spray drift potential [20]. Three types of nozzles were selected for testing: air-induction single flat fan, standard dual flat fan, and hollow cone at a pressure of 300 kPa. The application liquid consisted of glyphosate and a combination of glyphosate and adjuvant. Based on the conducted experiments, it was found that after adding the adjuvant to the liquid, the drift of the liquid was reduced by about 40% when using standard dual nozzles. On the other hand, in the case of using an air induction single nozzle, the addition of the adjuvant did not affect the drift, despite the increase in volume median diameter (VMD) value and improvement of droplet uniformity.

Many researchers have developed model-based computational fluid dynamics (CFD). Research has been conducted mainly in orchard crops as well as in greenhouses. CFD modeling has been the dominant technique in the design of air-assisted sprayers in recent years. Duga et al. and Salcedo et al. presented the results of research carried out in orchard crops. The authors presented a 2D and 3D computational model of fluid dynamics (CFD). The model took into account the tree habit and wind flow in the tree crown as well as the speed of the sprayer. On the basis of the model, the amount of liquid carried into the atmosphere (air drift) and into the soil (sedimentation drift) was calculated [21,22].

The aim of the research conducted by Gregorio et al. was to evaluate the drift reduction potential for hollow cone nozzles using light detection and ranging (LiDAR) technology. Based on the analysis of the results of this work, it was concluded that the LiDAR technique is an advantageous alternative in the evaluation of spray drift potential reduction [23,24].

Methods and algorithms of artificial intelligence are increasingly popular and very useful for solving various problems in agriculture. Some of these techniques have been employed for mathematical modeling of complex and nonlinear relationships, prediction, classification, and optimization. Thanks to artificial intelligence algorithms, complex agricultural ecosystems can be better described and understood. Wen et al. developed an unmanned aerial vehicles variable spray system based on an error back propagation artificial neural network (ANN) [25]. The ANN was trained to predict droplet deposition based on environment temperature, humidity, flight speed and altitude, wind speed, prescription value, nozzle pitch, and propeller pitch. The system, which is a combination of ANN (stable and reliable model with a prediction error of less than 20%) and multi-sensor for collecting real-time information about the spraying process, was successfully used for variable spray operation under different conditions. The ANN predictor model useful for spraying process optimization in precision agriculture was presented by Azizpanah et al. [26]. They developed accurate models for predicting the volumetric median diameter and drift phenomenon. The accuracy of the models described by the coefficient of determination (R^2) was higher than 90%. Yang et al. presented an approximate mathematical model of droplet drift for multirotor plant protection unmanned aerial vehicles [27]. In this model, a radial basis neural network was combined with computational fluid dynamics to better understand the influence of droplet size and windward airflow on the movement of droplets. In the work of Zhai et al. a precision farming system based on a combination of genetic algorithm and particle swarm optimization algorithm was proposed [28]. This system was validated through simulations of precise pesticide spraying. In these simulations, unmanned aerial vehicles were used as agents and were expected to complete the mission of precise pesticide spraying cooperatively. The aim of the mission was to optimize benefits and costs.

Many researchers have emphasized that the quality and effectiveness of the spraying procedure depends on the uniformity of the spray liquid fall, the coverage of the sprayed objects [29,30], and the application of the spray liquid [31]. Both the degree of coverage and the application of utility liquid can be used to assess changes in spraying technique, verify selected parameters of the sprayer, and assess the work of nozzles, depending on technical and technological factors [32,33]. Analysis of existing research results indicates that there is a relationship between the degree of coverage and biological effectiveness. According to some scientists, 30% covering of plants with liquid provides satisfactory efficacy (for most plant protection products). The assessment of the quality of sprayers and nozzles based on the coverage of sprayed objects is the fastest and simplest method [34]. The values of the degree of coverage are obtained after computer image analysis of the probes, which are most often water-sensitive papers [6,35,36].

However, the research does not show clearly whether this value applies to the average degree of coverage, calculated taking into account all components of the sprayed plants and in relation to systemic or contact pesticides.

According to the authors, the indicators presented above can serve as basic parameters for comparing the spraying equipment with various types of nozzles. Additional information useful in the assessment of the spraying procedure can be provided to the sprayer user by indicators proposed by the authors—average coverage of sprayed objects and coverage unevenness coefficient.

Therefore, the purpose of this research was to determine the impact of the type and size of the nozzles, liquid pressure, spray angle, and driving speed on the average degree of coverage and coverage unevenness coefficient, using various single and dual nozzles.

2. Materials and Methods

2.1. Experimental Set-Up

The following parameters of nozzles' work were used for the research:

- Pressure: 200, 300, and 400 kPa,
- Height of boom: 0.5 m,
- Spray angle, perpendicular to the ground: angled forward $+20°$, $+10°$, and $0°$ and backward $-10°$, and $-20°$
- Driving speed: 1.1, 2.2, and 3.3 m·s^{-1},
- Dose of liquid: 200 L·ha^{-1} (200 kPa), 240 L·ha^{-1} (300 kPa), 275 L·ha^{-1} (400 kPa))

Four types of nozzles were selected for testing: standard and air-induction single flat fan and anti-drift and air-induction dual flat fan. Standard nozzles are used in favorable weather conditions. A pre-orifice was applied to the anti-drift nozzle; thus, droplets would be larger and less prone to drift when compared with the standard nozzles. The largest, air-filled droplets were produced by the air-induction nozzles through the use of a Venturi air aspirator while reducing drift.

For each combination, the type of nozzle, and the pressure of the spray liquid, the droplet size was measured by laser diffraction method (Table 1). The Malvern Spraytech spray particle size analyzer was used for determining the droplet size distribution.

The tests were carried out under laboratory conditions. The test stand consisted of a device functioning as a self-propelled field sprayer and of artificial plants positioned along the machine's route (Figure 1).

Three artificial plants as three replicates were placed under a spray boom. Water sensitive papers were attached to artificial plants to form specific surfaces: the upper horizontal level surface, the bottom horizontal level surface, the vertical transverse leaving surface, and the vertical transverse approach surface. The scheme of the test stand was presented in our previous publication [37].

Table 1. The droplet size for each combination of the type of nozzle and the pressure.

Type of Nozzle	Nozzle Manufacturer	Pressure (kPa)	Flow Rate ($l \cdot min^{-1}$)	Droplet Size—VMD (μm)
single standard flat-fan—AXI 11002	Albuz	200	0.65	212
single standard flat-fan—AXI 11002	Albuz	300	0.79	193
single standard flat-fan—AXI 11002	Albuz	400	0.91	182
single air-induction flat-fan—AVI 11002	Albuz	200	0.65	554
single air-induction flat-fan—AVI 11002	Albuz	300	0.79	440
single air-induction flat-fan—AVI 11002	Albuz	400	0.91	382
dual anti-drift flat-fan—DGTJ 60 11002	TeeJet	200	0.65	299
dual anti-drift flat-fan—DGTJ 60 11002	TeeJet	300	0.79	264
dual anti-drift flat-fan—DGTJ 60 11002	TeeJet	400	0.91	249
dual air-induction flat-fan—AVI TWIN 11002	Albuz	200	0.65	543
dual air-induction flat-fan—AVI TWIN 11002	Albuz	300	0.79	436
dual air-induction flat-fan—AVI TWIN 11002	Albuz	400	0.91	384

VMD: volume median diameter.

Figure 1. Experimental set-up: 1—artificial plant, 2—spray boom, 3—metal track guide.

The degree of coverage was obtained based on a computer image analysis in Photoshop CC 2019 software. Water-sensitive papers change color from yellow to dark blue after contact with water, making it possible to analyze them. The dimensions of the water-sensitive papers are 26 mm × 76 mm, and three fragments with an area of 100 mm^2 were randomly selected for analysis. The degree of coverage was determined as the ratio of the surface covered with the liquid to the surface of the sampler.

$$P_{sp} = \frac{A_{pc}}{A_p} \cdot 100 \ [\%],\tag{1}$$

where: P_{sp}—degree of coverage (%), A_{pc}—surface covered with liquid (pixels), A_p—sampler surface (pixels)

The average degree of coverage was calculated as follows

$$\chi P_{sp} = \frac{P_{spg} + P_{spn} + P_{spo}}{n} \ [\%],$$ (2)

where χP_{sp}—average degree of coverage; P_{spg}—average degree of coverage of the upper horizontal level surface (%); P_{spn}—average degree of coverage of the vertical transverse approach surface (%); P_{spo}—average degree of coverage of the vertical transverse leaving surface (%); n—number of tests.

During the experiments, the bottom horizontal surface was not covered by liquid in any of the tests. Therefore, this surface was not taken into account in further analysis.

The coverage unevenness coefficient was calculated based on Equation (3) according to the formula [38]

$$\eta = \frac{\sqrt{\frac{1}{n-1} * \sum_{i=1}^{n} \left(P_{sp_i} - \chi P_{sp} \right)^2}}{\chi P_{sp}} \ [-]$$ (3)

where η—coverage unevenness coefficient (−); P_{sp}—degree of coverage of particular objects; χP_{sp}—average degree of coverage of all objects; n—number of tests.

2.2. Neural Network Models

Artificial neural networks are a very popular tool from artificial intelligence methods used for mathematical modeling, classification, clustering, and other tasks. ANNs are particularly useful when multidimensional, nonlinear relationships must be analyzed. ANNs are composed of very simple units called artificial neurons. Each neuron produces its output signal based on a vector of input signals, a vector of synaptic weights, and its activation function. The most popular are nonlinear activation functions such as sigmoid and hyperbolic tangent functions. In this research, a multilayer perceptron (MLP) as a neural network was used. MLP is a feedforward artificial neural network and consists of at least three layers: an input layer, one or more hidden layers, and an output layer. MLP networks are usually trained by an error backpropagation algorithm, which uses small incremental changes in connection weights in each iteration. After the training process (usually several thousand training cycles), the vector of connections weights changes from initial, random values to minimize the error between actually calculated and target output vector. The MLP with one hidden layer was used for this neural model's development. Four separate neural models were built, and for each model, the input signals were as follows: droplet size, driving speed, and spray angle. Two models were developed for single nozzles (with an average degree of coverage and coverage unevenness coefficient as an output signal). Analogous models were developed for the dual nozzles. To find the best MLP configuration for each model, 2000 ANNs were trained. The number of neurons in the hidden layer was changed from 10 to 40, different activation functions were used, and the matrix of initial synaptic weights was randomly generated. The experimental data were first normalized, and then the 270 data set was divided randomly into training, test, and validation sets at a 70:15:15 ratio. The accuracy of neural models was evaluated based on the coefficient of correlation (R), and the root mean squared error (RMSE), which are given by the following Equations (4) and (5)

$$R = \frac{\sum (Y_{meas} - \overline{Y}_{meas})\left(Y_{pred} - \overline{Y}_{pred}\right)}{\sqrt{\sum (Y_{meas} - \overline{Y}_{meas})^2 \sum \left(Y_{pred} - \overline{Y}_{pred}\right)^2}},$$ (4)

$$RMSE = \sqrt{\frac{1}{n} \sum_{i=1}^{n} \left(Y_{pred} - Y_{meas}\right)^2},$$ (5)

where: Y_{pred}—the absolute predicted value; $\overline{Y}_{pred}$—the average predicted value; Y_{meas}—the absolute measured (experimental) value; $\overline{Y}_{meas}$—the average measured value.

The accuracy of models is particularly important in the case of validation data set. One of the phenomena that can occur during the training process is overfitting. It occurs when the model is of great accuracy for the training data set (very best fit to the training data), but the accuracy for the validation data set is low. As a result, a neural network would not be able to generalize well to new data and is unsuitable for real-life applications.

Based on the best neural models, a sensitivity analysis was performed to indicate the contribution of the independent input variables in the models. For the simulations, the software Statistica v. 10 was used.

2.3. Optimization

The aim of the optimization process was to maximize the average degree of coverage and minimize the coverage unevenness coefficient at the same time. As an optimization method, a genetic algorithm (GA) implemented in the Excel 2013 Solver tool was chosen. The Excel Solver has been successfully applied for optimization procedures in prior literature. It was used by Barati in the estimation of nonlinear Muskingum routing parameters [39] and by Bhattacharjya for solving a groundwater flow inverse problem (estimation the unknown pumping rates of an aquifer, and estimation the aquifer transmissivity) [40]. The genetic algorithm is a search heuristic strongly inspired by nature, namely Charles Darwin's theory of natural evolution. GA works based on a population of individuals—potential problem solutions. Every individual is a vector of parameters known as genes. A set of genes is called a chromosome, which is an encoded form of a solution. Genes can be coded in various forms: as binary numbers, as real numbers, or as text. In each algorithm iteration, the tree operations: selection, crossover, and mutation are performed on individuals. The general rule of the algorithm is to assess each individual based on an objective function that represents the solution quality and to find the individual having the best objective function. This individual is considered as an optimal and, after decoding, is interpreted as a solution. In this work, the objective function was constructed as follows:

$$y = \chi P_{sp} + (2 - \eta), \tag{6}$$

The objective function is the sum of two terms, the degree of coverage, which must be maximized, and the expression $(2 - \eta)$, which is related to the coverage unevenness coefficient, and maximization of this expression means coverage unevenness coefficient minimization. The values of χP_{sp} and η were calculated based on neural models of the best accuracy. The aim of the optimization process was to find values of droplet size, driving speed, and spray angle, which maximizes the objective function. The optimization was performed separately for single and dual nozzles. Parameters of the algorithm were set as follows: population size—100; mutation rate—0.075; convergence—0.0001; random seed—0; and maximum time without improvement—30.

3. Results and Discussion

The test results are shown in Figures 2–7. Average coverage for various nozzles and spraying parameters is presented in Figures 2–4. An extremely important value of the presented results of the experiments is to emphasize the differences in both the values of the degree of coverage and the unevenness index.

Taking into account the average coverage values, it should be stated that the liquid pressure influences this parameter. Moreover, higher values of the average coverage were recorded for single standard and dual anti-drift nozzles.

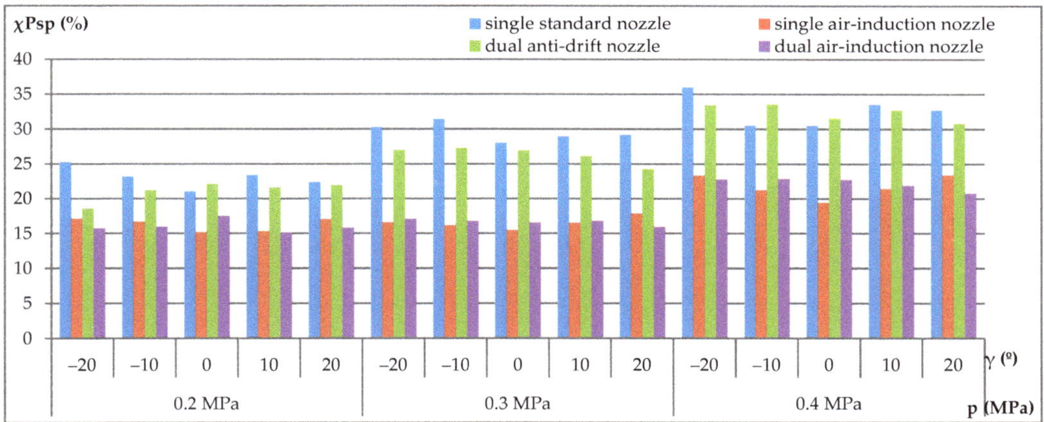

Figure 2. Average degree of coverage for selected nozzles at 1.1 m·s^{-1} driving speed.

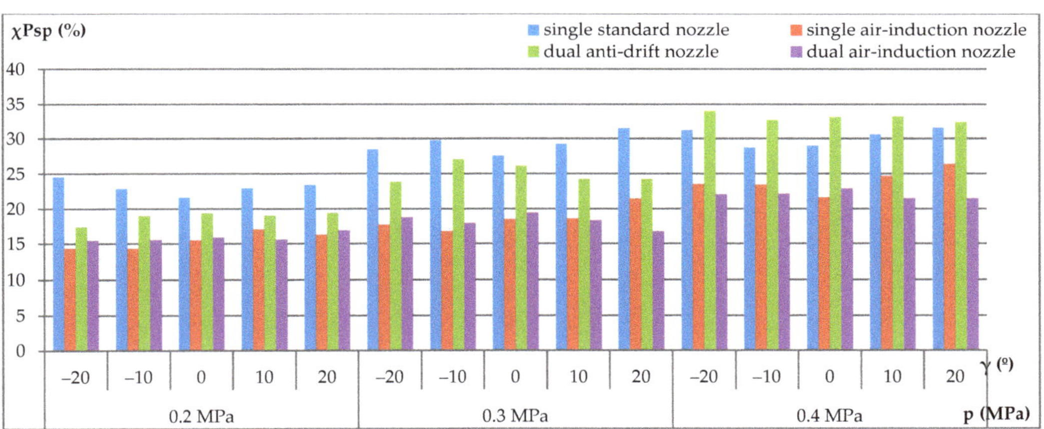

Figure 3. Average degree of coverage for selected nozzles at 2.2 m·s^{-1} driving speed.

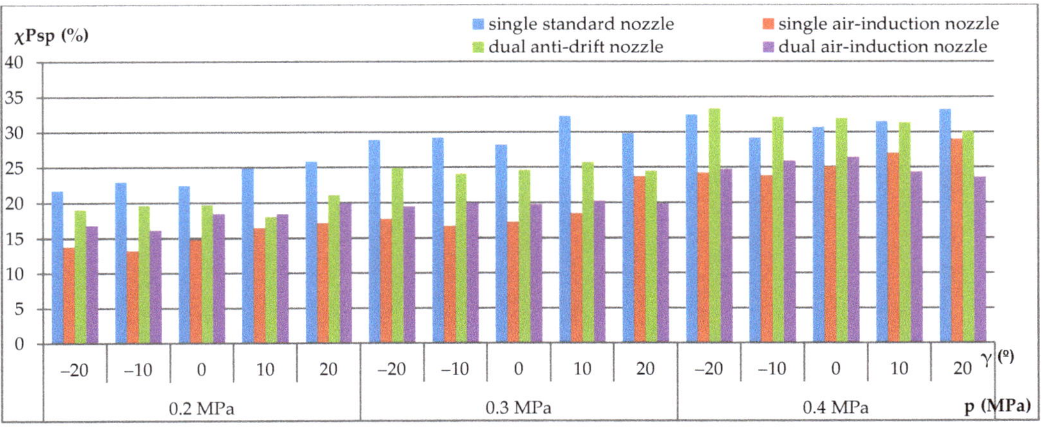

Figure 4. Average degree of coverage for selected nozzles at 3.3 m·s^{-1} driving speed.

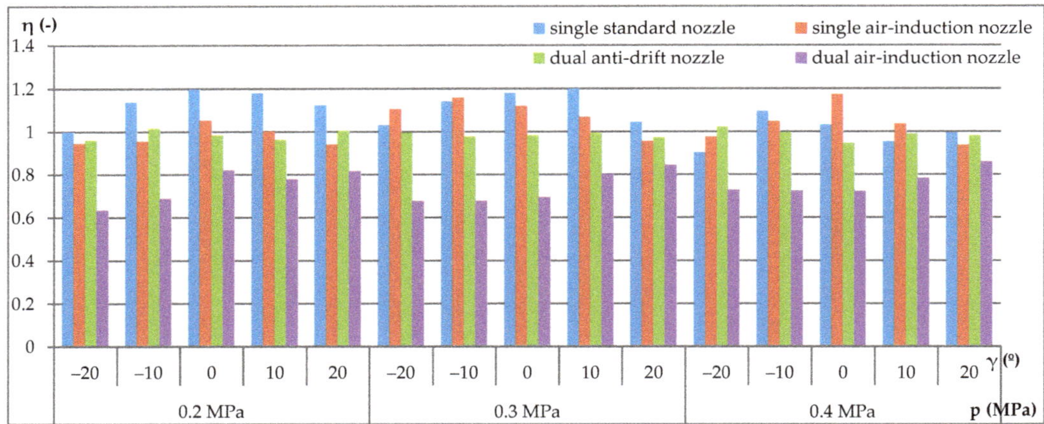

Figure 5. Coverage unevenness coefficient for selected nozzles at 1.1 m·s^{-1} driving speed.

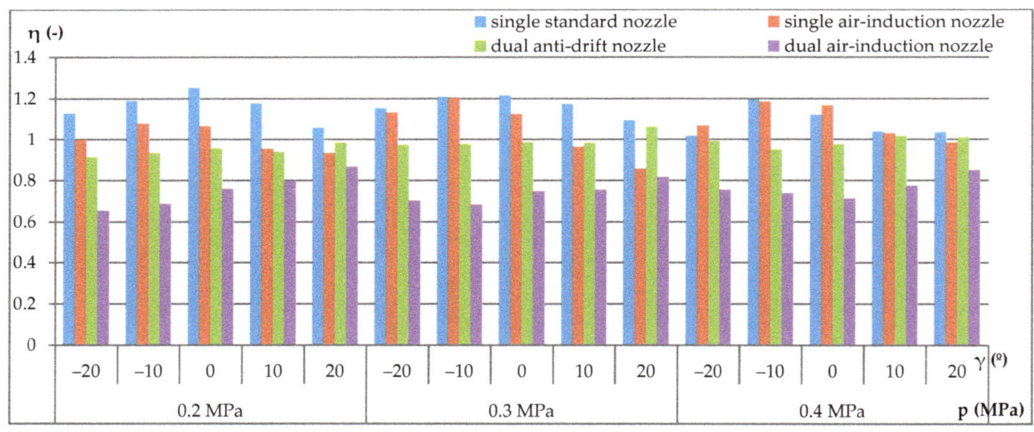

Figure 6. Coverage unevenness coefficient for selected nozzles at 2.2 m·s^{-1} driving speed.

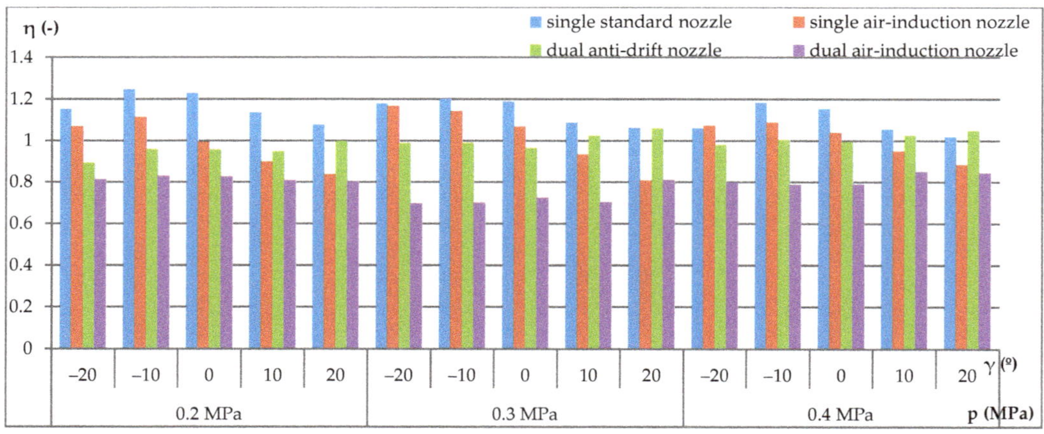

Figure 7. Coverage unevenness coefficient for selected nozzles at 3.3 m·s^{-1} driving speed.

Only a few studies have presented the on average degree of coverage results in the aspect of field crops. Average coverage studies were conducted by Qin et al. in the cultivation of cotton [41]. The authors of this study concluded that three factors, horizontal boom height, hang boom sprayer, and nozzles angle, were influenced by the average degree of coverage. Dereń et al. presented research results of the average total degree of coverage. This was based on experiments that found that higher values were obtained for dual flat fan nozzles [42]. In another study, the type of nozzles as well as the speed of sprayer and pressure of liquid influenced on the degree of coverage [43,44].

The results of the coverage unevenness coefficients are shown in Figures 5–7. The lowest values of this parameter, and thus the highest uniformity, are characteristic for dual air induction nozzles, regardless of the spraying conditions.

Analysis of the average coverage and uniformity of coverage were carried out, among others by Cai et al. in horticulture with the use of a fan sprayer equipped with a laser scanning system [45].

Based on the analysis of the experiments carried out by Musiu et al., it was found that reducing the liquid dose deteriorated the liquid coverage of plants but at the same time improved the homogeneity of distribution [33].

The four neural models were developed and then used for sensitivity analysis and the optimization process. The MLP structure and accuracy parameters for the best models obtained are presented in Table 2. The low values of residual mean square error (RMSE) error and high values of coefficient of correlation (R > 0.9) for the train, test, and validation data sets prove the high accuracy of all neural models.

Table 2. Error metrics of best model performances.

MLP Structure	Train		Test		Validation	
	R	**RMSE**	**R**	**RMSE**	**R**	**RMSE**
Single nozzles, average degree of coverage as an output parameter						
3-15-1	0.961	0.0020	0.968	0.0022	0.968	0.0019
Single nozzles, coverage uniformity coefficient as an output parameter						
3-36-1	0.968	0.0015	0.967	0.0019	0.933	0.0041
Dual nozzles, average degree of coverage as an output parameter						
3-17-1	0.959	0.0024	0.970	0.0021	0.976	0.0022
Dual nozzles, coverage uniformity coefficient as an output parameter						
3-11-1	0.979	0.0016	0.968	0.0027	0.992	0.0007

MLP: multilayer perceptron; RMSE: Residual mean square error.

Additionally, high values of R for the validation data set suggest that no overfitting effect occurred during the training process, and that models could be used in real-life applications.

Some mathematical models of the spraying process have been proposed in the literature. Baetens and coauthors proposed a 3D computational fluid dynamics model and a 2D diffusion-advection model for drift prediction [46,47]. The RTDrift model of spray drift was developed by Lebeau et al. [48]. These models included more input parameters than our model and took into account environmental parameters; however, their accuracy was lower. Li and coauthors proposed a prediction model of droplet coverage depending on droplet size, application distance, air delivery speed, and target leaf surface. The prediction accuracies of the model were 87.5%, 80%, and 100% for the three states of uniform, accumulation, and loss [31]. Slightly lower accuracies of neural models were obtained in our previous work, where the coverage of the three leaf surfaces was predicted based on droplet size, spray angle, and driving speed [49].

Based on the models presented in Table 2, a sensitivity analysis was performed to determine the parameters influencing the most and average degree of coverage and a coverage unevenness coefficient for single and dual nozzles. The results are presented in Figures 8 and 9. In the case of both parameters, the average degree of coverage and the coverage unevenness coefficient, the most influencing parameter is droplet size, which depends on nozzle type and pressure. The influence of driving speed and nozzle angular position is significantly lower. These results are in agreement with those presented for the coverage of the sprayed surfaces where the vertical transverse approach surface, the vertical transverse leaving surface, and the upper level surface were analyzed separately [49].

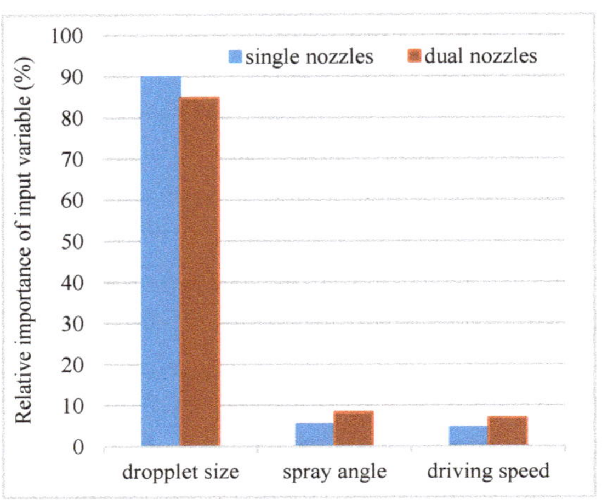

Figure 8. The relative importance of the input variables of the MLP model on the average degree of coverage.

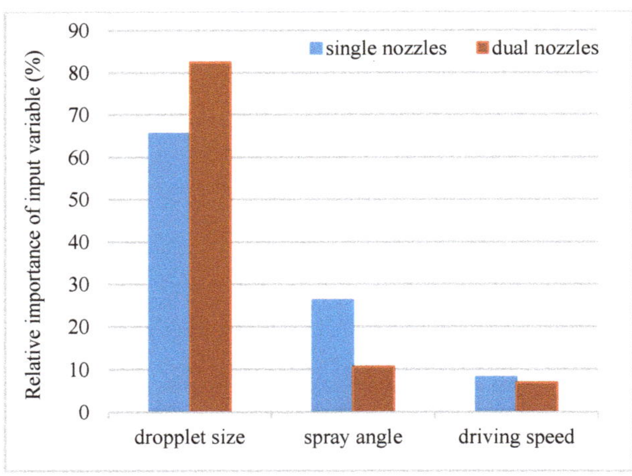

Figure 9. The relative importance of the input variables of the MLP model on the coverage unevenness coefficient.

The analysis of experimental data shows that in the case of single nozzles, the highest value of an average degree of coverage was observed for small droplet size (182.4 μm—produced by a single standard flat-fan nozzle with a pressure of 400 kPa), low driving speed (1.1 m·s^{-1}), and nozzle angular position of $-20°$. On the other hand, the lowest value of the coverage unevenness coefficient was produced by medium droplet size (400 μm—obtained from a single air-induction flat-fan nozzle with a pressure of 300 kPa), high driving speed

(3.3 m/s^{-1}), and nozzle angular position of 20°. Taking into account the average degree of coverage, a well known fact that the application of higher pressure results in a higher degree of coverage could be confirmed. At the same time, it can be concluded that standard nozzles have greater coverage unevenness compared to air induction nozzles. This is related to the range of coverage of individual horizontal and vertical surfaces. During the spraying process, there was significantly lower coverage on vertical surfaces and much higher coverage on horizontal surfaces obtained with single standard flat fan nozzles in comparison to single air induction flat fan nozzles. Therefore, greater uniformity of coverage was obtained when using single air induction flat fan nozzles.

The configuration of optimum parameters described above makes simultaneous optimization of the two indicators of spray process quality fairly difficult. As a result of the optimization carried out with the use of a genetic algorithm combined with neural models, the following optimum parameters of the spraying process were calculated: droplet size of 300 μm, driving speed of 1.1 m·s^{-1}, and nozzle angular position of −20°. For these parameters, the average degree of coverage equaled 35.54% (the range of experimental data was from 11.66% to 37.08%), and the coverage unevenness coefficient equaled 0.85 (the range of experimental data was from 0.80% to 1.25%). It can be stated that the spraying process parameters calculated in the optimization process give an average degree of coverage close to the maximum value and the coverage unevenness coefficient close to the minimum value.

In the case of dual nozzles, based on experimental data, it can be stated that the highest value of an average degree of coverage was observed for small droplet size (248.9 μm—produced by the dual anti-drift flat-fan nozzle with a pressure of 400 kPa), medium driving speed (2.2 m/s^{-1}), and nozzle angular position of −20°. The lowest value for the coverage unevenness coefficient was produced by a large droplet size (542.8 μm—obtained from the dual air-induction flat fan nozzle with a pressure of 200 kPa), low driving speed (1.1 m/s^{-1}), and nozzle angular position of −20°. When using dual flat fan nozzles, similar relationships were observed as in the case of using single flat fan nozzles. Based on the optimization process, the following optimum parameters of the spraying process could be proposed: droplet size of 352 μm, driving speed of 1.6 m/s^{-1}, and nozzle angular position of −20°. For these parameters, the average degree of coverage equaled 28.32% (the range of experimental data was from 13.81% to 35.31%), and the coverage unevenness coefficient equals 0.71 (the range of experimental data is from 0.63 to 1.06). It can be stated that for single nozzles, the spraying process parameters produced by optimization gave an average degree of coverage and coverage unevenness coefficient closer to optimum values than in the case of dual nozzles. The optimum spraying parameters obtained in this work generally correspond to these proposed in our previous work, where the coverage of three sprayed surfaces (the vertical transverse approach surface, the vertical transverse leaving surface, and the upper level surface) was optimized [49].

4. Conclusions

The findings provide a practical basis for the selection of appropriate nozzle parameters to ensure the highest uniformity with the best average liquid coverage. The main results of this study are as follows:

The highest value for an average degree of coverage was observed when applied with a single standard flat fan nozzle with a pressure of 400 kPa and a dual anti-drift flat fan nozzle with a pressure of 400 kPa.

The air induction flat fan nozzles were characterized by greater uniformity of coverage.

Based on the optimization results (maximization of coverage and minimization of unevenness), the two combinations of spraying conditions can be proposed. The first is single air induction nozzles using a pressure above 400 kPa, driving speed of 1.1 m/s^{-1} and spray angle, perpendicular to the ground of −20°. The second is dual air induction nozzles at pressure above 400 kPa, driving speed of 1.6 m·s^{-1} and spray angle, perpendicular to the ground of −20°.

Author Contributions: Conceptualization, B.C. and K.P.; methodology, B.C. and K.P.; software, B.C. and K.P.; validation, K.P.; formal analysis, B.C. and K.P.; investigation, B.C.; data curation, B.C. and K.P.; writing—original draft preparation, B.C. and K.P.; writing—review and editing, B.C. and K.P. All authors have read and agreed to the published version of the manuscript.

Funding: This research received no external funding.

Institutional Review Board Statement: Not applicable.

Informed Consent Statement: Not applicable.

Data Availability Statement: Not applicable.

Acknowledgments: The author would like to thank Antoni Szewczyk from the Institute of Agricultural Engineering, Wroclaw University of Environmental and Life Sciences, for his help in improving the methodological aspects of this research.

Conflicts of Interest: The authors declare no conflict of interest.

References

1. Bourodimos, G.; Koutsiaras, M.; Psiroukis, V.; Balafoutis, A.; Fountas, S. Development and Field Evaluation of a Spray Drift Risk Assessment Tool for Vineyard Spraying Application. *Agriculture* **2019**, *9*, 181. [CrossRef]
2. Ki-Hyun, K.; Ehsanul, K.; Shamin, J. Exposure to pesticides and the associated human health effects. *Sci. Total Environ.* **2016**, *575*, 252–535. [CrossRef]
3. Fargnoli, M.; Lombardi, M.; Puri, D.; Casorri, L.; Masciarelli, E.; Mandić-Rajčević, S.; Colosio, C. The Safe Use of Pesticides: A Risk Assessment Procedure for the Enhancement of Occupational Health and Safety (OHS) Management. *Int. J. Environ. Res. Public Health* **2019**, *16*, 310. [CrossRef]
4. Wilmart, O.; Legrève, A.; Scippo, M.L.; Reybroeck, W.; Urbain, B.; de Graaf, D.C.; Steurbaut, W.; Delahaut, P.; Gustin, P.; Kim Nguyen, B.; et al. Residues in Beeswax: A Health Risk for the Consumer of Honey and Beeswax? *J. Agric. Food Chem.* **2016**, *64*, 8425–8434. [CrossRef] [PubMed]
5. Machado, B.B.; Spadon, G.; Arruda, M.S.; Goncalves, W.N.; Carvalho, A.C.P.L.F.; Rodrigues, J.F., Jr. A smartphone application to measure the quality of pest control spraying machines via image analysis. *Assoc. Comput. Mach.* **2018**, 956–963. [CrossRef]
6. Nansen, C.; Ferguson, J.C.; Moore, J.; Groves, L.; Emery, R.; Garel, N.; Hewitt, A. Optimizing Pesticide Spray Coverage Using a Novel Web and smartphone Tool, SnapCard. *Agron. Sustain. Dev.* **2015**, *35*, 1075–1085. [CrossRef]
7. Li, W.; Xu, B.; Du, Y.; Mao, E.; Zhu, Z.; Li, Z. Auxiliary navigation system based on Baidu Map JavaScript API for high clearance sprayers. In Proceedings of the 2019 IEEE International Conference on Unmanned Systems and Artificial Intelligence (ICUSAI), Xi'an, China, 22–24 November 2019; pp. 160–165. [CrossRef]
8. Szwedziak, K.; Niedbała, G.; Grzywacz, Ż.; Winiarski, P.; Doležal, P. The Use of Air Induction Nozzles for Application of Fertilizing Preparations Containing Beneficial Microorganisms. *Agriculture* **2020**, *10*, 303. [CrossRef]
9. Berger, C.; Laurent, F. Trunk injection of plant protection products to protect trees from pests and diseases. *Crop Prot.* **2019**, *124*, 104831. [CrossRef]
10. Samseemoung, G.; Soni, P.; Suwan, P. Development of a Variable Rate Chemical Sprayer for Monitoring Diseases and Pests Infestation in Coconut Plantations. *Agriculture* **2017**, *7*, 89. [CrossRef]
11. Warneke, B.; Zhu, H.; Pscheidt, J.; Nackley, L. Canopy spray application technology in specialty crops: A slowly evolving landscape. *Pest. Manag. Sci.* **2020**. [CrossRef] [PubMed]
12. Otto, S.; Loddo, D.; Baldoin, C.; Zanin, G. Spray drift reduction techniques for vineyards in fragmented landscapes. *J. Environ. Manag.* **2015**, *162*, 290–298. [CrossRef]
13. Butler Ellis, M.C.; van de Zande, J.C.; van den Berg, F.; Kennedy, M.C.; O'Sullivan, C.M.; Jacobs, C.M.; Fragkoulis, G.; Spanoghe, P.; Gerritsen-Ebben, R.; Frewer, L.J.; et al. The BROWSE model for predicting exposures of residents and bystanders to agricultural use of plant protection products: An overview. *Biosyst. Eng.* **2017**, *154*, 92–104. [CrossRef]
14. Butler Ellis, M.C.; van den Berg, F.; van de Zande, J.C.; Kennedy, M.C.; Charistou, A.N.; Arapaki, N.S.; Butler, A.H.; Machera, K.A.; Jacobs, C.M. The BROWSE model for predicting exposures of residents and bystanders to agricultural use of pesticides: Comparison with experimental data and other exposure models. *Biosyst. Eng.* **2017**, *154*, 122–136. [CrossRef]
15. Kennedy, M.C.; Butler Ellis, M.C. Probabilistic modelling for Bystander and Resident exposure to pesticides using the Browse software. *Biosyst. Eng.* **2017**, *154*, 105–121. [CrossRef]
16. Tsaboula, A.; Papadakis, E.-N.; Vryzas, Z.; Kotopoulou, A.; Kintzikoglou, K.; Papadopoulou-Mourkidou, E. Environmental and human risk hierarchy of pesticides: A prioritization method, based on monitoring, hazard assessment and environmental fate. *Environ. Int.* **2016**, *91*, 78–93. [CrossRef] [PubMed]
17. Baldoin, C.; Friso, D. Assessment of the contribute of spray thickeners to the agro-chemical drift reduction using a mathematical model and a wind tunnel. *Appl. Math. Sci.* **2015**, *9*, 5603–5614. [CrossRef]
18. Friso, D.; Baldoin, C. Mathematical modeling and experimental assessment of agrochemical drift using a wind tunnel. *Appl. Math. Sci.* **2015**, *9*, 5451–5463. [CrossRef]

19. Friso, D.; Baldoin, C.; Pezzi, F. Mathematical modeling of the dynamics of air jet crossing the canopy of tree crops during pesticide application. *Appl. Math. Sci.* **2015**, *9*, 1281–1296. [CrossRef]
20. Griesang, F.; Decaro, R.; dos Santos, C.; Santos, E.; de Lima Roque, N.; da Costa Ferreira, M. How Much Do Adjuvant and Nozzles Models Reduce the Spraying Drift? Drift in Agricultural Spraying. *Am. J. Plant. Sci.* **2017**, *8*, 2785–2794. [CrossRef]
21. Duga, A.T.; Delele, M.A.; Ruysen, K.; Dekeyser, D.; Nuyttens, D.; Bylemans, D.; Nicolai, B.M.; Verboven, P. Development and validation of a 3D CFD model of drift and its application to air-assisted orchard sprayers. *Biosyst. Eng.* **2017**, *154*, 62–75. [CrossRef]
22. Salcedo, R.; Vallet, A.; Granell, R.; Garcera, C.; Molto, E.; Chueca, P. Eulerian-lagrangian model of the behaviour of droplets produced by an air-assisted sprayer in a citrus orchard. *Biosyst. Eng.* **2017**, *154*, 76–91. [CrossRef]
23. Gregorio, E.; Rocadenbosch, F.; Sanz, R.; Rosell-Polo, J. Eye-safe lidar system for pesticide spray drift measurement. *Sensors* **2015**, *15*, 3650–3670. [CrossRef] [PubMed]
24. Gregorio, E.; Torrent, X.; Planas, S.; Rosell-Polo, J. Assessment of spray drift potential reduction for hollow-cone nozzles: Part 2. LiDAR technique. *Sci. Total Environ.* **2019**, *687*, 967–977. [CrossRef]
25. Wen, S.; Zhang, Q.; Yin, X.; Lan, Y.; Zhang, J.; Ge, Y. Design of Plant Protection UAV Variable Spray System Based on Neural Networks. *Sensors* **2019**, *19*, 1112. [CrossRef]
26. Azizpanah, A.; Rajabipour, A.; Alimardani, R.; Kheiralipour, K.; Mohammadi, V. Precision spray modeling using image processing and artificial neural network. *Agric. Eng. Int. CIGR J.* **2015**, *17*, 65–74.
27. Yang, F.; Xue, X.; Cai, C.; Sun, Z.; Zhou, Q. Numerical Simulation and Analysis on Spray Drift Movement of Multirotor Plant Protection Unmanned Aerial Vehicle. *Energies* **2018**, *11*, 2399. [CrossRef]
28. Zhai, Z.; Martínez Ortega, J.-F.; Lucas Martínez, N.; Rodríguez-Molina, J.A. Mission Planning Approach for Precision Farming Systems Based on Multi-Objective Optimization. *Sensors* **2018**, *18*, 1795. [CrossRef] [PubMed]
29. Malneršič, A.; Dular, M.; Širok, B.; Oberti, R.; Hočevar, M. Close-range air-assisted precision spot-spraying for robotic applications: Aerodynamics and spray coverage analysis. *Biosyst. Eng.* **2016**, *146*, 216–226. [CrossRef]
30. Marian, O.; Muntean, M.; Drocaş, I.; Ranta, O.; Molnar, A.; Catunescu, G.; Bărbieru, V. Assessment Method of Coverage Degree for Pneumatic Sprayers Used in Orchards. *Agric. Agric. Sci. Procedia* **2016**, *10*, 47–54. [CrossRef]
31. Li, J.; Cui, H.; Ma, Y.; Xun, L.; Li, Z.; Yang, Z.; Lu, H. Orchard Spray Study: A prediction model of droplet deposition states on leaf surfaces. *Agronomy* **2020**, *10*, 747. [CrossRef]
32. Liao, J.; Hewitt, A.J.; Wang, P.; Luo, X.W.; Zang, Y.; Zhou, Z.Y.; Lan, Y.; O'Donnell, C. Development of droplet characteristics prediction models for air induction nozzles based on wind tunnel tests. *Int. J. Agric. Biol. Eng.* **2019**, *12*, 1–6. [CrossRef]
33. Musiu, E.; Lijun, Q.L.; Wu, Y. Spray deposition and distribution on the targets and losses to the ground as affected by application volume rate, airflow rate and target position. *Crop. Prot.* **2019**, *116*, 170–180. [CrossRef]
34. Legleiter, T.R.; Johnson, W.G. Herbicide coverage in narrow row soybean as influenced by spray nozzle design and carrier volume. *Crop. Prot.* **2016**, *83*, 1–8. [CrossRef]
35. Özlüoymak, Ö.B.; Bolat, A. Development and assessment of a novel imaging software for optimizing the spray parameters on water-sensitive papers. *Comput. Electron. Agric.* **2020**, *168*, 105104. [CrossRef]
36. Lipiński, A.J.; Lipiński, S. Binarizing water sensitive papers—How to assess the coverage area properly? *Crop. Prot.* **2020**, *127*, 104949. [CrossRef]
37. Dereń, K.; Cieniawska, B.; Szewczyk, A.; Sekutowski, T.; Zbytek, Z. Average liquid coverage depending on the type of the nozzle, spraying parameters and characteristics of the sprayed objects. *J. Res. Appl. Agric. Eng.* **2017**, *62*, 22–26.
38. *ISO 5682-3, Equipment for Crop Protection—Spraying Equipment—Part 3: Test Methods for Volume/Hectare Adjustment Systems of Agricultural Hydraulic Pressure Sprayers*; ISO: Geneva, Switzerland, 1997.
39. Barati, R. Application of excel solver for parameter estimation of the nonlinear Muskingum models. *KSCE J. Civ. Eng.* **2013**, *17*, 1139–1148. [CrossRef]
40. Bhattacharjya, R.K. Solving groundwater flow inverse problem using spreadsheet solver. *J. Hydrol. Eng.* **2011**, *16*, 472–477. [CrossRef]
41. Qin, W.C.; Xue, X.Y.; Cui, L.F.; Zhou, Q.; Xu, Z.F.; Chang, F.L. Optimization and test for spraying parameters of cotton defoliant sprayer. *Int. J. Agric. Biol. Eng.* **2016**, *9*, 63–72. [CrossRef]
42. Dereń, K.; Szewczyk, A.; Sekutowski, T.R. The effect of the type of preparation with the content of nano-copper and copper on the coverage of winter rape plants. *J. Agric. Eng.* **2018**, *63*, 51–55.
43. Drocas, I.; Marian, O.; Ranta, O.; Molnar, A.; Muntean, M.; Cătunescu, G. Study on determining the degree of coverage when performing phytosanitary treatments using water sensitive paper. *Lucr. Ştiinţifice Agron.* **2014**, *57*, 159–163.
44. Bolat, A.; Özlüoymak, Ö. Evaluation of performances of different types of spray nozzles in site-specific pesticide spraying. *Semin. Ciências Agrárias.* **2020**, *41*, 1199–1212. [CrossRef]
45. Cai, J.C.; Wang, X.; Gao, Y.Y.; Yang, S.; Zhao, C.J. Design and performance evaluation of a variable-rate orchard sprayer based on a laser-scanning sensor. *Int. J. Agric. Biol. Eng.* **2019**, *12*, 51–57. [CrossRef]
46. Baetens, K.; Ho, Q.T.; Nuyttens, D.; De Schampheleire, M.; Melese Endalew, A.; Hertog, M.; Nicolaï, B.; Ramon, H.; Verboven, P. A validated 2-D diffusion- advection model for prediction of drift from ground boom sprayers. *Atmos. Environ.* **2009**, *43*, 1674–1682. [CrossRef]
47. Baetens, K.; Nuyttens, D.; Verboven, P.; De Schampheleire, M.; Nicolaï, B.; Ramon, H. Predicting drift from field spraying by means of a 3D computational fluid dynamics model. *Comput. Electron. Agric.* **2007**, *56*, 161–173. [CrossRef]

48. Lebeau, F.; Verstraete, A.; Stainier, C.; Destain, M.-F. RTDrift: A real time model for estimating spray drift from ground applications. *Comput. Electron. Agric.* **2011**, *77*, 161–174. [CrossRef]
49. Cieniawska, B.; Pentoś, K.; Łuczycka, D. Neural modeling and optimization of the coverage of the sprayed surface. *Bull. Pol. Acad. Sci. Tech. Sci.* **2020**, *68*, 601–608. [CrossRef]

Article

The Relationship between Soil Electrical Parameters and Compaction of Sandy Clay Loam Soil

Katarzyna Pentoś [1,*], Krzysztof Pieczarka [1] and Kamil Serwata [2]

[1] Institute of Agricultural Engineering, Wrocław University of Environmental and Life Sciences, 37b Chełmońskiego Street, 51-630 Wrocław, Poland; krzysztof.pieczarka@upwr.edu.pl
[2] Agri Solutions Sp. z o.o., Ligota Wielka 34, 56-400 Oleśnica, Poland; k.serwata@agrisolutions.eu
*   Correspondence: katarzyna.pentos@upwr.edu.pl; Tel.: +48-71-320-5970

Abstract: Soil spatial variability mapping allows the delimitation of the number of soil samples investigated to describe agricultural areas; it is crucial in precision agriculture. Electrical soil parameters are promising factors for the delimitation of management zones. One of the soil parameters that affects yield is soil compaction. The objective of this work was to indicate electrical parameters useful for the delimitation of management zones connected with soil compaction. For this purpose, the measurement of apparent soil electrical conductivity and magnetic susceptibility was conducted at two depths: 0.5 and 1 m. Soil compaction was measured for a soil layer at 0–0.5 m. Relationships between electrical soil parameters and soil compaction were modelled with the use of two types of neural networks—multilayer perceptron (MLP) and radial basis function (RBF). Better prediction quality was observed for RBF models. It can be stated that in the mathematical model, the apparent soil electrical conductivity affects soil compaction significantly more than magnetic susceptibility. However, magnetic susceptibility gives additional information about soil properties, and therefore, both electrical parameters should be used simultaneously for the delimitation of management zones.

Keywords: apparent soil electrical conductivity (ECa); magnetic susceptibility (MS); EM38; neural networks

Citation: Pentoś, K.; Pieczarka, K.; Serwata, K. The Relationship between Soil Electrical Parameters and Compaction of Sandy Clay Loam Soil. *Agriculture* **2021**, *11*, 114. https://doi.org/10.3390/agriculture11020114

Academic Editor: Sebastian Kujawa and Gniewko Niedbała

Received: 28 December 2020
Accepted: 28 January 2021
Published: 1 February 2021

Publisher's Note: MDPI stays neutral with regard to jurisdictional claims in published maps and institutional affiliations.

1. Introduction

One of the objectives of precision agriculture is optimizing production by increasing yield or by cost reduction due to optimizing the use of natural resources, and improving soil quality [1]. Another important aspect is also environmental impact reduction, e.g., by the variable-rate application of fertilizers or reduction of the application of plant protection products. For this purpose, many new technologies have been developed and used in practice. One available technique popular in precision agriculture is the delimitation of management zones. Management zones are areas of the field with similar characteristics, such as soil texture and chemical composition [2]. The optimization of crop production requires knowledge of many soil properties. These properties are very often determined by expensive and time-consuming methods. Delimitation of management zones allows for a reduction in sample numbers involved in defining soil properties. However, this process is difficult due to the high soil spatial variability and complex combination of factors that may influence soil properties [3].

Soil spatial variations may be analyzed based on many techniques such as apparent soil electrical conductivity (EC_a) maps, crop yield maps and soil organic matter maps [4–6]. The electrical soil parameters are influenced by many physical-chemical features of soil. The apparent soil electrical conductivity is defined as the soil capacity for conducting electric current. EC_a reflects some soil properties such as water content, bulk density, salinity, clay content, texture, organic matter content, size and distribution of pores, depth to contrasting soil layers, and organic carbon content [7–9]. Magnetic susceptibility (MS) expresses the level of magnetization of a soil in reaction to an applied magnetic field. Magnetite and maghemite as ferromagnetic minerals are the main determinants of soil magnetic susceptibility [10]. MS is strongly affected by soil drainage class, texture, clay content, organic matter, and organic carbon [11–14].

Geospatial measurements of apparent soil electrical conductivity and magnetic susceptibility are reliable, quick, easy, nondestructive, and economic; therefore, they are recognized as a valuable mapping tool indicating soil productivity [15,16]. Furthermore, the important feature of EC_a is its temporal stability, which has been demonstrated in prior literature [17,18]. Many studies have investigated the potential of soil electrical parameters in the delimitation of management zones. Serrano et al. (2014) reported the potential usability of EC_a measurement using the Veris 2000 XA sensor for monitoring soil texture, moisture content, organic matter, pH, and phosphorus soil content [19]. Pedrera-Parrilla et al. (2016) delimited three management zones with similar soil conditions based on EC_a measurement [1]. They found an exponential relationship between spatially averaged soil water content and EC_a for water contents lower than 0.11 kg·kg^{-1}. Peralta et al. (2013) reported that EC_a variability can be explained by some soil properties, namely clay content, soil organic matter, cation exchange capacity, and soil water content [15]. de Souza Bahia et al. (2017) reported a high correlation between MS and clay, iron oxides, carbon, and nitrogen, and revealed that MS is a good predictor of key properties of soils under study [20]. Magnetic susceptibility was used by some researchers for soil mapping according to the degree of contamination with heavy metals or organic compounds [21,22]. Ayoubi et al. (2019) reported a strong relationship between magnetic susceptibility and concentrations of chromium, iron, nickel, zinc, cobalt, and manganese [10]. It can be stated that the apparent electrical conductivity of soil has become one of the most popular tools for characterizing the spatial variability of soil parameters. Magnetic susceptibility is most often used for soil mapping in the context of environmental pollution.

Soil compaction may result from natural causes such as soil parameters, freezing, and drying or from mechanical operations with the use of heavy wheeled machines. As a soil compaction influences the yield, the delimitation of management zones connected with this parameter is of great importance. Soil compaction affects the water infiltration rate, soil air permeability, nitrogen availability, root length, root penetration, and rooting depth [23,24]. Soil compaction mapping is necessary, e.g., for a variable-depth tillage technology that plays an important role in applications of precision agriculture.

Artificial neural networks (ANNs) are very useful tools for solving prediction problems in agriculture, and there are many applications of neural network modelling in state-of-the-art literature. The advantage of ANNs is the ability to produce high-quality prediction models of complex and nonlinear relationships [25–27].

Generally, for the delimitation of management zones, a single electrical parameter (EC_a or MS) is used. EC_a is more popular and is connected with many physical-chemical properties of soil. However, there is still a lack of investigations regarding relationships between EC_a and MS with soil compaction. Thus, the objectives of this study are: (1) To assess whether field-scale EC_a and MS geospatial measurements are potential estimators of soil compaction and (2) to indicate the optimum set of electrical soil parameters for the delimitation of management zones connected with soil compaction.

2. Materials and Methods

2.1. Experimental Data Acquisition

The 12.86 ha (1.286×10^5 m^2) large experimental area is located in Poland, Lower Silesia province, Olesnica district ($51°10'13.5''$ N and $17°27'6.6''$ E). Soil sampling was carried out during the summer period of 2019. In the 2018 growing season, the area was grown with winter oilseed rape, and after combine harvesting, the field was cultivated with the universal cultivator Horsch Terrano with a depth of 0.01 m. Thereafter, until the measurement procedure, no cultivation was carried out. Soil was classified as a sandy clay loam, according to the Polish Soil Classification System and USDA Soil Taxonomy [28]. The physical characterization of soil is shown in Table 1.

Table 1. Soil texture parameters.

Texture	Diameter [mm]	Percentage
skeletans	>2	1.2
sand	2–0.05	57.3
silt	0.05–0.002	18.4
clay	<0.002	24.3
soil texture	Sandy clay loam (SCL)	

The apparent soil electrical conductivity and magnetic susceptibility were measured by electromagnetic induction with the a Geonics EM38-MK2 Ground Conductivity Meter at two depths: 0.5 and 1 m. For the vertical dipole orientation used during measurements, the effective depth is 1.5 m for a coil distance of 1.0 m and 0.75 m for a coil distance of 0.5 m. The position of EM38 was registered by GPS. The technical parameters of the EM38-MK2 meter are detailed in Table 2.

Table 2. Technical parameters of the EM38-MK2 meter.

Measured quantities	1: Apparent conductivity in millisiemens per meter $(mS \cdot m^{-1})$ 2: In-phase ratio of the secondary to primary magnetic field in parts per thousand (ppt)
Intercoil spacing	1 and 0.5 m
Operating frequency	14.5 kHz
Measuring range	Conductivity: $1000\ mS \cdot m^{-1}$ In-phase: $\pm$ 28 ppt for 1 m separation In-phase: $\pm$ 7 ppt for 0.5 m separation
Measurement resolution	$\pm 0.1\%$ of full scale
Measurement accuracy	$\pm 5\%$ at $30\ mS \cdot m^{-1}$
Noise levels	Conductivity: $0.5\ mS \cdot m^{-1}$; in-phase: 0.02 ppt

The soil compaction was measured for a soil layer 0–0.5 m with a Eijkelkamp Penetrologger with GPS. In measurements, a cone with an angle of $60°$ and a base of $0.0001\ m^2$ was used. Penetration speed was equal to $0.03\ m \cdot s^{-1}$. Soil moisture was measured within a short time period after the EM38 survey and was equal to 30%.

After the measurement of electrical parameters and analysis of results, the five management zones were delimited (depicted in different colors in Figure 1). Then, the measurement of soil compaction was conducted every 30 m (measurement points are depicted in Figure 1). Based on GPS information of the penetrologger, coordinates were assigned to each measurement point. Thereafter, with the use of the least squares method, the nearest point of EC_a and MS measurement was connected to each point of the soil compaction measurement.

2.2. Artificial Neural Networks

An artificial neural network is a computational tool modelled on the biological functioning of the human nervous system. An ANN is made up of fundamental information-processing units called artificial neurons. Artificial neurons are arranged in layers. At least two layers are obligatory in the ANN structure—input and output layer. In this work, the two types of ANNs are used for modelling relationships under study—multilayer perceptron (MLP) and radial basis function (RBF) neural network. The MLP is a feedforward neural network with one or two hidden layers of neurons placed between the input and output layer. The MLP is trained with the use of a backpropagation algorithm. An RBF network is a feedforward neural network with one hidden layer. Neurons in a hidden layer are RBF nonlinear activation units, usually with Gaussian functions [29]. The output signal of the RBF network is calculated by a linear neuron. RBF neural networks are considered a tool with better approximation abilities and faster learning speed than other types of ANNs [30].

Figure 1. The map of experimental area with management zones depicted in different colors and soil compaction measurement points.

As a result of measurements, a set of 154 data was obtained. Outliers were removed from the data set. For neural modelling, this data set was randomly divided into training and validation sets in the proportion of 80:20. Prior to the neural models training process, data were normalized into a range <0;1>. In this research, both MLP and RBF neural models were developed in the Statistica v. 13 software. The input vector was composed of a certain combination of apparent soil electrical conductivity and magnetic susceptibility measured at the depth of 0.5 and 1 m. The form of input vectors determined the number of nodes in the input layer (2 or 4). For each model, the group of 2000 neural structures was trained. Each structure was characterized by an ANN type (MLP or RBF), the number of neurons in the hidden layer (from 10 to 40), the initial connection weight matrix, and in the case of the MLP network—transfer functions of neurons in hidden and output layers (hyperbolic tangent, sigmoidal, and exponential). After the training and testing process, the neural model of the best accuracy was indicated based on the coefficient of correlation (R) between experimental data and data calculated by the neural model for the validation data set. As an output model parameter, a compaction of a certain soil layer was used.

For the indication of the optimal vector of input parameters, the determination of the contribution of independent input variables in an ANN model was necessary. For this purpose, the sensitivity analysis implemented in the Statistica v. 13 environment was employed. This method can be used for the MLP neural network to provide information about the relative importance of the input model parameters. The sensitivity analysis is based on replacing the values of each input variable by its mean values. The mean values are calculated using the training data set. Afterward, an error ratio is calculated as follows:

$$R_e = \frac{E_{ch}}{E_{oryg}} \tag{1}$$

where R_e is an error ratio, E_{ch} is a network error with a certain input changed by its mean value, and E_{oryg} is a network error with the input with its original value. Based on the error ratio R_e, the percentages of influence of the input parameters of an ANN model on its output is calculated. This method of sensitivity analysis was used, e.g., in the modelling and analysis of the structural damage after an earthquake [31].

3. Results and Discussion

The statistics of the experimental data (electrical and physical soil properties) are presented in Table 3.

Table 3. Statistics of experimental data.

The Parameter	Minimum	Maximum	Mean	Standard Deviation
apparent soil electrical conductivity 0.5 m [mS·m^{-1}]	0.39	15.23	5.10	3.05
magnetic susceptibility 0.5 m [-]	−0.11	−0.01	−0.09	0.02
apparent soil electrical conductivity 1 m [mS·m^{-1}]	0.00	36.95	8.18	6.77
magnetic susceptibility 1 m [-]	−4.18	0.16	−2.09	0.70
soil compaction (depth 0–0.1 m) [MPa]	0.17	1.03	0.40	0.17
soil compaction (depth 0–0.2 m) [MPa]	0.36	1.60	0.86	0.24
soil compaction (depth 0–0.3 m) [MPa]	0.42	2.04	1.15	0.28
soil compaction (depth 0–0.4 m) [MPa]	0.57	2.20	1.39	0.28
soil compaction (depth 0–0.5 m) [MPa]	0.65	2.20	1.41	0.28
soil compaction (depth 0.1–0.2 m) [MPa]	0.48	2.20	1.28	0.37
soil compaction (depth 0.2–0.3 m) [MPa]	0.42	3.41	1.71	0.51
soil compaction (depth 0.3–0.4 m) [MPa]	0.89	3.28	2.04	0.51
soil compaction (depth 0.4–0.5 m) [MPa]	0.17	3.39	1.14	0.58

As shown in Table 3, the apparent soil electrical conductivity is higher for the measurement depth of 1 m than for 0.5 m. The magnetic susceptibility is negative when measured at a depth of 0.5 m and changes its values from negative to positive, when measured at a depth of 1 m. Soil compaction values increase with the measurement depth when the soil layer being measured is between 0 and a certain depth. When the measurement is carried out for a soil layer of 0.1 m, the soil compaction increases with the increase in soil layer depth up to 0.4 m. The mean soil compaction of the soil layer between 0.4 and 0.5 m is similar to 0.1–0.2 m soil layer compaction. Maps of spatial variability for selected soil parameters presented in Table 3 are depicted in Figure 2.

Maps presented in Figure 2 confirm the similarity in the spatial distribution pattern for EC_a and MS, as well as for soil compaction measured at depths of 0–0.5 and 0.4–0.5 m.

An EM38-MK2 meter provides the measurement of two electrical soil parameters—apparent soil electrical conductivity and magnetic susceptibility. Generally, the apparent soil electrical conductivity is the parameter used for the delimitation of management zones. In order to establish the optimum set of parameters for the delimitation of management zones connected with soil compaction, the group of neural models was developed. As input model parameters, the following vectors were used: [EC_a 0.5 m; EC_a 1 m], [EC_a 0.5 m; MS 0.5 m], [EC_a 0.5 m; MS 0.5 m; EC_a 1 m; MS 1 m], and [MS 0.5 m; MS 1 m]. As an output parameter, the soil compaction of various soil layer depths was used. The type, structure, and quality of the best neural models are summarized in Table 4.

In the process of the development of neural models, the two neural network types were used, namely the multilayer perceptron and radial basis function neural network. The data presented in Table 4 suggest that the better neural network type for modelling the relationships between soil electrical parameters and soil compaction was the RBF neural network. The number of neurons in the hidden layer varies significantly, for MLP models from 11 to 39 and for RBF models from 10 to 36.

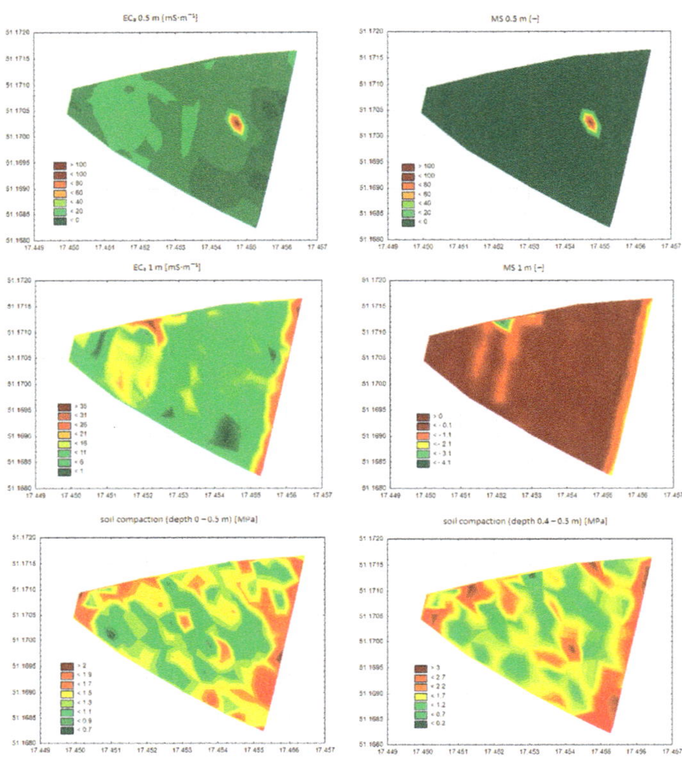

Figure 2. Maps of spatial variability for apparent soil electrical conductivity, magnetic susceptibility, soil compaction (0–0.5 m), and soil compaction (0.4–0.5 m).

Table 4. Neural models characteristics.

Input Parameters	Output Parameter	ANN Type	ANN Structure	R
EC_a 0.5; MS 0.5; EC_a 1; MS 1	soil compaction (0–0.5 m)	MLP	4-17-1	0.769
EC_a 0.5; MS 0.5; EC_a 1; MS 1	soil compaction (0.4–0.5 m)	RBF	4-14-1	0.826
EC_a 0.5; MS 0.5	soil compaction (0–0.5 m)	MLP	2-20-1	0.877
EC_a 0.5; MS 0.5	soil compaction (0.4–0.5 m)	MLP	2-19-1	0.846
EC_a 0.5; MS 0.5	soil compaction (0–0.4 m)	RBF	2-17-1	0.700
EC_a 0.5; MS 0.5	soil compaction (0–0.3 m)	RBF	2-24-1	0.446
EC_a 0.5; MS 0.5	soil compaction (0–0.2 m)	RBF	2-10-1	0.521
EC_a 0.5; MS 0.5	soil compaction (0–0.1 m)	RBF	2-20-1	0.615
EC_a 0.5; MS 0.5	soil compaction (0.3–0.4 m)	RBF	2-33-1	0.662
EC_a 0.5; MS 0.5	soil compaction (0.2–0.3 m)	MLP	2-12-1	0.594
EC_a 0.5; MS 0.5	soil compaction (0.1–0.2 m)	RBF	2-36-1	0.476
EC_a 0.5; EC 1	soil compaction (0–0.5 m)	RBF	2-27-1	0.759
EC_a 0.5; EC 1	soil compaction (0.4–0.5 m)	RBF	2-11-1	0.732
EC_a 0.5; EC 1	soil compaction (0–0.4 m)	RBF	2-28-1	0.656
EC_a 0.5; EC 1	soil compaction (0–0.3 m)	RBF	2-21-1	0.517
EC_a 0.5; EC 1	soil compaction (0–0.2 m)	RBF	2-15-1	0.433
EC_a 0.5; EC 1	soil compaction (0–0.1 m)	MLP	2-16-1	0.501
EC_a 0.5; EC 1	soil compaction (0.3–0.4 m)	RBF	2-21-1	0.648
EC_a 0.5; EC 1	soil compaction (0.2–0.3 m)	RBF	2-27-1	0.470
EC_a 0.5; EC 1	soil compaction (0.1–0.2 m)	RBF	2-12-1	0.471
MS 0.5; MS 1	soil compaction (0–0.5 m)	RBF	2-10-1	0.725
MS 0.5; MS 1	soil compaction (0.4–0.5 m)	MLP	2-39-1	0.790

The input parameters: Apparent soil electrical conductivity, 0.5 m; magnetic susceptibility, 0.5 m; apparent soil electrical conductivity, 1 m; and magnetic susceptibility, 1 m are marked in the table as EC_a 0.5, MS 0.5, EC_a 1, and MS 1, respectively.

Taking into account the fact that electrical parameters were measured at depths of 0.5 and 1 m, it can be expected that a better quality of models is obtained when the output model parameter is the soil compaction measured at a depths of 0–0.5 and 0.4–0.5 m. It was confirmed by the results summarized in Table 4. For an apparent soil electrical conductivity of 0.5 m and magnetic susceptibility of 0.5 m, as well as for apparent soil electrical conductivities of 0.5 and 1 m as input model parameters, the nine neural models were developed with soil compaction measured for each soil layer described in the section 'Materials and Methods'. The model quality increases when soil compaction is measured at greater depths. Only when the output model parameter is the soil compaction measured for a soil layer of 0–0.5 and 0.4–0.5 m is the correlation coefficient R for the validation data set greater than 0.7. The high R value calculated for the validation data set proves the model usefulness for real-life applications and means that no overfitting effect occurred during the training process. Based on the results described above, for magnetic susceptibility, 0.5 and 1 m, as well as for a combination of all electrical parameters as input vector components, only soil compaction measured for soil layers of 0–0.5 and 0.4–0.5 m was taken into account as an output parameter of neural models. In the case of magnetic susceptibility, 0.5 and 1 m as input model parameters, the values of correlation between predicted and experimental soil compaction equal 0.725 for the soil layer at 0–0.5 m and 0.790 for the soil layer at 0.4–0.5 m. When all electrical parameters were input vector components, the R values were 0.769 for the soil layer at 0–0.5 m and 0.826 for the soil layer at 0.4–0.5 m. The highest quality of models was obtained for an apparent soil electrical conductivity at 0.5 m and magnetic susceptibility at 0.5 m as input model parameters. This means that both electrical parameters play important roles in the delimitation of management zones. A slightly lower quality of models was calculated in the case of these two electrical parameters measured at depths of 0.5 and 1 m. It can be assumed that the quality of these models could increase for soil compaction measured at a depth between 0.5 and 1 m. However, soil compaction measured in deeper soil layers is of no practical relevance, because of limitations in rooting depth, as well as characteristics of the working resistance of soil-working machines.

Based on the model of the relationship between apparent soil electrical conductivity and magnetic susceptibility measured at depths of 0.5 and 1 m, and soil compaction 0–0.5 m, the sensitivity analysis was performed in order to determine the average variable relative importance. The results are depicted in Figure 3.

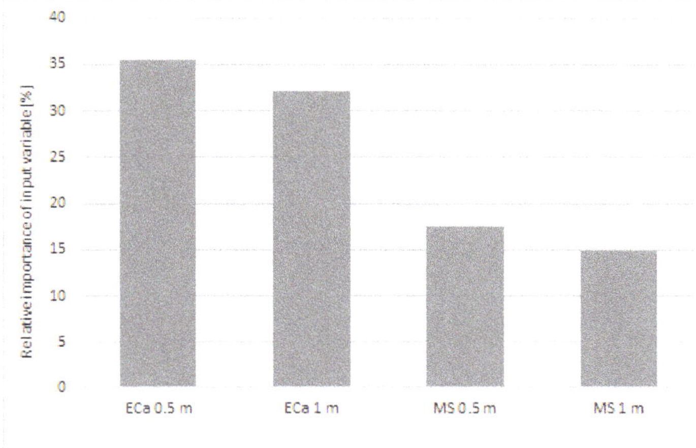

Figure 3. The relative importance of input variables of multilayer perceptron (MLP) model on soil compaction (0–0.5 m).

As illustrated in Figure 3, apparent soil electrical conductivity affects soil compaction about twice as much as magnetic susceptibility. Electrical parameters measured at a depth of 0.5 m influence the compaction of the soil layer at 0–0.5 m slightly more than electrical parameters measured at a depth of 1 m. Soil compaction is affected by many physical, chemical and biological properties of soils, namely water content, organic matter content, mineralogy, soil structure and texture, and particle-size distribution [32]. These parameters also influence soil electrical parameters. Therefore, a close relation between soil compaction and both soil apparent electrical conductivity and magnetic susceptibility can be expected. The correlation between the compaction and electrical resistivity of noncohesive soils was investigated by Kowalczyk et al. (2014) [33]. Al-Gaadi et al. (2012) proved the possibility of assessing soil compaction by EC_a measured by a EM38-MK2 device [34]. The level of correlation between EC_a and soil compaction depends on the EM38-MK2 orientation (vertical or horizontal), moisture content, and the height of the device above the ground. Keskin et al. (2011) found a negative correlation between tillage depth, which depends on soil compaction and soil electrical conductivity [35]. The correlation coefficient ranged from R = −0.56 to R = −0.84 depending, inter alia, on the depth of electrical conductivity measurement. The authors stated that the electrical conductivity measurement system can be used for tillage depth determination.

Electrical soil parameters are increasingly popular for the delimitation of management zones in precision agriculture. Some measuring instruments were developed for this purpose. Some of them measure only one parameter (apparent soil electrical conductivity or magnetic susceptibility). Some instruments allow the measurement of both those parameters (e.g., Geonics EM38). Therefore, various surveys have been presented in the literature for the delimitation of management zones. Guo et al. (2019) presented a geostatistical method combined with a fuzzy clustering algorithm for determination of the spatiotemporal variation of soil salinity based on EC_a measurement conducted with the EM38 tool [36]. Machado et al. (2015) reported a relatively high linear correlation (R = 0.778) between EC_a measured by the EMP-400 profiler sensor at 15 kHz and cation exchange capacity. They found a lower correlation between EC_a and other soil parameters, namely organic matter, potassium, and base saturation [37]. Valente et al. (2014) indicated EC_a as an important tool for defining management zones due to its significant correlation with the micronutrients manganese, zinc and copper, as well as with macronutrients potassium and phosphorus [6]. They also found a strong correlation of EC_a with the remaining phosphorus. It proves the usability of EC_a measurement in precision fertilizer application. Grimley et al. (2004) used only magnetic susceptibility for the delineation of hydric soils in the midwestern USA. They found the critical MS values for differentiation of silty and sandy surface soil textures [11]. Soderstrom et al. (2013) used EC_a, MS, and gamma ray data for mapping soil properties. They reported that EC_a and MS were better indicators of soil properties under study than gamma ray measurements [38]. Jordanova et al. (2013) suggested that the combination of EC_a and magnetic soil parameters can improve the process of soil classification by decreasing the nonunique interpretation of the relationship between apparent conductivity and certain soil physical properties [39].

Artificial neural networks have been employed by many researchers for prediction of soil properties and, based on results, ANNs have been introduced as an efficient model of relationships under study. In our research, the correlation coefficient R does not exceed 0.877, which is not a very high correlation when compared with other neural prediction models evaluated by authors [40,41]. It can be assumed that in addition to the soil parameters reflected by EC_a and MS, other soil features influence the soil compaction. Similar or lower neural models' quality was reported in a prior literature for ANNs as a tool for the prediction of soil properties. Khanbabakhani et al. (2020) indicated ANNs as an efficient model of the relationship between soil texture and longitude, altitude, elevation, and slope percentage. They compared an ANN model with three other interpolation methods including inverse distance weighting, kriging, and co-kriging used for soil texture mapping. Based on statistical indices, the efficiency of the ANN model was better than other methods. However, the correlation of the model was less than 0.5 [42]. A feed-forward back-propagation network with one hidden layer was used by Aitkenhead et al. (2015) as a predictive model of soil organic matter [43]. They calculated different qualities of models depending on values of organic matter in the training data set. For a full-organic-matter-content-range model, they reported R = 0.93, and for a small-organic-matter-content model (less than 20%), R = 0.82. Tasan et al. (2020) used a MLP neural network for the prediction of soil moisture described by two parameters, namely field capacity and permanent wilting point [44]. Among the ANN models developed in this research, a model with four input parameters (sand, clay, percent calcium carbonate, and cation exchange capacity) yielded the greatest quality (R = 0.88 for field capacity and R = 0.81 for permanent wilting point as output parameter of model).

4. Conclusions

In recent literature, there is a lack of scientific papers regarding relationships between electrical parameters and soil compaction. Therefore, in this research, two parameters: EC_a and MS, as potential estimators of soil compaction were investigated. The spatial variability maps present similar patterns for apparent soil electrical conductivity and magnetic susceptibility, both measured at a depth of 0.5 m. A similar pattern was also found for soil compaction measured at a depth of 0–0.5 m, and soil compaction measured at a depth of 0.4–0.5 m. The neural models of high quality were developed for prediction of soil compaction measured for soil layers at 0–0.5 and 0.4–0.5 m based on electrical parameters of the soil. The RFB neural network turned out to be a better prediction tool for this purpose. In the mathematical model, the apparent soil electrical conductivity affects soil compaction significantly more than magnetic susceptibility. In spite of this fact, MS provides additional information about soil properties. Therefore, MS should be used as a complementary parameter to EC_a in the process of the delimitation of management zones. It can be concluded that the measurement of EC_a and MS can be considered a relatively inexpensive, easy, and fast technique with the potential to make significant contributions to the identification and prediction of soil compaction. This technique is promising for real-life applications, e.g., for variable-depth tillage technology based on soil compaction mapping that can reduce fuel consumption.

Author Contributions: Conceptualization, K.P. (Katarzyna Pentoś), K.P. (Krzysztof Pieczarka) and K.S.; data curation, K.P. (Katarzyna Pentoś), K.P. (Krzysztof Pieczarka) and K.S.; formal analysis, K.P. (Katarzyna Pentoś) and K.P. (Krzysztof Pieczarka); investigation, K.P. (Katarzyna Pentoś), K.P. (Krzysztof Pieczarka) and K.S.; methodology, K.P. (Katarzyna Pentoś), K.P. (Krzysztof Pieczarka) and K.S.; writing—original draft, K.P. (Katarzyna Pentoś); writing—review and editing, K.P. (Katarzyna Pentoś) and K.P. (Krzysztof Pieczarka); visualization, K.P. (Katarzyna Pentoś) and K.P. (Krzysztof Pieczarka). All authors have read and agreed to the published version of the manuscript.

Funding: This research received no external funding.

Institutional Review Board Statement: Not applicable.

Informed Consent Statement: Not applicable.

Data Availability Statement: Data are available by contacting the authors.

Conflicts of Interest: The authors declare no conflict of interest.

References

1. Pedrera-Parrilla, A.; Brevik, E.C.; Giraldez, J.V.; Vanderlinden, K. Temporal stability of electrical conductivity in a sandy soil. *Int. Agrophys.* **2016**, *30*, 349–357. [CrossRef]
2. Moral, F.J.; Terron, J.M.; da Silva, J.R.M. Delineation of management zones using mobile measurements of soil apparent electrical conductivity and multivariate geostatistical techniques. *Soil. Till. Res.* **2010**, *106*, 335–343. [CrossRef]
3. Brevik, E.C.; Calzolari, C.; Miller, B.A.; Pereira, P.; Kabala, C.; Baumgarten, A.; Jordan, A. Soil mapping, classification, and pedologic modeling: History and future directions. *Geoderma* **2016**, *264*, 256–274. [CrossRef]
4. Dalchiavon, F.C.; Carvalho, M.D.E.; Montanari, R.; Andreotti, M. Strategy of specification of management areas: Rice grain yield as related to soil fertility. *Rev. Bras. Cienc. Solo* **2013**, *37*, 45–54. [CrossRef]

5. Fleming, K.L.; Heermann, D.F.; Westfall, D.G. Evaluating soil color with farmer input and apparent soil electrical conductivity for management zone delineation. *Agron. J.* **2004**, *96*, 1581–1587. [CrossRef]

6. Valente, D.S.M.; De Queiroz, D.M.; Pinto, F.D.D.; Santos, F.L.; Santos, N.T. Spatial variability of apparent electrical conductivity and soil properties in a coffee production field. *Eng. Agric.* **2014**, *34*, 1224–1233. [CrossRef]

7. Bertermann, D.; Schwarz, H. Laboratory device to analyse the impact of soil properties on electrical and thermal conductivity. *Int. Agrophys.* **2017**, *31*, 157–166. [CrossRef]

8. Kuhn, J.; Brenning, A.; Wehrhan, M.; Koszinski, S.; Sommer, M. Interpretation of electrical conductivity patterns by soil properties and geological maps for precision agriculture. *Precis. Agric.* **2009**, *10*, 490–507. [CrossRef]

9. Martinez, G.; Vanderlinden, K.; Ordonez, R.; Muriel, J.L. Can Apparent Electrical Conductivity Improve the Spatial Characterization of Soil Organic Carbon? *Vadose Zone J.* **2009**, *8*, 586–593. [CrossRef]

10. Ayoubi, S.; Adman, V.; Yousefifard, M. Use of magnetic susceptibility to assess metals concentration in soils developed on a range of parent materials. *Ecotox. Environ. Safe.* **2019**, *168*, 138–145. [CrossRef]

11. Grimley, D.A.; Arruda, N.K.; Bramstedt, M.W. Using magnetic susceptibility to facilitate more rapid, reproducible and precise delineation of hydric soils in the midwestern USA. *Catena* **2004**, *58*, 183–213. [CrossRef]

12. Siqueira, D.S.; Marques, J.; Matias, S.S.R.; Barron, V.; Torrent, J.; Baffa, O.; Oliveira, L.C. Correlation of properties of Brazilian Haplustalfs with magnetic susceptibility measurements. *Soil Use Manag.* **2010**, *26*, 425–431. [CrossRef]

13. Maher, B.A. Magnetic properties of modern soils and Quaternary loessic paleosols: Paleoclimatic implications. *Palaeogeogr. Palaeocl.* **1998**, *137*, 25–54. [CrossRef]

14. Yang, P.G.; Yang, M.; Mao, R.Z.; Byrne, J.M. Impact of Long-Term Irrigation with Treated Sewage on Soil Magnetic Susceptibility and Organic Matter Content in North China. *Bull. Environ. Contam. Tox.* **2015**, *95*, 102–107. [CrossRef]

15. Peralta, N.R.; Costa, J.L.; Balzarini, M.; Angelini, H. Delineation of management zones with measurements of soil apparent electrical conductivity in the southeastern pampas. *Can. J. Soil. Sci.* **2013**, *93*, 205–218. [CrossRef]

16. Molin, J.P.; de Castro, C.N. Establishing management zones using soil electrical conductivity and other soil properties by the fuzzy clustering technique. *Sci. Agric.* **2008**, *65*, 567–573. [CrossRef]

17. De Caires, S.A.; Wuddivira, M.N.; Bekele, I. Spatial analysis for management zone delineation in a humid tropic cocoa plantation. *Precis. Agric.* **2015**, *16*, 129–147. [CrossRef]

18. Li, Y.; Shi, Z.; Li, F. Delineation of site-specific management zones based on temporal and spatial variability of soil electrical conductivity. *Pedosphere* **2007**, *17*, 156–164. [CrossRef]

19. Serrano, J.; Shahidian, S.; da Silva, J.M. Spatial and Temporal Patterns of Apparent Electrical Conductivity: DUALEM vs. Veris Sensors for Monitoring Soil Properties. *Sensors* **2014**, *14*, 10024–10041. [CrossRef]

20. De Souza Bahia, A.S.R.; Marques, J.; La Scala, N.; Cerri, C.E.P.; Camargo, L.A. Prediction and Mapping of Soil Attributes using Diffuse Reflectance Spectroscopy and Magnetic Susceptibility. *Soil Sci. Soc. Am. J.* **2017**, *81*, 1450–1462. [CrossRef]

21. Wang, X.S. Magnetic properties and heavy metal pollution of soils in the vicinity of a cement plant, Xuzhou (China). *J. Appl. Geophys.* **2013**, *98*, 73–78. [CrossRef]

22. Sarris, A.; Kokinou, E.; Aidona, E.; Kallithrakas-Kontos, N.; Koulouridakis, P.; Kakoulaki, G.; Droulia, K.; Damianovits, O. Environmental study for pollution in the area of Megalopolis power plant (Peloponnesos, Greece). *Environ. Geol.* **2009**, *58*, 1769–1783. [CrossRef]

23. Tan, X.; Chang, S.X.; Kabzems, R. Soil compaction and forest floor removal reduced microbial biomass and enzyme activities in a boreal aspen forest soil. *Biol. Fert. Soils* **2008**, *44*, 471–479. [CrossRef]

24. Kristoffersen, A.O.; Riley, H. Effects of soil compaction and moisture regime on the root and shoot growth and phosphorus uptake of barley plants growing on soils with varying phosphorus status. *Nutr. Cycl. Agroecosys.* **2005**, *72*, 135–146. [CrossRef]

25. Niedbała, G.; Nowakowski, K.; Rudowicz-Nawrocka, J.; Piekutowska, M.; Weres, J.; Tomczak, R.J.; Tyksiński, T.; Álvarez, P.A. Multicriteria Prediction and Simulation of Winter Wheat Yield Using Extended Qualitative and Quantitative Data Based on Artificial Neural Networks. *Appl. Sci.* **2019**, *9*, 2773. [CrossRef]

26. Niedbała, G.; Piekutowska, M.; Weres, J.; Korzeniewicz, R.; Witaszek, K.; Adamski, M.; Pilarski, K.; Czechowska-Kosacka, A.; Krysztofiak-Kaniewska, A. Application of Artificial Neural Networks for Yield Modeling of Winter Rapeseed Based on Combined Quantitative and Qualitative Data. *Agronomy* **2019**, *9*, 781. [CrossRef]

27. Wang, L.; Wang, P.X.; Liang, S.L.; Zhu, Y.C.; Khan, J.; Fang, S.B. Monitoring maize growth on the North China Plain using a hybrid genetic algorithm-based back-propagation neural network model. *Comput. Electron. Agric.* **2020**, *170*. [CrossRef]

28. Distribution, P.S. Particle size distribution and textural classes of soils and mineral materials-classification of Polish Society of Soil Science (2009) Annals of. *Soil Sci.* **2009**, *60*, 5–16.

29. Faris, H.; Aljarah, I.; Mirjalili, S. Evolving radial basis function networks using moth-flame optimizer. In *Handbook of Neural Computation*; Academic Press: Cambridge, MA, USA, 2017; pp. 537–550. [CrossRef]

30. Ahmadian, A.S. Numerical Modeling and Simulation. In *Numerical Models for Submerged Breakwaters*; Ahmadian, A.S., Ed.; Butterworth-Heinemann: Oxford, UK, 2016; pp. 109–126.

31. Hadzima-Nyarko, M.; Nyarko, E.K.; Moric, D. A neural network based modelling and sensitivity analysis of damage ratio coefficient. *Expert Syst. Appl.* **2011**, *38*, 13405–13413. [CrossRef]

32. Nawaz, M.F.; Bourrie, G.; Trolard, F. Soil compaction impact and modelling. A review. *Agron. Sustain. Dev.* **2013**, *33*, 291–309. [CrossRef]

33. Kowalczyk, S.; Maslakowski, M.; Tucholka, P. Determination of the correlation between the electrical resistivity of non-cohesive soils and the degree of compaction. *J. Appl. Geophys.* **2014**, *110*, 43–50. [CrossRef]
34. Al-Gaadi, K. Employing Electromagnetic Induction Technique for the Assessment of Soil Compaction. *Americ. J. Agric. Biol. Sci.* **2012**, *7*, 425–434.
35. Keskin, S.; Khalilian, A.; Han, Y.J.; Dodd, R. Variable-depth Tillage based on Geo-referenced Soil Compaction Data in Coastal Plain Soils. *Int. J. Appl. Sci. Tech.* **2011**, *1*, 22–32.
36. Guo, Y.; Zhou, Y.; Zhou, L.Q.; Liu, T.; Wang, L.G.; Cheng, Y.Z.; He, J.; Zheng, G.Q. Using proximal sensor data for soil salinity management and mapping. *J. Integr. Agric.* **2019**, *18*, 340–349. [CrossRef]
37. Machado, F.C.; Montanari, R.; Shiratsuchi, L.S.; Lovera, L.H.; Lima, E.D. Spatial dependence of electrical conductivity and chemical properties of the soil by electromagnetic induction. *Rev. Bras. Cienc. Solo* **2015**, *39*, 1112–1120. [CrossRef]
38. Soderstrom, M.; Eriksson, J.; Isendahl, C.; Araujo, S.R.; Rebellato, L.; Schaan, D.P.; Stenborg, P. Using proximal soil sensors and fuzzy classification for mapping Amazonian Dark Earths. *Agric. Food Sci.* **2013**, *22*, 380–389. [CrossRef]
39. Jordanova, D.; Jordanova, N.; Werban, U. Environmental significance of magnetic properties of Gley soils near Rosslau (Germany). *Environ. Earth Sci.* **2013**, *69*, 1719–1732. [CrossRef]
40. Pentos, K.; Pieczarka, K.; Lejman, K. Application of Soft Computing Techniques for the Analysis of Tractive Properties of a Low-Power Agricultural Tractor under Various Soil Conditions. *Complexity* **2020**, *2020*. [CrossRef]
41. Cieniawska, B.; Pentos, K.; Luczycka, D. Neural modeling and optimization of the coverage of the sprayed surface. *Bull. Pol. Acad. Sci. Tech.* **2020**, *68*, 601–608. [CrossRef]
42. Khanbabakhani, E.; Torkashvand, A.M.; Mahmoodi, M.A. The possibility of preparing soil texture class map by artificial neural networks, inverse distance weighting, and geostatistical methods in Gavoshan dam basin, Kurdistan Province, Iran. *Arab. J. Geosci.* **2020**, *13*. [CrossRef]
43. Aitkenhead, M.J.; Donnelly, D.; Sutherland, L.; Miller, D.G.; Coull, M.C.; Black, H.I.J. Predicting Scottish topsoil organic matter content from colour and environmental factors. *Eur. J. Soil Sci.* **2015**, *66*, 112–120. [CrossRef]
44. Tasan, S.; Demir, Y. Comparative Analysis of MLR, ANN, and ANFIS Models for Prediction of Field Capacity and Permanent Wilting Point for Bafra Plain Soils. *Commun. Soil Sci. Plan.* **2020**, *51*, 604–621. [CrossRef]

Article

Evaluation of Convolutional Neural Networks' Hyperparameters with Transfer Learning to Determine Sorting of Ripe Medjool Dates

Blanca Dalila Pérez-Pérez [1], Juan Pablo García Vázquez [1,*] and Ricardo Salomón-Torres [2,*]

[1] Facultad de Ingeniería, Universidad Autónoma de Baja California (UABC), Mexicali 21100, Mexico; dalila.perez9@uabc.edu.mx

[2] Unidad Académica San Luis Río Colorado, Universidad Estatal de Sonora (UES), Sonora 83500, Mexico

* Correspondence: pablo.garcia@uabc.edu.mx (J.P.G.V.); ricardo.salomon@ues.mx (R.S.-T.); Tel.: +52-686-191-2431 (J.P.G.V.); +52-653-534-4255 (R.S.-T.)

Abstract: Convolutional neural networks (CNNs) have proven their efficiency in various applications in agriculture. In crops such as date, they have been mainly used in the identification and sorting of ripe fruits. The aim of this study was the performance evaluation of eight different CNNs, considering transfer learning for their training, as well as five hyperparameters. The CNN architectures evaluated were VGG-16, VGG-19, ResNet-50, ResNet-101, ResNet-152, AlexNet, Inception V3, and CNN from scratch. Likewise, the hyperparameters analyzed were the number of layers, the number of epochs, the batch size, optimizer, and learning rate. The accuracy and processing time were considered to determine the performance of CNN architectures, in the classification of mature dates' cultivar Medjool. The model obtained from VGG-19 architecture with a batch of 128 and Adam optimizer with a learning rate of 0.01 presented the best performance with an accuracy of 99.32%. We concluded that the VGG-19 model can be used to build computer vision systems that help producers improve their sorting process to detect the Tamar stage of a Medjool date.

Keywords: *Phoenix dactylifera* L.; Medjool dates; image classification; convolutional neural networks; deep learning; transfer learning

Citation: Pérez-Pérez, B.D.; García Vázquez, J.P.; Salomón-Torres, R. Evaluation of Convolutional Neural Networks' Hyperparameters with Transfer Learning to Determine Sorting of Ripe Medjool Dates. *Agriculture* **2021**, *11*, 115. https://doi.org/10.3390/agriculture11020115

Academic Editors: Sebastian Kujawa and Gniewko Niedbała
Received: 31 December 2020
Accepted: 25 January 2021
Published: 1 February 2021

Publisher's Note: MDPI stays neutral with regard to jurisdictional claims in published maps and institutional affiliations.

1. Introduction

The date palm fruit (*Phoenix dactylifera* L.) is a berry composed of a fleshy mesocarp, covered by a thin epicarp and an endocarp covering all of its seed [1]. The name of this fruit is "date," which comes from the Greek word "Daktylos," which means "finger" [2]. This fruit has been the primary source of food in several countries in the Middle East, playing an essential role in the economy, society, and environment [3].

This fruit's growth presents a progressive maturity level in four stages known by their Arabic names: Kimri, Khalal, Rutab, and Tamar. At its first stage of growth (Kimri), the fruit is small, green, and with a hard texture. In its second stage (Khalal), the fruit reaches its maximum size and changes it is green color to yellow or red. In the third stage (Rutab), the fruit is losing weight and moisture, turning the fruit into a brown color. In the last stage (Tamar), the fruit is ripe and ready to be harvested [4].

According to Food and Agriculture Organization of the United Nations data, the world's largest date producers are Egypt, Saudi Arabia, Iran, Algeria, and Iraq, producing 66% of the world production in 2018 [5]. However, despite not being a native crop of the American continent, the date has also become a priority fruit for cultivation in southern California and Arizona in the United States and northwestern Mexico, where high-quality dates, such as Medjool cultivar, are grown [6].

The date palm producers face several challenges concerning harvesting, sorting, and packaging because they are mainly performed manually [7]. Therefore, many employers are hiring for these activities that involve long working hours. People perform repetitive tasks, causing mistakes in the correct inspection of the fruit's quality attributes, such as color (maturity level), size, and texture.

Particularly in the Medjool date harvesting process, fruit pickers shake the palm bunch so that the ripe dates fall into containers. This can cause the ripe fruit to suffer damage to its texture or that fruits in other ripening stages are also harvested. The dates are placed in trays, where the immature ones will be extracted and grouped in other trays to dry them in the sun until they reach their full maturity. In contrast, the minute or damaged ones are commonly separated to develop date by-products or for animal consumption. Finally, the Medjool date sorting (which has the required degree of maturity) is packed.

To automate the processes related to harvesting, sorting, and packaging of dates, recently, there has been interest in exploring convolutional neural networks (CNNs) [8,9]. CNN has shown exceptional accuracy for classifying fruits and vegetables, considering several quality parameters, such as color (maturity level), shape, texture, and size [10–14].

Regarding dates, we identified that some studies use machine learning algorithms and image processing techniques to sort among date palm fruit or to detect among their different maturity stages [15–17]. Further, there are research works that propose using CNNs [8,9,18]. However, these studies do not present models to detect the maturity stage of the Medjool date.

The main contribution of this article was the identification of the hyperparameters that best influenced the training of a CNN architecture that transfers learning to Medjool's mature date sorting. To achieve it, we performed a comparison of the performance of eight CNN architectures. Two versions of the CNN architecture are called the Visual Geometric Group (VGG) from Oxford University, VGG-16 and VGG-19. Three versions of the CNN architecture are called Residual Network from Microsoft research, ResNet-50, ResNet-101, and ResNet-152. WE also looked at AlexNet, Inception Version 3, and a CNN from scratch. The hyperparameters analyzed were the number of layers, the number of epochs, the batch size, optimizer, and learning rate.

2. Materials and Methods

2.1. Image Acquisition

The images corresponding to ripe and unripe dates in trays were taken in September 2020, during the first round of harvest of Medjool dates in the plantation located in Colonia La Herradura (32°36′56″ N, 115°15′36″ W) in the Mexicali Valley, Mexico. The acquisition of images was made with three different cameras, using natural light between 8:00 a.m. and 2:00 p.m. We used a Canon EOS Rebel T6 of 18 megapixels and the cameras of the smartphones Samsung, SM-N950F and SM-N960F, which have a dual camera of 12 megapixels.

2.2. Image Data Set

The image data set contained 1002 images in JPG format, which are of different sizes (5184 × 3456, 4449 × 3071, and 4376 × 3375 pixels). The network architectures were trained with JPG images because they are fed with low-quality images in real scenarios. We refer to low-quality as images with blur, noise, contrast, or compression. We considered that if you are trained in architecture with this type of image, the system will be able to classify the Medjool date in images with these features. Further, a study shows that convolutional neural networks are minimally affected in their performance by using JPG format [19].

The image data set was distributed as follows: 501 images of ripe dates and 501 images of unripe dates on trays (Figure 1). The dates in trays were previously classified as ripe or unripe by expert people.

Figure 1. Example of dataset images. (**A–C**) Trays with ripe dates. (**D–F**) Trays with unripe dates.

2.3. Convolutional Neural Networks

The convolutional neural network (CNN) is a type of artificial neural network, where neurons correspond to receptive fields similar to the neurons in the primary visual cortex of a biological brain [20]. Also, CNN is identical to ordinary neural networks such as multilayer perceptron. They are composed of neurons that have weights and biases that can learn. Each neuron receives some input, performs a scalar product, and then applies an activation function [21]. The CNN as a multilayer perceptron has a loss or cost function on the last layer, which will be fully connected.

Figure 2 presents a CNN structure, which consists of three blocks. The first is the input, an image. Next, we can see the block of feature extraction, which consists of convolutional and pooling layers. Finally, the third block is of classification, which consists of fully connected layers and softmax. The structure of the convolutional network changes as the number of convolution and pooling layers increases.

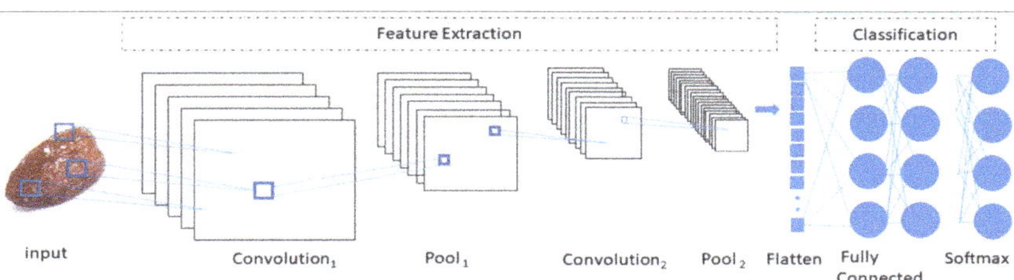

Figure 2. Representation of a basic convolutional neural network (CNN).

The main difference between convolutional neural networks from ordinary neural networks is that they explicitly assume that the inputs are images [21], allowing specific encoding properties in the architecture, gaining efficiency and reducing the number of parameters in the network. In this way, CNNs can model complex variations and behaviors, giving quite accurate predictions. This study considered eight CNNs' architectures: VGG-16, VGG-19, Inception V3, ResNet-50, ResNet-101, ResNet-152, AlexNet, and one CNN from scratch.

2.3.1. VGG-16 and VGG-19 Architectures

VGG is the abbreviation for the Visual Geometric Group [22]. The VGG model was developed by Simonay and Zimmerman [23]. VGG uses 3×3 convolutional layers stacked on top of each other in increasing depth. The reduction of volume size is handled by max pooling. Two fully connected layers, each with 4096 nodes, are then followed by a softmax classifier [23]. The number 16 or 19 is the layer of networks considered deep.

2.3.2. Inception V3 Architecture (GoogLeNet V3)

This architecture was born with the name of GoogLeNet, but subsequent updates have been called Inception vN, where N refers to the version number put out by Google [22]. The basic module of Inception [24] consists of four branches concatenated in parallel: a 1×1 kernel convolution, followed by two 3×3 convolutions; a 1×1 convolution, followed by a 3×3 convolution; a pooling, followed by a 1×1 convolution; and finally a 1×1 convolution. Inception consists of 10 modules, although these modules are going slightly as the net gets deeper. Five of the modules are changed with the purpose of reducing the computational cost by replacing the $n \times n$ convolutions with two convolutions, a 1×7 followed by a 7×1. Two last modules replace the last two convolutions: 3×3 of the first branch with two convolutions each and one 1×3 followed by another 3×1, this time in parallel. In total, Inception V3 has 42 layers with parameters.

2.3.3. ResNet-50, ResNet-101, and ResNet-152 Architectures (Residual Neural Network)

ResNet [25] does not have a fixed depth and depends on the number of consecutive modules used. However, increasing the network's depth to obtain a greater precision makes the network more difficult to optimize. ResNet addresses this problem by adjusting a residual application in place of the original and adding several connections between layers. These new connections skip several layers and perform an identity or a 1×1 convolution. The base block of this network is called the residual block. When the network has 50 or more layers, it is composed of three sequential convolutions, a 1×1, a 3×3, and a 1×1, and a connection that links the input of the first convolution to the output of the third convolution. This study used three models with this architecture, ResNet-50, ResNet-101, and ResNet-152, which are composed of 50, 101, and 152 layers, respectively.

2.3.4. AlexNet

This architecture consists of five convolutional layers and three fully connected layers. Some convolution layers are followed by max-pooling layers (1, 2, and 5 layers). The Rectified Linear Unit (ReLU) nonlinearity is applied to the output of every convolutional and fully connected layer. The fully connected layers have 4096 neurons each [26]. To avoid data over-adjustment, a regularization method is used, known as a dropout, which consists of "turning off" neurons with a predetermined probability during training.

2.3.5. CNN from Scratch

The CNN that we built from scratch was composed of four alternate convolutional and max-pooling layers, followed by a dropout after every other convolutional and pooling pair. After the last grouping layer, we attached a fully connected layer with 256 neurons, another dropout layer, and, finally, a softmax classification layer for our classes. The loss function was the cross entropy since it is useful with convolutional neural networks, most significantly for purposes of image classification [27]. In order to compare the performance of a network that learns from scratch against other architectures that start from transfer learning, a convolutional network was trained from scratch.

2.3.6. CNNs' Optimization Techniques and Hyperparameters

Techniques

All the above networks were too deep to train them from scratch with our data set. Therefore, we used transfer learning, which consists of taking the features learned in other contexts and using them in a new and similar problem [28]. Transfer learning is usually done for tasks where the data set has too little data to train a full-scale model from scratch. This was our case since we only had 1002 Medjool date images.

Transfer learning is commonly used in two ways: (1) pretraining model, which consists of using a pretrained model that replace its last layers with others, so that the characteristics are of the new data set and (2) convolutional network tuning, which is a strategy to tune the weights of the layers using backward propagation.

For this study, the application of transfer learning was the pretraining model. We used the pre-trained networks with ImageNet, which is a large visual database designed for use in visual object recognition [26]. We removed the final classification layer, the neuron softmax layer at the end, which corresponds to ImageNet, and instead replaced it with a new softmax layer for our image data set. A summary of the utilized CNN architectures is shown in Table 1.

Table 1. Characteristics of the CNNs' architectures used in this study.

Network	Depth (Hidden Layers)	Image Size	Parameters
CNN from scratch	24	224 × 224	1,209,058
VGG-16	16	224 × 224	134,268,738
VGG-19	19	224 × 224	143,667,240
ResNet-50	50	224 × 224	23,591,810
ResNet-101	101	224 × 224	42,662,274
ResNet-152	152	224 × 224	58,375,042
Inception v3	48	299 × 299	21,806,882
AlexNet	8	227 × 227	56,328,962

Hyperparameters

Hyperparameters are variables that define the structure of a convolutional network as well as allow it to be trained [29]. These hyperparameters are learning rate, epochs, optimizer, batch size, number of layers, and activation functions, among others, which can be adjusted to make CNN more efficient. In this study, we changed the values of the hyperparameters optimizer, learning rate, batch, and epochs. Our CNN used an optimizer Adaptive Moment Estimation (Adam) and Stochastic Gradient Descent (SGD) since those are well-known optimizers, which have good performance to classify images in CNN [30]. The learning rates for the optimizers were 0.01 and 0.001. The batch size value was 64 and 128, the epochs were 25 and 400, and the number of layers depended on the CNN architecture used (Table 1).

2.4. Experimental Framework

To implement and evaluate the CNN architectures presented in Section 2.3, we used the Google Colab cloud service based on Jupyter's Notebooks, which allows the free use of Google's GPUs or TPUs, with the libraries Scikit-learn, PyTorch, TensorFlow, Keras, and OpenCV [31]. The hardware specifications used in this experiment were GPU: Nvidia-Tesla-T4; CPU: Intel(R) Xeon(R) CPU @ 2.20 GHz; RAM: ~12.78 GB available; Hard Disk: ~32.20 GB available; and the software specifications were and Operating System: 18.04.5 LTS (Bionic Beaver) with the libraries Keras 1.0.8 and Tensorflow 1.15 as a back-end.

2.5. Performance Evaluation

The accuracy is the metric used to evaluate the classification performance of the architectures proposed in this paper. This metric calculates the percentage of samples that are correctly classified, and it is represented in the next equation:

$$\text{Accuracy} = \frac{\text{tp} + \text{tn}}{\text{tp} + \text{tn} + \text{fp} + \text{fn}} \tag{1}$$

where tp represents true positives, those that belonged to the class and were correctly classified in that class; tn represents true negatives, those that did not belong to the class and were correctly classified in another class; fp represents false positives, those that did not belong to the class and were wrongly assigned to the class; and finally, fn represents false negatives, those that belonged to the class and were mistakenly classified in another class.

3. Results

Using the Adam parameter as an optimizer, it can be observed in Table 2 that for the evaluation with 25 epochs, the highest performance percentage was for VGG-16 (96.63% and 95.27%), with a learning rate (0.001), and for VGG-19 (93.92% and 97.30%), with a learning rate (0.01). The lowest performance was for AlexNet (64.19%) and ResNet-152 (64.17%), for a learning rate (0.001), and CNN from scratch (46.62% and 53.38%), with a learning rate (0.01). On the other hand, for 400 epochs, the highest percentage was Inception V3 (98.65%) and VGG-19 (98.75%), both for a learning rate (0.001) and for Inception V3 (98.65%) and VGG-19 (99.32%), with a learning rate (0.01). Likewise, the lowest performance was for ResNet-101 and ResNet-152 (both with 80.41%) and ResNet-101 (79.05%), for a learning rate (0.001) and, finally, AlexNet (67.57%) and CNN from scratch (43.24%), both with a learning rate (0.01). It can also be observed that the two best results were for VGG-19 (99.32% and 98.65%) for a batch (128), followed by Inception V3 (98.65%) for both batches (64); all these for 400 epochs.

Table 2. Accuracy evaluation of eight CNN architectures, changing the values of the hyperparameters of a batch, learning rate, and epochs, using the Adaptive Moment Estimation (Adam) parameter as an optimizer.

Parameters	CNN from Scratch	VGG-16	VGG-19	ResNet-50	ResNet-101	ResNet-152	AlexNet	Inception V3
Epochs	25	25	25	25	25	25	25	25
	400	400	400	400	400	400	400	400
Batch = 64, Optimizer = Adam, Learning Rate = 0.001								
Accuracy	93.24	96.63	90.54	68.92	71.62	74.32	64.19	96.62
(%)	94.59	96.62	95.95	81.08	80.41	80.41	88.51	98.65
Time	9	24	14	11	14	25	11	12
(min)	16	40	43	33	41	61	19	131
Batch = 128, Optimizer = Adam, Learning Rate = 0.001								
Accuracy	85.13	95.27	87.84	70.95	70.27	64.17	75.00	93.24
(%)	93.92	97.29	98.65	83.11	75.67	81.08	85.81	95.27
Time	11	12	5	13	7	9	11	13
(min)	12	34	46	45	16	54	19	48
Batch = 64, Optimizer = Adam, Learning Rate = 0.01								
Accuracy	46.62	85.81	93.92	83.78	65.54	75.68	85.81	95.95
(%)	95.27	95.95	96.62	86.49	79.05	84.46	67.57	98.65
Time	10	12	12	10	11	16	16	11
(min)	14	43	45	34	45	65	18	47
Batch = 128, Optimizer = Adam, Learning Rate = 0.01								
Accuracy	53.38	97.29	97.30	84.46	76.35	66.89	89.19	93.92
(%)	43.24	96.62	99.32	84.46	79.73	81.76	87.16	95.95
Time	11	12	13	12	11	15	11	13
(min)	16	38	43	33	46	60	18	44

Regarding the time parameter in Table 2, CNN from scratch had the lowest values for processing time. However, some values were higher than those reported by ResNet-50, ResNet-101, ResNet-152, and AlexNet architectures. Likewise, the highest processing times in 25 epochs were for ResNet-152 (25 min) and Inception V3 (13 min), with a learning rate (0.001), and for ResNet-152 and AlexNet (16 min) and ResNet-152 (15 min), for a learning rate (0.01). For 400 epochs, the highest process time was for Inception V3 (131 min) and ResNet-152 (54 min), both for a learning rate (0.001) and ResNet-152 (65 and 60 min), with a learning rate (0.01). The ResNet-152 architecture was the CNN that required the most processing time on its network for most hyperparameters. The highest processing times were not associated with high or low accuracy.

Table 3 reveals that using the Stochastic Gradient Descent (SGD) parameter as an optimizer, for an evaluation with 25 epochs, the highest performance percentage was for VGG-19 (87.16%) and VGG-16 (87.16%), with a learning rate (0.001), and for Inception V3 (92.56% and 91.89%), with a learning rate (0.01). While the lowest performance was for AlexNet (52.70%) and CNN from scratch (51.35%), for a learning rate (0.001), and for ResNet-50 and ResNet-152 (both with 45.94%) and ResNet-50 (45.94%), with a learning rate (0.01). On the other hand, for 400 epochs, the highest percentage was obtained by Inception V3 (95.94%) and CNN from scratch (94.59%), both for a learning rate (0.001), and VGG-19 (94.59%) and Inception V3 (95.27%), with a learning rate (0.01). Likewise, the lowest performance was obtained by AlexNet (56.08% and 60.81%), for a learning rate (0.001), and, finally, ResNet-50 (50% and 52.03%) with a learning rate (0.01). It can also be observed that the two best CNN architectures turned out to be CNN from scratch (94.59%) and Inception V3 (95.27%) for a batch (128), followed by Inception V3 (95.94%) and VGG-19 (94.59%) for a batch (64).

Table 3. Accuracy evaluation of eight CNNs' architectures, changing the batch's hyperparameters values, learning rate, and epochs, using the Stochastic Gradient Descent (SGD) parameter as an optimizer.

Parameters	CNN from Scratch	VGG-16	VGG-19	ResNet-50	ResNet-101	ResNet-152	AlexNet	Inception V3
Epochs	25	25	25	25	25	25	25	25
	400	400	400	400	400	400	400	400
Batch = 64, Optimizer = SGD, Learning Rate = 0.001								
Accuracy	54.05	86.49	87.16	66.89	53.38	54.05	52.70	86.50
(%)	93.24	93.24	91.21	75.68	75.00	64.19	56.08	95.94
Time	12	14	13	11	12	13	11	13
(min)	16	41	46	36	47	58	19	42
Batch = 128, Optimizer = SGD, Learning Rate = 0.001								
Accuracy	51.35	87.16	85.81	68.92	53.37	53.37	53.38	79.05
(%)	94.59	90.54	92.57	71.62	74.32	64.87	60.81	93.24
Time	10	23	13	12	12	13	11	14
(min)	28	50	48	32	44	115	18	43
Batch = 64, Optimizer = SGD, Learning Rate = 0.01								
Accuracy	83.11	88.51	77.10	45.94	46.62	45.94	53.38	92.56
(%)	89.86	92.57	94.59	50.00	66.89	65.54	83.78	93.92
Time	10	12	12	11	8	14	11	12
(min)	15	45	45	32	44	58	31	44
Batch = 128, Optimizer = SGD, Learning Rate = 0.01								
Accuracy	79.72	78.37	79.73	45.94	46.62	53.37	54.05	91.89
(%)	91.26	90.54	88.51	52.03	56.08	64.19	80.41	95.27
Time	11	11	12	11	69	13	34	12
(min)	16	41	44	53	44	54	21	42

Table 3 shows that, for the time parameter, there was no defined pattern to identify the architecture that presented the lowest processing time in all its hyperparameters. Low values mostly appeared for CNN from scratch. However, the lowest value was for the ResNet-101 model with 8 min, in epochs (25), batch (64), and learning rate (0.01). Likewise, the accuracy of CNN from scratch was better than that reported by ResNet-50, ResNet-101,

ResNet-152, and AlexNet architectures. The highest processing times in 25 epochs was for VGG-16 (14 and 23 min), with a learning rate (0.001), and for ResNet-152 (14 min) and ResNet-101 (69 min), for a learning rate (0.01). For 400 epochs, the highest process time was for ResNet-152 (58 and 115 min) for a learning rate (0.001) and (58 and 54 min), with a learning rate (0.01). Finally, ResNet-52 architecture required the most processing time for most hyperparameters. The highest processing times were not associated with high or low accuracy.

4. Discussion

Convolutional Neural Networks (CNNs) are used in several agriculture areas such as leaf and plant disease detection, land cover classification, crop type classification, plant recognition, segmentation of root and soil, crop yield estimation, fruit counting, obstacle detection in row crops and grass mowing, and identification of weeds, to mention a few [32,33]. For example, in Mohanty et al. [34], they presented the training of CNN architectures AlexNet and Google Net with a PlanVillage image data set to detect 26 types of diseases in 14 kinds of crops. Their results showed an accuracy of 99.35% to identify healthy and diseased plants. Meanwhile, Rahnemonfar and Sheppard [35] proposed using the CNN architectures' inception and Residual Networks (ResNet) architectures to estimate the yield of a tomato plant using synthetic images. Their results indicated that, with 91% accuracy, they can evaluate the yield.

Another example was presented in [36], where authors proposed training several convolutional networks to identify four fruits (mango, orange, apple, and banana). They were classified into two categories: fresh and rotten. The best performing models were Inception version 3 and the Visual Geometric Group of 16-layer (VGG-16) architectures, which received the learning transfer. Their results showed identification and classification percentages of 90% accuracy. A similar study was presented in [13], where the use of a VGG-16 network to classify vegetables and fruits was proposed. A total of 26 categories were classified: pumpkin, celery, cauliflower, pineapple, pomegranate, grapefruit, banana, cucumber, broccoli, onion, carrot, etc. The authors claimed to have 95.6% accuracy in classifying these fruits and vegetables. Regarding dates, we identified research works that proposed using CNNs to sort among dates or to detect among their different maturity stages [8,9,16].

Currently, determining the stage of maturity in the Medjool date using traditional image processing and machine learning methods is complicated. This is because these methods are trained to extract features in various cultivars such as their appearance, color (associated with the maturity stages), shape, and texture [7,16]. However, there are no studies where a feature extraction or predictive model for sorting Medjool dates that we are aware of. Furthermore, recent models cannot determine sorting Medjool because this cultivar is harvested, sorted, packaged, and consumed in its Tamar stage.

To contribute with a model that may be useful in sorting the Medjool date through images, we compared the performance of eight CNN architectures in this study. Additionally, some hyperparameters' values were modified, and transfer learning was used to identify and propose the use of CNN with the best precision.

As shown in Table 2, our findings indicated that when we use an Adam optimizer, the VGG architectures show the best accuracy, with the VGG-19 model that reached the highest percentage of accuracy with 99.32%. Likewise, the ResNet and CNN from scratch architectures showed the lowest performance percentages; the CNN from a scratch model achieved the most insufficient precision, with 43.24%. The highest average percentage generated among the eight architectures was 89.53%, using the combination of batch (64), learning rate (0.01), and epochs (400), with an average time of 48.71 min, while the lowest was 75%, combining a batch (64), learning rate (0.001), and epochs (25), with an average time of 12.25 min.

Likewise, Table 3 indicates that no architecture showed the best accuracy when we used an SGD parameter as an optimizer. However, the ResNet-50 architecture showed the lowest performance percentages, with batch (64 and 128) and learning rate (0.001). The highest percentage generated among the eight architectures was 80.57%, using the combination of batch (64), learning rate (0.001), and epochs (400), with an average time of 38.13 min, while the lowest was 66.21%, combining a batch (128), learning rate (0.01), and epochs (25), with an average time of 21.63 min.

It was noticeable that, if the number of epochs for all models was increased, the percentage of accuracy and required processing time also increased. Likewise, we observed that the highest processing times corresponded to the ResNet-152 architecture, which could be associated with the fact that this architecture had the highest number of layers. However, none of its precision was higher than 85% performance.

The optimizer can help us minimize the error function that allows us to conform to the training set examples. In this study, the accuracy was higher for Adam than for SGD.

Several studies have focused on identifying the CNN that offers the best precision for selecting dates from cultivars in their various stages of maturity [8,9,18]. However, there are currently no reported studies that use any CNN to classify the date cultivar Medjool.

Table 4 compares similar studies to ours, where CNNs' architectures with the best performance have been reported. Nasiri et al., 2019 [9], only worked on VGG-16 with two hyperparameters, obtaining the highest accuracy of 96.98%. Likewise, Altaheri et al., 2019 [8], worked on two CNN with transfer learning and fine-tuning, modifying three hyperparameters twice, obtaining the highest percentage for VGG-16 with an accuracy of 97.25%. Faisal et al., 2020 [16], compared four CNNs' performances, evaluating four hyperparameters, resulting in ResNet as the best model, with an accuracy of 99.01%. Finally, our study evaluated eight CNNs' performances, using transfer learning and modifying four hyperparameters twice, resulting in the VGG-19 model with the highest performance, with 99.32% accuracy.

Table 4. Comparison of studies that report CNN architectures in the detection of various stages of maturity in the date palm fruit.

Reference	Date Palm Cultivar	Maturity Stages	Number of Images (Dataset)	CNN Architectures	Hyperparameters	Best Accuracy
Nasiri et al., 2019 [9]	Shahani	Khalal, Rutab, Tamar, and defective date	+1300 images	VGG-16	Epochs = 15 Batch = 32	VGG-16 96.98%
Altaheri et al., 2019 [8]	Barhi, Khalas, Meneifi, Naboot Saif and Sullaj	Immature-1, Immature-2, pre-Khalal, Khalal, Khalal-with-Rutab, pre-Tamar, and Tamar.	8072 images	AlexNet, VGG-16, Transfer learning and Fine-Tuning	Epochs = 50 and 200 Batch = 32 and 128 Learning rate = 0.0001, 0.0002	VGG-16 97.25%
Faisal et al., 2020 [18]	Barhi, Khalas, Meneifi, Naboot Saif, and Sullaj	Immature, Khalal, Khalal with Rutab, Pre-Tamar, and Tamar	8079 images	ResNet, VGG-19, Inception V3, NASNet and support vector machine (SVM) (regression and linear)	Epochs = 50 Batch = 16 Optimizer = Adam Learning rate = 0.0001	ResNet 99.01%
This Study	Medjool	Ripe and unripe	1002 images	VGG-16, VGG-19, Inception V3, ResNet-50, ResNet-101, ResNet-152, AlexNet, CNN from scratch	Epochs = 25 and 400 Batch = 64 and 128 Optimizers = Adam, Stochastic Gradient Descent Learning rate = 0.001, 0.01	VGG-19 99.32%

One aspect to consider in this comparison is that the Medjool date is only consumed in its Tamar stage. Therefore, this study only used two stages for its sorting. The number of images was lower compared to the rest of the studies. However, in our work, the percentage of accuracy was higher due to the application of transfer learning and modification in various hyperparameters, which influence architectures' performance [37,38].

In our study, resulting from choosing the hyperparameters epochs (400), batch (128), optimizer (Adam), and a learning rate (0.01), we identified that VGG-19 architecture had the best performance. Likewise, this architecture could be included as part of the software that controlled a robotic mechanism to support the date palm farmer in an automated system of sorting ripe fruits.

5. Conclusions

This study evaluated the precision and processing time of eight CNN architectures. Seven of them were pretrained by an extensive image database designed for object recognition (ImageNet). These models were named VGG-16, VGG-19, Inception V3, ResNet-50, ResNet-101, ResNet-152, and AlexNet, which received transfer learning when their last classification layer was replaced. Additionally, a model that learns from scratch was used, that is, without obtaining learning.

All CNN architectures were evaluated by modifying the epochs, batch, optimizer, and learning rate hyperparameters since these parameters have been reported to have positive effects on the performance of convolutional networks. The results indicated that the CNN with the best performance for the sorting Medjool date was the architecture of the VGG group, which used the Adam optimizer. From these architectures, the VGG-19 model was the one that reported the best accuracy, with 99.32%. Likewise, the ResNet group architectures were the ones that reported the lowest performance using the same optimizer, the ResNet-152 model, which reported the most insufficient accuracy, with 64.17%. The use of the SGD optimizer did not have a significant effect on obtaining high accuracies.

Finally, it will be necessary to continue working on the best accuracy and the shortest processing time, with the modification of other hyperparameters. The inclusion in the evaluation of different fruit attributes, such as its size, gives it a high commercial value. It is essential in the packing process of this fruit.

Author Contributions: Conceptualization, B.D.P.-P. and J.P.G.V.; methodology, J.P.G.V. and R.S.-T.; software, B.D.P.-P.; validation, B.D.P.-P., J.P.G.V., and R.S.-T.; formal analysis, R.S.-T. and J.P.G.V.; investigation, J.P.G.V. and R.S.-T.; resources, B.D.P.-P.; data curation, B.D.P.-P.; writing, original draft preparation, J.P.G.V. and R.S.-T.; writing, review and editing, R.S.-T.; visualization, B.D.P.-P. All authors contributed to its editing and approved the final draft. All authors have read and agreed to the published version of the manuscript.

Funding: This research received no external funding.

Institutional Review Board Statement: Not applicable.

Informed Consent Statement: Not applicable.

Data Availability Statement: Dataset is available on https://data.mendeley.com/datasets/872xk9 npmz/1.

Acknowledgments: We would like to thank CONACyT for the scholarship granted to the first author (CVU-409617). We would like to thank Ramiro Quiroz, of the company Palmeras RQ. We accessed his plantation to take the photographs. We thank Emmanuel Santiago Durazo for his contributions to the work done.

Conflicts of Interest: The authors declare no conflict of interest.

References

1. Salomon-Torres, R.; Sol-Uribe, J.A.; Salasab, B.; García, C.; Krueger, R.; Hernández-Balbuena, D.; Norzagaray-Plasencia, S.; García-Vázquez, J.P.; Ortiz-Uribe, N. Effect of Four Pollinating Sources on Nutritional Properties of Medjool Date (*Phoenix dactylifera* L.) Seeds. *Agriculture* **2020**, *10*, 45. [CrossRef]
2. Chao, C.T.; Krueger, R.R. The Date Palm (*Phoenix dactylifera* L.): Overview of Biology, Uses, and Cultivation. *HortScience* **2007**, *42*, 1077–1082. [CrossRef]
3. Radwan, E.; Radwan, E. The current status of the date palm (Phoenix dactylifera) and its uses in the Gaza Strip, Palestine. *Biodiversitas J. Biol. Divers.* **2017**, *18*, 1047–1061. [CrossRef]
4. Aleid, S.M. Dates. In *Tropical and Subtropical Fruits: Postharvest Physiology, Processing and Packaging*; Siddid, M., Ahmed, J., Lobo, M.G., Ozadali, F., Eds.; Wiley: New York, NY, USA, 2012; pp. 179–202.
5. Food and Agriculture Organization. FAOSTAT. Available online: http://www.fao.org/faostat/en/#data/QL (accessed on 11 December 2020).
6. Ortiz-Uribe, N.; Salomon-Torres, R.; Krueger, R. Date Palm Status and Perspective in Mexico. *Agriculture* **2019**, *9*, 46. [CrossRef]
7. Altaheri, H.; Alsulaiman, M.; Muhammad, G.; Amin, S.U.; Bencherif, M.; Mekhtiche, M. Date fruit dataset for intelligent harvesting. *Data Brief* **2019**, *26*, 104514. [CrossRef] [PubMed]

8. Altaheri, H.; Alsulaiman, M.; Muhammad, G. Date Fruit Classification for Robotic Harvesting in a Natural Environment Using Deep Learning. *IEEE Access* **2019**, *7*, 117115–117133. [CrossRef]

9. Nasiri, A.; Taheri-Garavand, A.; Zhang, Y.-D. Image-based deep learning automated sorting of date fruit. *Postharvest Biol. Technol.* **2019**, *153*, 133–141. [CrossRef]

10. Guo, Z.; Zhang, M.; Lee, D.-J.; Simons, T. Smart Camera for Quality Inspection and Grading of Food Products. *Electronics* **2020**, *9*, 505. [CrossRef]

11. Sa, I.; Ge, Z.; Dayoub, F.; Upcroft, B.; Perez, T.; McCool, C. DeepFruits: A Fruit Detection System Using Deep Neural Networks. *Sensors* **2016**, *16*, 1222. [CrossRef]

12. Steinbrener, J.; Posch, K.; Leitner, R. Hyperspectral fruit and vegetable classification using convolutional neural networks. *Comput. Electron. Agric.* **2019**, *162*, 364–372. [CrossRef]

13. Zeng, G. Fruit and vegetables classification system using image saliency and convolutional neural network. In Proceedings of the 2017 IEEE 3rd Information Technology and Mechatronics Engineering Conference (ITOEC), Chongqing, China, 3–5 October 2017; IEEE: Piscataway, NJ, USA, 2017; pp. 613–617.

14. Rudnik, K.; Michalski, P. A Vision-Based Method Utilizing Deep Convolutional Neural Networks for Fruit Variety Classification in Uncertainty Conditions of Retail Sales. *Appl. Sci.* **2019**, *9*, 3971. [CrossRef]

15. Muhammad, G. Date fruits classification using texture descriptors and shape-size features. *Eng. Appl. Artif. Intell.* **2015**, *37*, 361–367. [CrossRef]

16. Haidar, A.; Dong, H.; Mavridis, N. Image-based date fruit classification. In Proceedings of the 2012 IV International Congress on Ultra Modern Telecommunications and Control Systems, St. Petersburg, Russia, 3–5 October 2012; IEEE: Piscataway, NJ, USA, 2012; pp. 357–363.

17. Manickavasagan, A.; Al-Mezeini, N.; Al-Shekaili, H. RGB color imaging technique for grading of dates. *Sci. Hortic.* **2014**, *175*, 87–94. [CrossRef]

18. Faisal, M.; Albogamy, F.; ElGibreen, H.; Algabri, M.; AlQershi, F.A. Deep Learning and Computer Vision for Estimating Date Fruits Type, Maturity Level, and Weight. *IEEE Access* **2020**, *8*, 206770–206782. [CrossRef]

19. Dodge, S.; Karam, L. Understanding how image quality affects deep neural networks. In Proceedings of the 2016 Eighth International Conference on Quality of Multimedia Experience (QoMEX), Lisbon, Portugal, 6–8 June 2016; IEEE: Piscataway, NJ, USA, 2016; pp. 1–6.

20. Geirhos, R.; Janssen, D.H.; Schütt, H.H.; Rauber, J.; Bethge, M.; Wichmann, F.A. Comparing deep neural networks against humans: Object recognition when the signal gets weaker. *arXiv* **2017**, arXiv:1706.06969.

21. Feng, J.; Darrell, T. Learning the Structure of Deep Convolutional Networks. In Proceedings of the 2015 IEEE International Conference on Computer Vision (ICCV), Santiago, Chile, 7–13 December 2015; IEEE: Piscataway, NJ, USA, 2015; pp. 2749–2757.

22. Zaccone, G.; Karim, M.R. *Deep Learning with TensorFlow: Explore Neural Networks and Build Intelligent Systems with Python*; Packt Publishing Ltd.: Birmingham, UK, 2018.

23. Simonyan, K.; Zisserman, A. Very deep convolutional networks for large-scale image recognition. *arXiv* **2014**, arXiv:1409.1556.

24. Szegedy, C.; Vanhoucke, V.; Ioffe, S.; Shlens, J.; Wojna, Z. Rethinking the inception architecture for computer vision. In Proceedings of the IEEE Conference on Computer Vision and Pattern Recognition, Las Vegas, NV, USA, 27–30 June 2016; pp. 2818–2826.

25. He, K.; Zhang, X.; Ren, S.; Sun, J. Deep residual learning for image recognition. In Proceedings of the IEEE Conference on Computer Vision and Pattern Recognition, Las Vegas, NV, USA, 27–30 June 2016; pp. 770–778.

26. Krizhevsky, A.; Sutskever, I.; Hinton, G.E. Imagenet classification with deep convolutional neural networks. In *Advances in Neural Information Processing Systems*; Curran Associates, Inc.: New York, NY, USA, 2012; pp. 1097–1105.

27. Zhu, Q.; He, Z.; Zhang, T.; Cui, W. Improving Classification Performance of Softmax Loss Function Based on Scalable Batch-Normalization. *Appl. Sci.* **2020**, *10*, 2950. [CrossRef]

28. Pan, S.J.; Yang, Q. A Survey on Transfer Learning. *IEEE Trans. Knowl. Data Eng.* **2009**, *22*, 1345–1359. [CrossRef]

29. Han, J.-H.; Choi, D.-J.; Park, S.-U.; Hong, S.-K. Hyperparameter Optimization Using a Genetic Algorithm Considering Verification Time in a Convolutional Neural Network. *J. Electr. Eng. Technol.* **2020**, *15*, 721–726. [CrossRef]

30. Bera, S.; Shrivastava, V.K. Analysis of various optimizers on deep convolutional neural network model in the application of hyperspectral remote sensing image classification. *Int. J. Remote. Sens.* **2019**, *41*, 2664–2683. [CrossRef]

31. Bisong, E. Google Colaboratory. In *Building Machine Learning and Deep Learning Models on Google Cloud Platform*; Apress: Berkeley, CA, USA, 2019; pp. 59–64.

32. Kamilaris, A.; Prenafeta-Boldú, F.X. A review of the use of convolutional neural networks in agriculture. *J. Agric. Sci.* **2018**, *156*, 312–322. [CrossRef]

33. Kamilaris, A.; Prenafeta-Boldú, F.X. Deep learning in agriculture: A survey. *Comput. Electron. Agric.* **2018**, *147*, 70–90. [CrossRef]

34. Mohanty, S.P.; Hughes, D.P.; Salathé, M. Using Deep Learning for Image-Based Plant Disease Detection. *Front. Plant. Sci.* **2016**, *7*, 1419. [CrossRef]

35. Rahnemoonfar, M.; Sheppard, C. Deep Count: Fruit Counting Based on Deep Simulated Learning. *Sensors* **2017**, *17*, 905. [CrossRef] [PubMed]

36. Ashraf, S.; Kadery, I.; Chowdhury, A.A.; Mahbub, T.Z.; Rahman, R.M. Fruit Image Classification Using Convolutional Neural Networks. *Int. J. Softw. Innov.* **2019**, *7*, 51–70. [CrossRef]

37. Kandel, I.; Castelli, M. The effect of batch size on the generalizability of the convolutional neural networks on a histopathology dataset. *ICT Express* **2020**, *6*, 312–315. [CrossRef]
38. Motta, D.; Santos, A.; Álisson, B.; Machado, B.A.S.; Ribeiro-Filho, O.G.V.; Camargo, L.O.A.; Valdenegro-Toro, M.A.; Kirchner, F.; Badaró, R. Optimization of convolutional neural network hyperparameters for automatic classification of adult mosquitoes. *PLoS ONE* **2020**, *15*, e0234959. [CrossRef]

Article

Mapping Paddy Rice Using Weakly Supervised Long Short-Term Memory Network with Time Series Sentinel Optical and SAR Images

Mo Wang [1,2], **Jing Wang** [3,*] **and Li Chen** [1,2]

[1] Agricultural Information Institute, Chinese Academy of Agricultural Sciences, Beijing 100081, China; wangmo@caas.cn (M.W.); chenli02@caas.cn (L.C.)
[2] Key Laboratory of Agricultural Big Data, Ministry of Agriculture and Rural Affairs, Beijing 100081, China
[3] China Center for Information Industry Development, Beijing 100086, China
* Correspondence: wangj.13b@igsnrr.ac.cn

Received: 4 October 2020; Accepted: 16 October 2020; Published: 19 October 2020

Abstract: Rice is one of the most important staple food sources worldwide. Effective and cheap monitoring of rice planting areas is demanded by many developing countries. This study proposed a weakly supervised paddy rice mapping approach based on long short-term memory (LSTM) network and dynamic time warping (DTW) distance. First, standard temporal synthetic aperture radar (SAR) backscatter profiles for each land cover type were constructed on the basis of a small number of field samples. Weak samples were then labeled on the basis of their DTW distances to the standard temporal profiles. A time series feature set was then created that combined multi-spectral Sentinel-2 bands and Sentinel-1 SAR vertical received (VV) band. With different combinations of training and testing datasets, we trained a specifically designed LSTM classifier and validated the performance of weakly supervised learning. Experiments showed that weakly supervised learning outperformed supervised learning in paddy rice identification when field samples were insufficient. With only 10% of field samples, weakly supervised learning achieved better results in producer's accuracy (0.981 to 0.904) and user's accuracy (0.961 to 0.917) for paddy rice. Training with 50% of field samples also presented improvement with weakly supervised learning, although not as prominent. Finally, a paddy rice map was generated with the weakly supervised approach trained on field samples and DTW-labeled samples. The proposed data labeling approach based on DTW distance can reduce field sampling cost since it requires fewer field samples. Meanwhile, validation results indicated that the proposed LSTM classifier is suitable for paddy rice mapping where variance exists in planting and harvesting schedules.

Keywords: paddy rice mapping; dynamic time warping; LSTM; weakly supervised learning; cropland mapping

---

1. Introduction

Rice is an important staple food source for more than half of the global population [1]. In Asia in particular, the rice planting area was 146 million ha in 2018, accounting for 88% of the world's total (FAOSTAT, 2020). The demand for rice over the next 30 years is anticipated to increase by 90% in Asia [2]. Monitoring rice planting is of great importance to global food security and informed policymaking.

Remote sensing has been providing timely and efficient means for agricultural monitoring for decades. Paddy rice is the only crop that grows under wetland conditions [3]. Its growing cycle comprises three phases: vegetative phase, reproductive phase, and ripening phase, which are detailly divided into 10 growth stages. The growing cycle of paddy rice features unique optical reflectance and radar scattering characteristics compared to other crops due to the abundant water conditions in the early stages. Many study cases have employed optical (e.g., Landsat, Moderate Resolution Imaging Spectroradiometer (MODIS), SPOT, and Sentinel-2), microwave (e.g., RADARSAT, Phased Array Type L-band Synthetic Aperture Radar (PALSAR), and Sentinel-1), or a combination of the two remote sensing data sources to map rice-growing area at regional or continental scales. Dong and Xiao [4] sorted out rice mapping approaches into four categories, i.e., (1) reflectance data and image statistic-based approaches, (2) vegetation index (VI) data and enhanced image statistic-based approaches, (3) VI or radar backscatter-based temporal analysis approaches, and (4) phenology-based approaches through remote sensing recognition of key growth phases. The essence of rice mapping approaches is classification models using a single (category 1 and 2) or multiple temporal (category 3 and 4) remote sensing images. A single temporal image has its limitations in discriminating paddy rice from other land cover types due to similar remote sensing signals at certain growing stage and mixed pixels [5]. Approaches based on multi-temporal and phenology-based classifiers are drawing more research interests for paddy rice mapping and other cropland mappings [5–9].

Flooding and transplanting signals are unique features for paddy rice. They are often used as key signatures to identify the rice field. Xiao et al. [7,10] generated rice maps for China and Southeast Asia with time-series MODIS data by identifying flooding and transplanting phases using the relationship between normalized difference vegetation index (NDVI) (enhanced vegetation index (EVI)) and land surface water index (LSWI). Yin et al. [11] applied a time series vegetation index to detect flooding and transplanting phases using fusion data from MODIS and Landsat images. Besides vegetation index and water index, synthetic aperture radar (SAR) datasets were also used to detect flooding and transplanting signals. Torbick et al. [12] and Zhang et al. [13] used PALSAR data to identify rice growing stages. Other phenology-based approaches with SAR data sources include European Remote Sensing Satellite-1 (ERS-1) [14] and RADARSAT-2 [15]. More recently, the novel European Space Agency (ESA) Sentinel-1 has been frequently used as a SAR data source for paddy rice mapping [8,9,16,17], as it provides higher spatial resolution (10 m) and more frequent revisit (6 days). It should be noted that optical remote sensing data is often limited to construct an ideal time series dataset due to cloud cover in rice-growing regions. Therefore, radar remote sensing (e.g., SAR) is more suited for time series analysis since it is independent of weather conditions.

Many multi-temporal or phenology-based paddy rice mapping studies treated all rice fields as synchronized in phenology and input multi-temporal data into classifiers simultaneously. Such a manner ignores the planting schedule difference of farmers and variation in rice-growing periods of different rice cultivars. For example, Le Toan et al. [18] observed a shift of up to 6–8 weeks between fields within a region in Indonesia. This could lead to classification error for regional rice mapping with medium and high spatial resolution images. Some studies [19–21] attempted to address this problem with dynamic programming algorithms, such as dynamic time warping (DTW). The authors classified cropland on the basis of the DTW distance of NDVI from each pixel to samples using a threshold or a simple classifier. To obtain a complete NDVI time series dataset, they either had to reconstruct the data from cloud cover or use existing data products, which often are not available. The reconstruction of data for cloudy days could introduce uncertainties to the final time series dataset. Radar remote sensing data, on the other hand, has greater potential in finding phenological alignment using dynamic programming algorithms. Yet, few cropland mapping studies exist that have used such a strategy, except for Li and Bijker [22] who focused on vegetable classification.

Being consistent with general land cover classification development, various supervised and unsupervised classification algorithms have been applied for paddy rice mapping, such as maximum likelihood [23,24], support vector machines (SVM) [25–27], random forest (RF) [27,28], and decision trees (DT) [29]. In recent years, deep learning methods have shown great capability in fields such as computer vision and remote sensing image classification [30,31]. Convolutional neural network (CNN) and recurrent neural network (RNN) are the most adopted algorithms for remote sensing classification tasks, including rice mapping. Compared to shallow learning methods (such as SVM, RF, and DT), deep learning models rely more on abundant training samples to make reliable predictions and to avoid over-fitting. Kussul et al. [31] presented one of the early attempts of using deep learning models for cropland classification. The authors proposed a 1-D CNN model from multiple Sentinel-1 and Landsat images sensed in the growing season. A total of 547 polygons of training samples were selected by extensive field surveys. Zhang et al. [32] devised a CNN-based approach that classifies small patches for each pixel with fused multi-temporal Landsat and MODIS NDVI data, along with a synthetic phenological variable. They acquired training samples on the basis of existing land use/land cover (LULC) maps and Google Earth. Sun et al. [33] proposed a long short-term memory (LSTM) recurrent neural network (RNN) model for yearly cropland classification of North Dakota, USA, using Landsat imagery. More than 10 million samples were selected to train the LSTM model on the basis of the Cropland Data Layer (CDL). All these deep learning-based cropland classification approaches rely on either a trustworthy labeled training data source or a laborious field survey, which might not be available or may be too expensive for some projects.

With concerns of variance in rice cultivation schedule and abundant training samples needed for deep learning models, we propose a paddy rice mapping approach that labels weak samples using DTW distance with time series SAR data and then feeds them into an LSTM network classifier. First, standard SAR backscatter profiles for each land cover type were constructed on the basis of a relatively small number of paddy rice pixel samples. The polarization of SAR data was carefully selected for that which best reflected flooding and transplanting signals of rice growing stages. DTW distances from each pixel of the study area to standard profiles were then computed. The *k*-nearest pixels were selected as weak samples. Using field samples and weak samples as a training set, we trained an LSTM model using time series multispectral and SAR bands as input. This weakly supervised deep learning approach could benefit from the strong learning ability of RNN models on time series data with less field sampling cost.

2. Materials and Methods

2.1. Study Area and Materials

2.1.1. Study Area

The study area is Tongxiang County of China's Zhejiang Province. It was chosen because it is one of China's major rice-growing regions. The county is located in the Hangjiahu Plain of East China, with an area of 727 km^2 (Figure 1). The annual number of days $\geq$10 °C is 217–222. The accumulated temperature in degree-days is 4772–4960 °C, featured with a subtropical monsoon climate. Rainfall and river network of the region provide adequate irrigation for single/double-cropping systems. Motivated by the agricultural economy, only single-cropping of rice is practiced in Tongxiang. Rice growing season of the study area is from May to November. However, rice harvest could happen as early as in October and as late as in December.

The study area meets the requirements to test our method for several reasons. First, it represents the typical geography of rice cultivation regions in East China with small to medium field size. Second, the weather of the region is an example of a scenario wherein optical remote sensing data are often not applicable due to cloud cover during the rice-growing season. Third, the size of the region is fit for rice mapping with high to medium resolution images that are to be trained with deep learning models.

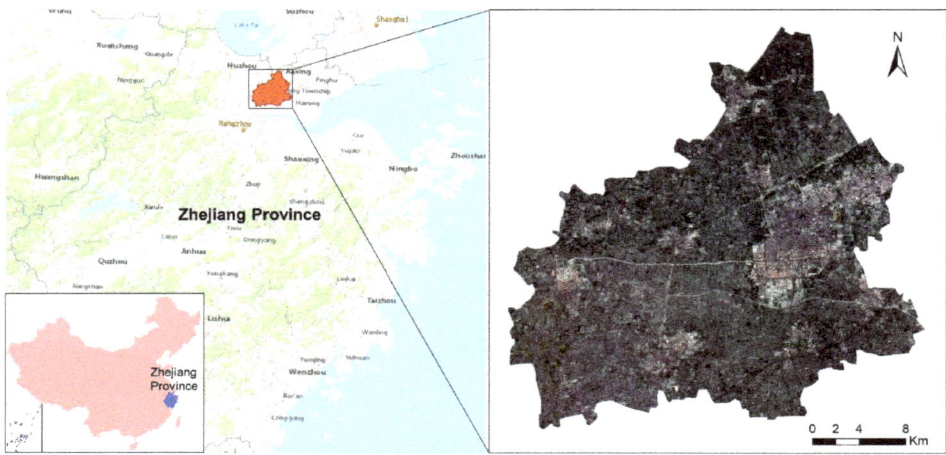

Figure 1. Location of the study area.

2.1.2. Datasets and Preprocessing

Time Series SAR Data

Time series SAR data were used to compute similarities between pixels to land cover samples on the basis of sequence alignment. Time coverage of the SAR data was from April to December, which was extended by a month for the beginning and the end of the period compared with the normal cultivation period (May to November). This was because abnormal cultivation schedules exist for some farmers, for example, early rice planting (e.g., before May) and late harvest (e.g., after November). A total of 23 images of Sentinel-1A Level-1 Ground Range Detected (GRD) product with an interval of 12 days from April to December 2018 were accessed from ESA Copernicus Open Access Hub (https://scihub.copernicus.eu/). Sentinel-1B data product was omitted since it was not provided by ESA for the study area. The images were sensed with C-band in interferometric wide (IW) mode, and in dual of vertical transmitted; vertical received (VV); and vertical transmitted, horizontally received (VH) channels. Attributes of SAR images are listed in Table 1. The radar backscatter data were used to detect key phenology stages of paddy rice, especially flooding signals of early growing stages. According to Clement et al. [34], VV polarization slightly outperforms VH polarization while delineating flooding. Hence, VV bands were chosen to construct the SAR time series.

Several preprocessing steps were carried out using Sentinel Toolboxes software SNAP. First, subsets of original images were generated by cropping extend to the study area. The images were then calibrated to sigma naught ($\sigma0$) backscatter coefficients in a linear unit. Terrain correction was applied subsequently using Range Doppler Terrain Correction with Shuttle Radar Topography Mission (STRM) 3 sec grid. We generated 10×10 m pixel spacing images, which were reprojected to the Universal Transverse Mercator (UTM) zone 51 N. Then, a multi-temporal speckle filtering was applied on the terrain-corrected images using a 7×7 gamma map filter with 3 looks to reduce the noise in the images. Finally, the linear intensity values of the VV bands were converted to the dB unit.

Sentinel-2 Time Series Images

Sentinel-2 multispectral images covering rice growing season were obtained from ESA Copernicus Open Access Hub. Screened by cloud cover, only 11 cloud-free dates were usable. Top-of-atmosphere level 1C data products of those dates were downloaded. Some preprocessing steps were also required for the level 1C product. First, atmospheric, terrain, and cirrus correction were performed using the Sen2Cor v2.8 tool provided by ESA to achieve bottom-of-atmosphere level 2A product. Among the

13 spectral bands of Sentinel-2 images, we selected 10 bands (B2-B8, B8a, B11, and B12) as input features, including visible, near-infrared (NIR), and short-wave infrared (SWIR) bands. The reason to include SWIR bands is that they are sensible to leaf water content [35], which could help differentiate rice fields from other crops. B2-B4 and B8 were in a resolution of 10 m. The remaining 6 bands (20 m resolution) were resampled to 10 m by nearest neighbor interpolation using the processing module in SNAP. Table 2 lists the attributes of used Sentinel-2 images.

Table 1. Attributes of synthetic aperture radar (SAR) images.

SAR Images	
Sensor	Sentinel-1A
Data level	Level-1 ground range detected
Spatial resolution	10 m
Wavelength	C band
Polarization	VV
Pass	Ascending
Acquisition mode	IW
Acquisition date (23 successive images with 12-day interval)	4 April 2018 16 April 2018 28 April 2018 24 December 2018

Table 2. Attributes of optical images.

Multispectral Optical Images	
Sensor	Sentinel-2A, Sentinel-2B
Data level	Level-1C
Spatial resolution	10 m, 20 m
Band	B2, B3, B4, B5, B6, B7, B8, B8a, B11, B12
Acquisition date (11 images)	9 April 2018 19 April 2018 4 May 2018 9 May 2018 18 June 2018 18 July 2018 23 July 2018 28 July 2018 7 August 2018 1 October 2018 10 November 2018

Field Sampling Data

Field sampling was conducted to collect samples belonging to broad land cover categories, i.e., paddy rice fields, water bodies (rivers, lakes, and reservoirs), construction (buildings, roads, bare surfaces, and paved surfaces), vegetation (forest, shrubs, orchards, and other planted area), and other croplands. A total of 506 sample sites (points of interest, POIs) were selected through scattering the study area for the five land cover types. The sample sites were selected in a homogeneous spatial distribution principle to cover cultivation variation between sub-regions. A geo-referenced field photo was taken for each POI in November 2018. Locations of geo-referenced photos were converted into kmz format and imported to Google Earth. Polygons of sample sites (areas of interest, AOIs) were then delineated with the assistance of Google Earth and reference photos. Boundaries of polygons can be distinguished with high-resolution satellite images of 2018 from Google Earth. The land cover type of an AOI was assigned with the land cover type of the field photo where it was taken (as shown in Figure 2).

Figure 2. Illustration of the field sampling process. Image (**a**) (paddy rice field) and (**b**) (other croplands) are example photos of points of interest (POIs). Their locations are shown in the Google Earth image on the left. Boundaries of fields (**a**,**b**) were delineated with the Google Earth image. Likewise, areas of interest (AOIs) for the other three land cover types were also created. Note that the dates of the Google Earth image and the field photos are not the same but were taken in the same year.

A total number of 506 AOIs (one for each sample site) were created and then rasterized into a GeoTIFF image with 10 m resolution. Final AOIs include rice field (4569 pixels), water body (4805 pixels), vegetation (4372 pixels), construction (6094 pixels), and other croplands (4592 pixels).

2.2. Methodology

We integrated time series optical and radar remote sensing data to best capture phenological characteristics of paddy rice. The time series data were to be fed into a RNN to cope with non-linear relationship and temporal correlation. The framework of this study is shown in Figure 3. First, a labeling strategy was developed. Following the preprocessing steps, standard SAR backscatter profiles for each land cover type were created with a relatively small number of field samples. Pairwise DTW distances to the standard SAR backscatter profile for each land cover type were computed for the whole study area. Top-*k* nearest pixels were then screened out as weakly labeled samples with the highest confidence. A specifically designed LSTM network was trained with integrated input features of optical and SAR data, using field samples (supervised) or optionally combined with DTW-labeled samples (weakly supervised). Finally, a supervised or weakly supervised approach was to be adopted on the basis of experiment results to generate a paddy rice map.

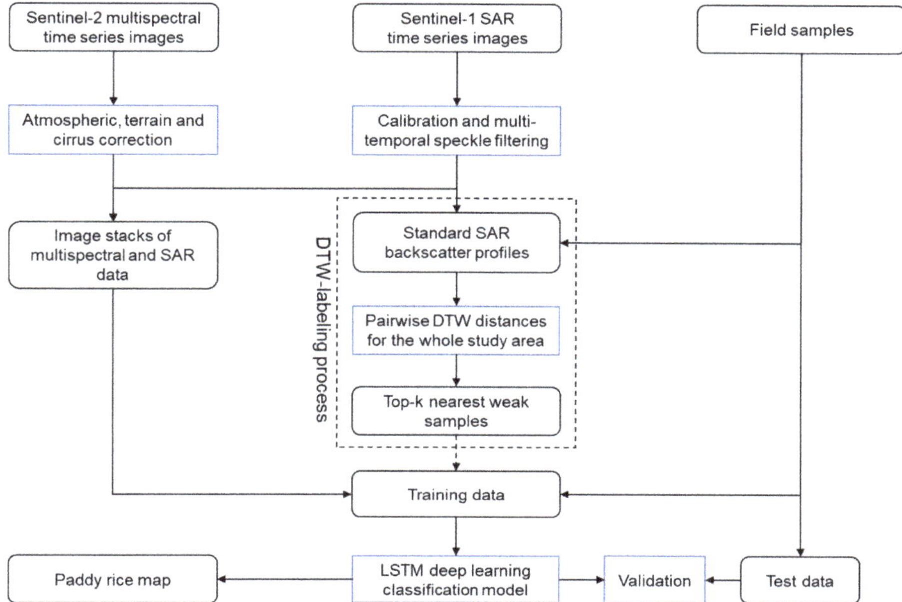

Figure 3. The framework of mapping paddy rice with dynamic time warping (DTW)-labeling strategy and weakly supervised classification with long short-term memory (LSTM) deep learning classifier.

2.2.1. Standard Time Series SAR Backscatter Profiles

To simulate the scenario wherein only a small number of field samples are available, we took 10% of field sample pixels of each land cover type as a base to compute standard SAR backscatter profiles. SAR backscatter values of each date were extracted from the time series of the 23 Sentinel-1 SAR backscatter coefficient images from April to December. For each date and land cover type, the standard SAR backscatter value was computed by averaging on the selected base samples (Figure 4). The five land cover types had different levels of SAR backscatter in dB. Waterbody had a distinctly low radar backscatter profile with litter temporal variance. In contrast, construction had the highest radar backscatter value. Vegetation and cropland were similar in the radar backscatter level. Compared to other cropland, rice fields showed a greater variance due to temporal change in the structure of the canopy and that of the underlying soil surface.

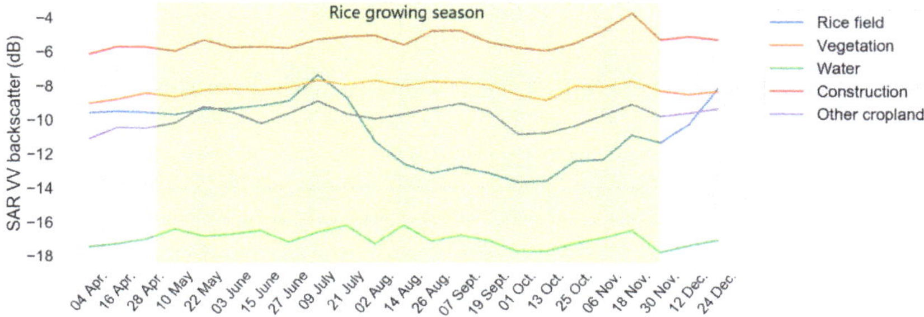

Figure 4. Time series SAR backscatter profiles for the five land cover types.

2.2.2. DTW Distance-Based Sampling

The dynamic time warping (DTW) algorithm uses dynamic programming techniques to find the optimal alignment and the minimal distance of two temporal sequences. The algorithm was originally developed for speech recognition to cope with different speaking speeds [36], and was introduced in time series remote sensing image analysis in recent years [19,20,37,38]. The alignment of two sequences is through being warped non-linearly by stretching or shrinking the time dimension. In our case, DTW can deal with temporal distortion in rice-growing phases of different rice fields or rice varieties. Mathematical principle and a detailed procedure to implement DTW was elaborated by Berndt and Clifford [39]. The basic idea is illustrated in Figure 5, where A and B are two SAR backscatter time series of rice field samples. The DTW algorithm finds the best alignment that allows minimum accumulated distance for each time step. In this example, the planting date of B was later than that of A. DTW algorithm aligns the phases between these two SAR time series by shifting with the corresponded time gap. As a result, the similarity measure in the SAR time series curve is justified, regardless of planting date difference and rice varieties.

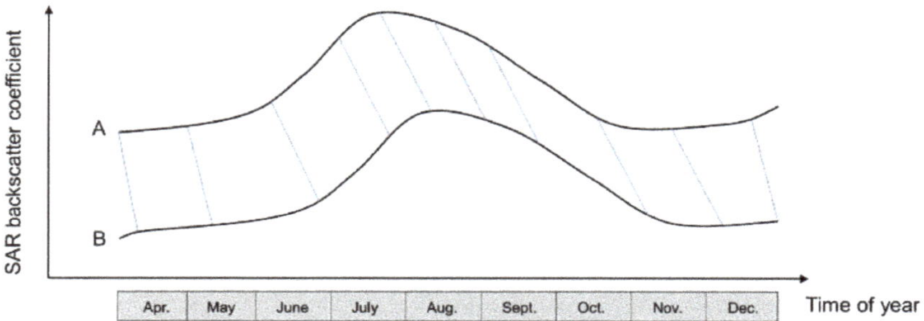

Figure 5. Illustration of DTW alignment between two time series SAR backscatter curves.

For each land cover type, pairwise DTW distances to its standard SAR time series were computed. The shorter the distance, the more likely a pixel belongs to the corresponding land cover type. Top-k nearest strategy was applied to select weak samples, which means they had the highest confidence belonging to a certain land cover type. The decision on the number of screened samples was made on the basis of the rough size of the rice planting area in the study area. A larger number of samples may enhance the training of the deep learning model. However, it could bring more uncertainty in the sample label, i.e., lower sample quality. The number of labeled samples should be a small portion of the total rice planting area. In practice, a DTW Python module by Pierre Rouanet (https://github.com/pierre-rouanet/dtw) was adopted.

2.2.3. LSTM Deep Learning Classifier

LSTM [40] is an RNN architecture designed to avoid the long-term dependency problem. Compared to feed-forward neural networks, RNN has a feedback connection that deals with temporal dynamic behavior. RNNs use an internal state (memory) from the previous time step as one of the inputs for the current time step. Traditional RNN often encounters vanishing gradient problem while processing long-sequence data. LSTM, on the other hand, can avoid this problem by adding gates to control the cell state. A common LSTM unit has an input gate, an output gate, and a forget gate. A standard RNN unit and LSTM unit are illustrated in Figure 6.

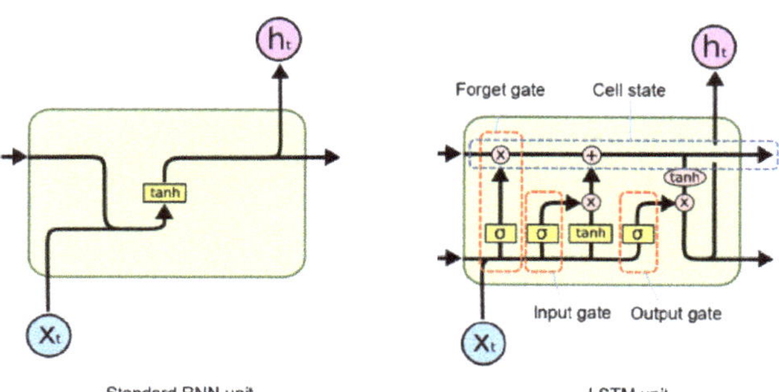

Figure 6. The architecture of standard recurrent neural network (RNN) cell and LSTM cell [41].

Gates in LSTM cells use a sigmoid activation function to decide whether to keep a feature. "0" means the gates are blocking inputs. "1" means gates are allowing inputs to pass through it. Equations of the three gates are

$$i_t = \sigma(w_i[h_{t-1}, x_t] + b_i) \tag{1}$$

$$f_t = \sigma\big(w_f[h_{t-1}, x_t] + b_f\big) \tag{2}$$

$$o_t = \sigma(w_o[h_{t-1}, x_t] + b_o) \tag{3}$$

where i_t, f_t, and o_t denote input gate, forget gate, and output gate, respectively; σ denotes sigmoid function; and w_x, h_{t-1}, x_t, and b_x denote the weight for respective gate neurons, the output of the previous LSTM block, input at the current timestamp, and biases for the respective gates, respectively.

The cell state and final output are defined by equations

$$\widetilde{c_t} = tanh(w_c[h_{t-1}, x_t] + b_i) \tag{4}$$

$$c_t = f_t \times c_{t-1} + i_t \times \widetilde{c_t} \tag{5}$$

$$h_t = o_t \times tanh(c_t) \tag{6}$$

where c_t and $\widetilde{c_t}$ denote cell state and candidate for cell state at time t, and h_t denotes the final output (hidden state) of the LSTM cell.

Considering the large span of the time series data, we developed an LSTM network to identify paddy rice from other land cover types. The core of the model is a stack of two LSTM layers, one fully connected layer, and two dropout layers, as shown in Figure 7. Output layers used the softmax activation function to predict the likelihood that a pixel belongs to one land cover type. Categorical cross-entropy was chosen as the loss function. The choice was made on the basis of several concerns. First, it has been proven that the use of cross-entropy losses made great performance improvements in models with softmax outputs [42]. Second, the training samples for each class were fairly balanced, which makes the network unbiased towards each class. Therefore, other loss functions were neglected, for instance, focal loss, which is suitable to deal with the imbalance problem in datasets. Parameters used to train the network are shown in Table 3. The model was implemented with Python 3.7 language and TensorFlow library.

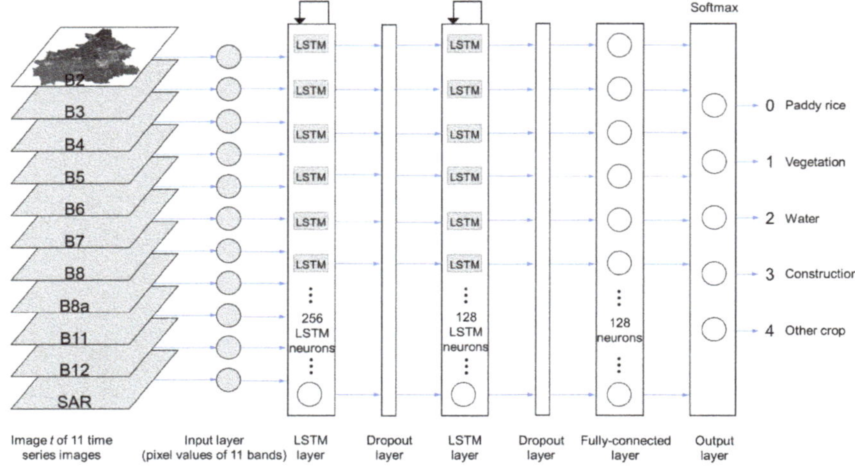

Figure 7. LSTM network structure. The inputs of the classifier are pixel values of 11 bands (10 multispectral bands and 1 SAR band). At time t, the classifier is fed with the band values of image t, and the LSTM layers receive cell state and hidden state from the previous timestamp $t - 1$. The final prediction is made by the output layer of the last timestamp.

Table 3. Parameters of proposed LSTM classifier.

Learning Rate	Dropout Rate	Batch Size	Loss Function	Optimizer
0.001	0.5	64	Categorical cross-entropy	Adam algorithm

The input multispectral features (10 bands) of the classification model were carefully selected from optical imagery, as described within Section 2.2.2. Meanwhile, SAR backscatter signals provided valuable information to discriminate paddy rice fields. The SAR backscatter coefficient was therefore added as an input feature. It should be noted that the dates of Sentinel-1 SAR images did not coincide with that of Sentinel-2 optical images. The SAR images with the closest sensing date to corresponding optical images were selected. This strategy should be feasible since the time difference between the two data sources was minor (less than 5 days). The SAR data were normalized to the same range (0, 1) as optical bands. The final input features comprised 10 optical bands and 1 SAR band.

2.2.4. Experiment Design

We designed 3 experiment schemes on the basis of the number of field samples. In each scheme, supervised and weakly supervised learning were both to be tested. Scheme 1 and 2 are two comparative experiments that were to be validated by a fixed test set with a sufficient sample number. First, the test set was randomly selected from the rest of the field samples, apart from those for computing standard SAR backscatter profiles. The sample number of the test set accounted for 50% of all field samples. In scheme 1, 10% of field samples and 5000 DTW-labeled samples each type were used for training. The 10% was the same set of samples for computation of the standard SAR backscatter profiles in Section 3.1. Scheme 1 is designed to test for the case that only a small number of field samples were available. In scheme 2, training samples were the rest of the half field samples, apart from the test set plus 2000 DTW-labeled samples of each type. Scheme 3 utilized 80% of field samples and 2000 DTW-labeled samples of each type for training. It was designed to compare weakly supervised with supervised learning in a condition of relatively adequate field samples. The details of experiment schemes are listed:

- Scheme 1: Supervised learning training on 10% of field samples compares with weakly supervised learning training on (10% of field samples + 5000 DTW-labeled samples for each land cover type).
- Scheme 2: Supervised learning training on 50% of field samples compares with weakly supervised learning training on (50% of field samples + 2000 DTW-labeled samples for each land cover type).
- Scheme 3: Supervised learning training on 80% of field samples compares with weakly supervised learning training on (80% of field samples + 2000 DTW-labeled samples for each land cover type).

3. Results

3.1. DTW Distance-Based Sampling Results

Using the SAR time series of each pixel in the study area and the five standard SAR profiles, we generated five pairwise DTW distance maps (Figure 8). Each map depicted the likelihood that image pixels belong to the corresponding land cover type. The lower the value, the more likely the pixel belonged to the corresponding land cover type. For water body and construction ((b) and (d), respectively, in Figure 8), their DTW distance maps show visibly different patterns than the other three. This can be explained by the SAR backscatter differences between vegetation and non-vegetation. In some parts of the map, DTW distances for paddy rice ((a) in Figure 8) and other crops ((e) in Figure 8) show a similar pattern of low value. Those pixels are likely to be agricultural land whose SAR profile lies between the standard paddy rice profile and other-crop profile. This ambiguity was avoided by the top-k nearest strategy to pick out the weak samples with the highest confidence.

For each land cover type, the top-k (5000 or 2000 for experiments) pixels with the lowest DTW distance were selected as weak samples. No conflict of labels was found in the selected weak samples since a relatively small amount (to the pixel number of the whole study area) of weak samples were picked.

3.2. LSTM Classification Results

We evaluated the experiments with four metrics, namely, overall accuracy (OA), producer's accuracy (PA) for paddy rice, user's accuracy (UA) for paddy rice, and Kappa coefficient. UA evaluates how often the class in the classification map will be present on the ground, and PA evaluates how often real land cover types in the study area are correctly shown on the classification map. They can be derived from the confusion matrix. PA and UA for paddy rice were of most concern since our goal was solely focusing on paddy rice mapping. The evaluation of three experimental schemes is shown in Table 4 (PA and UA for paddy rice are emphasized in bold and in tables thereafter).

Paddy rice PA and UA for supervised learning in scheme 1 were 0.904 and 0.917, respectively, whereas for weakly supervised learning were 0.981 (improved by 0.77) and 0.961 (improved by 0.44), respectively. However, supervised learning surpassed weakly supervised in OA (0.937 to 0.854) and Kappa (0.921 to 0.817), which indicated that weak samples introduced errors in land cover types other than paddy rice. Further analysis of classification results for scheme 1 was listed in Tables 5 and 6. Compared with supervised learning, PA and OA for paddy rice were notably improved. Meanwhile, PA and OA for vegetation and other crops were depressed. The expanded weak samples might introduce errors in labels for vegetation and other crops since they could have similar time series SAR profiles (Figure 4). Scheme 1 proves that with a small number of field samples for training, the proposed weakly supervised approach had evident improvement in paddy rice identification, regardless of possibly worse identification of other land cover types.

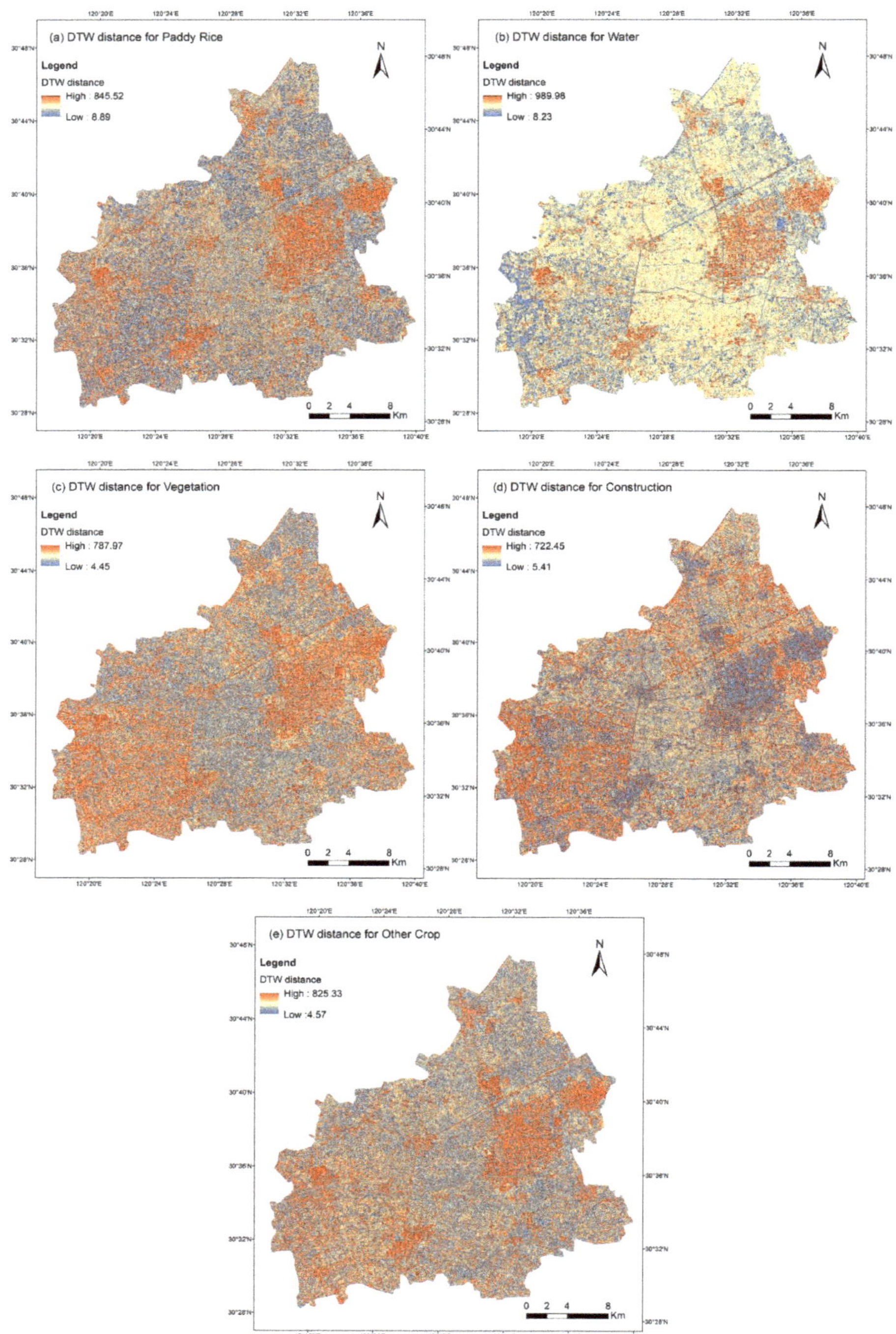

Figure 8. DTW distances of SAR time series to standard SAR backscatter profiles. (**a–e**) are five DTW distance maps for each land cover type respectively.

Table 4. Evaluation of three experiment schemes.

	Experiment	OA	Paddy Rice PA	Paddy Rice UA	Paddy Rice Support	Kappa
Scheme 1	Supervised	0.937	**0.904**	**0.917**	2281	0.921
	Weakly supervised	0.854	**0.981**	**0.961**	2281	0.817
Scheme 2	Supervised	0.973	**0.968**	**0.972**	2281	0.966
	Weakly supervised	0.986	**0.985**	**0.993**	2281	0.982
Scheme 3	Supervised	0.982	**0.988**	**0.987**	913	0.978
	Weakly supervised	0.989	**0.996**	**0.984**	913	0.986

Table 5. Confusion matrix for supervised learning in scheme 1.

	Paddy Rice	Vegetation	Water	Construction	Other Crops
Paddy rice	2092	54	0	23	112
Vegetation	42	2276	3	4	66
Water	5	0	2118	24	0
Construction	3	0	21	3015	2
Other crops	173	158	7	72	1885
PA	**0.904**	0.915	0.986	0.961	0.913
UA	**0.917**	0.948	0.986	0.99	0.821
OA			0.937		
Kappa			0.921		

Table 6. Confusion matrix for weakly supervised learning in scheme 1.

	Paddy Rice	Vegetation	Water	Construction	Other Crops
Paddy rice	2193	12	0	6	70
Vegetation	6	2066	0	58	261
Water	6	0	2091	16	34
Construction	0	50	2	2742	247
Other crops	30	775	39	163	1288
PA	**0.981**	0.712	0.981	0.919	0.678
UA	**0.961**	0.864	0.974	0.902	0.561
OA			0.854		
Kappa			0.817		

Scheme 2 used more field samples for training than scheme 1. The experiment results also showed improvement with weakly supervised learning in paddy rice PA and UA, although not as much as scheme 1. PA was improved by 0.017 from 0.968 to 0.985, and UA improved by 0.021 from 0.972 to 0.993. This can be explained by the classification model. The proposed classification model was fairly fitted by training with 50% of field samples. No evident improvement was observed in scheme 3, while 80% of field samples were used for training. Paddy rice PA was slightly improved. Considering the relatively small sample number of the test set, this improvement was neglectable.

3.3. Paddy Rice Map

On the basis of the model evaluation, we implemented the weakly supervised approach in scheme 3 to generate a paddy rice map (Figure 9). The resulting map shows that the majority of paddy rice cultivation was in the north and west of the county. The total area of mapped paddy rice fields was 13,998.8 ha, slightly less than the government-published figure of 14,666 ha. With closer examination of four selected areas (Figure 10), we found that the resulting map depicts paddy rice fields with fair accuracy. Meanwhile, the precise boundaries of contiguous rice fields failed to be delineated due to the resolution of the data source. Boundaries between rice fields are generally 1–3 m in width, and hence are unfeasible to be mapped with 10-m resolution images. The narrow boundaries were also counted as paddy rice fields in the produced map.

As we were focusing on paddy rice mapping, the classification accuracy of other land cover types was less regarded. With visual interpretation, water and construction type were well identified, while vegetation and other crops were mixed up in some areas. In some cases, these two broad categories could have a similar time series pattern in terms of spectral and radar signals.

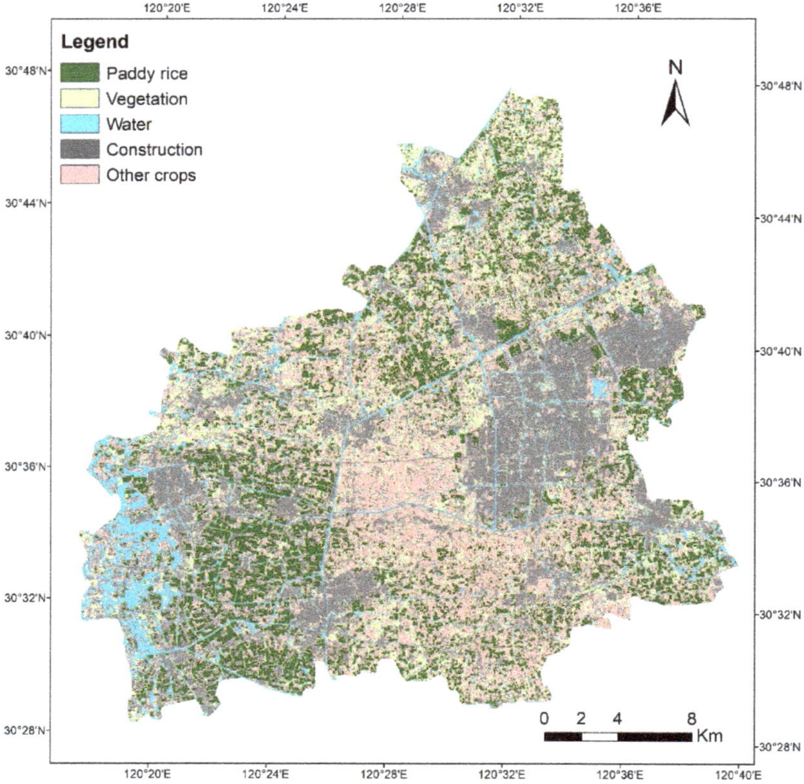

Figure 9. Classification map for paddy rice with weakly supervised LSTM deep learning classifier.

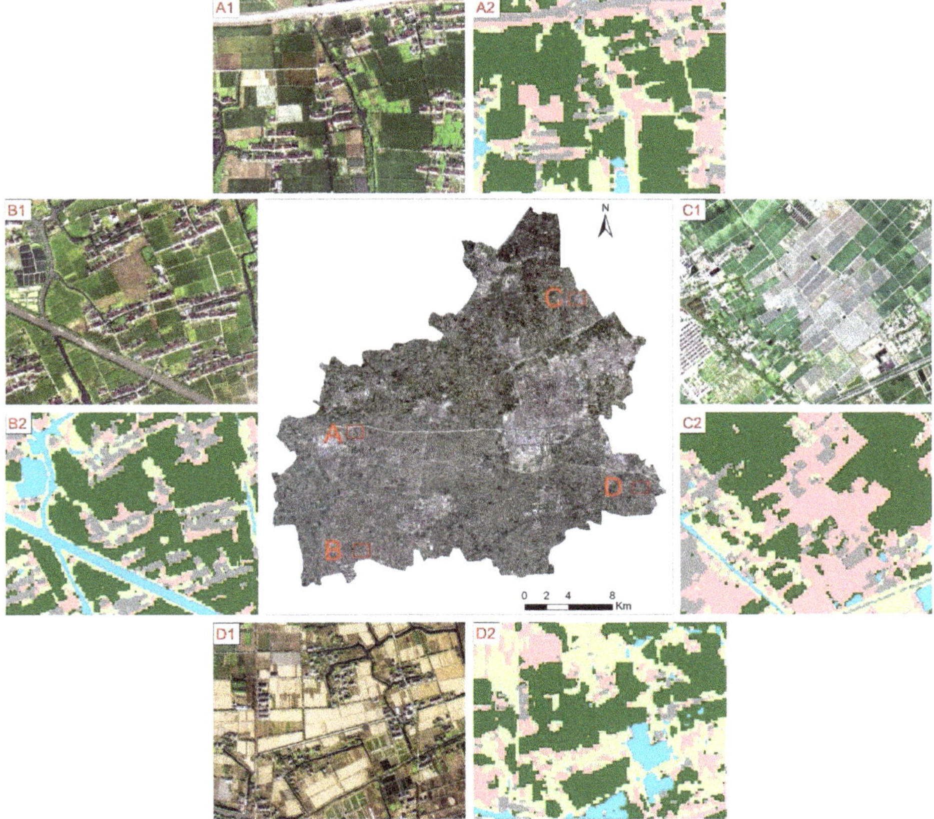

Figure 10. Classification result inspection on selected areas with high-resolution satellite images. **A1** is the high-resolution remote sensing image and **A2** is its classification result, vice versa for **B–D**. Images (**A1,B1**) were sensed in May 2018, (**C1**) was sensed in July 2018, and (**D1**) was sensed in December 2018. By visual interpretation, paddy rice fields in (**A1,B1,C1**) are displayed in green, while in (**D1**), paddy rice fields are displayed in yellow since the crop was harvested.

4. Discussion

This study mapped paddy rice fields using a weakly supervised LSTM deep learning network with time series optical and SAR remote sensing data. The motivation of this study was to show that deep learning classifiers have a strong capability in crop mapping, but often require a larger number of training samples. To solve this problem, a DTW distance-based sampling method was developed to deal with the time alignment in rice cultivation and label weak samples. Experiments on different numbers of training samples showed that the DTW distance-based sampling improved classification results for paddy rice, whereas the field samples were insufficient.

Previous studies, e.g., Zhong et al. [43] and Sun et al. [33], have applied RNNs in crop mapping, although not specifically for rice mapping. The effectiveness of using RNNs to classify croplands has been proven. However, the experiment results of those studies are not comparable to our study since different datasets and experiment schemes were used. Those studies relied on high-quality field survey data to provide a large number of training samples. Our experiments showed that with weakly supervised learning, only 10% of field samples as training data achieved comparable results (PA 0.981,

UA 0.961) to supervised learning on 50% of field samples (PA 0.968, UA 0.972), and weakly supervised learning on 50% field samples (PA 0.985, UA 0.993) had similar performance to supervised learning on 80% field samples (PA 0.988, UA 0.987). Although in scheme 1, weakly supervised learning with weak samples impaired the differentiation of vegetation and other crops, the goal of paddy rice mapping was promoted. The results imply that DTW-labeled paddy rice, water, and construction samples had high confidence and contributed significantly to the training of the RNN classifier.

Many factors could have an impact on the identification and mapping of paddy rice with multi-temporal remote sensing data. First, the resolution of satellite images creates the bottleneck of the precision of paddy rice mapping. The average size of paddy rice fields determines the appropriate image resolution to be used. The typical length of crop field sides is tens of meters in the study area. The boundaries between crop fields are narrow. Among the 11 input feature bands, 5 of them were in 10-m resolution and 6 of them were resampled to 10 m from 20 m. A mixed pixel problem still exists for small land parcels such as boundaries between croplands. Image segmentation on high-resolution images as supplementary data is a possible solution to alleviate this problem. The second factor is the continuity of time series data. Optical remote sensing data were critically confined by weather conditions in the study area. Sentinel-2 images of only 11 dates (less than a quarter of sensed images) from April to November were qualified for generating cloud-free data. With more complete time series images, the RNN classifier would yield more reliable results. This is a common problem for many studies using optical remote sensing data. The third factor is feature selection. Multi-temporal profile analysis of paddy rice widely utilizes temporal analysis of vegetation indexes (VIs) and water indexes (WIs) [19,44,45]. Some evident characteristics can be observed with VIs or WIs at a certain phase of paddy rice growth. For instance, in the heading phase, normalized difference vegetation index (NDVI) and enhanced vegetation index (EVI) show an increase and decrease variation, respectively, with a peak; land surface water index (LSWI) shows a jump in the transplanting phase due to water inundation [46]. As for the deep learning classifier in this study, we avoided VIs or WIs as input features due to the following concern. The range of VIs or WIs is different from that of spectral bands. If the VIs or WIs are rescaled to the range of spectral reflectance, the principle of the indexes is no longer valid. On the other hand, the neural networks would learn an implicit representation of raw input data, which is analogous to manually created features such as VIs and WIs.

Flooding signals are key features at early growing phases of paddy rice to differentiate from other crop types. Before transplanting, radar backscatter is low due to flooded fields. Radar backscatter increases during the vegetative phase, reaching a maximum at the heading stage [47]. We utilized the SAR backscatter coefficient as an input to the LSTM classifier. However, the power of SAR data as an input feature to identify the rice field was restrained by the number of valid optical images. Only a small portion of SAR images was used.

Most studies on rice mapping using multi-temporal data ignored planting schedule differences. Inputting all multi-temporal features as a whole into shallow machine learning classifiers, such as SVM and random forest, could impair the classifier's predictive ability. RNNs have strong adaptability to avoid this problem through learned memory control. Accordingly, many land cover classification studies using deep learning classifiers achieved better results compared to shallow classifiers. In the meantime, a larger number of training samples were required for the training of deep learning models. The proposed labeling method was especially designed for paddy rice, whose radar signal pattern through the growing season is unique. The experiment was conducted for a single-cropping regime in East China. However, the DTW-labeling strategy is valid for the multiple-cropping regime and other regions of the world, since it only relies on all-weather SAR remote sensing data. If applied to a multiple-cropping regime, each cropping regime should be categorized as an independent land cover type. For instance, single/double/triple-cropping rice fields should be separate land cover types. With enough labeled samples, state-of-the-art deep learning classifiers (non-sequential or sequential as in this study) could be chosen to adapt to a specific rice-growing region in the world.

5. Conclusions

This study proposed a paddy rice mapping approach based on a weakly supervised LSTM network with DTW distance-based sampling strategy. The purpose of this study was to enable a good fit for deep learning classifiers in the circumstance that field samples were not sufficient or to reduce the cost of field sampling. Apart from field samples, a larger number of weak samples were labeled on the basis of DTW distance to the standard SAR profiles of each land cover type. A specifically designed LSTM classifier was trained on combined optical spectral bands and SAR backscatter data as input features. A paddy rice map was finally generated with the best training scheme.

A few conclusions can be made on the basis of experiment results. First, the weakly supervised approach has a positive impact on paddy rice mapping when limited field samples are available. The DTW distance-based sampling strategy effectively reduced field sampling costs since fewer field samples were required. Second, the LSTM network classifier achieved high precision on paddy rice mapping. The mechanism of LSTM networks is feasible for paddy rice mapping where variance exists in terms of planting and harvesting schedules.

Author Contributions: Conceptualization, M.W.; data curation, L.C.; methodology, M.W.; project administration, J.W.; resources, L.C.; software, M.W.; writing—original draft, M.W.; writing—review and editing, J.W. and L.C. All authors have read and agreed to the published version of the manuscript.

Funding: This research was supported by the China Central Public-Interest Scientific Institution Basal Research Fund (no. Y2020PT14), Basal Research Fund of AII CAAS (no. JBYW-AII-2020-17), Open Research Fund Program of Key Laboratory of Agricultural Big Data, Ministry of Agriculture and Rural Affairs, and Special Project of National Science and Technology Library (no. 2020QBW008).

Conflicts of Interest: The authors declare no conflict of interest.

References

1. Sarris, A. Rice in Global Markets. In Proceedings of the FAO Rice Conference 2004, Rome, Italy, 12–13 February 2004.
2. Muthayya, S.; Sugimoto, J.D.; Montgomery, S.; Maberly, G.F. An overview of global rice production, supply, trade, and consumption. *Ann. N. Y. Acad. Sci.* **2014**, *1324*, 7–14. [CrossRef]
3. Bouman, B. How much water does rice use. *Management* **2009**, *69*, 115–133.
4. Dong, J.; Xiao, X. Evolution of regional to global paddy rice mapping methods: A review. *ISPRS J. Photogramm. Remote Sens.* **2016**, *119*, 214–227. [CrossRef]
5. Jin, C.; Xiao, X.; Dong, J.; Qin, Y.; Wang, Z. Mapping paddy rice distribution using multi-temporal Landsat imagery in the Sanjiang Plain, northeast China. *Front. Earth Sci.* **2016**, *10*, 49–62. [CrossRef] [PubMed]
6. Zhong, L.; Gong, P.; Biging, G.S. Efficient corn and soybean mapping with temporal extendability: A multi-year experiment using Landsat imagery. *Remote Sens. Environ.* **2014**, *140*, 1–13. [CrossRef]
7. Xiao, X.; Boles, S.; Liu, J.; Zhuang, D.; Frolking, S.; Li, C.; Salas, W.; Moore, B., III. Mapping paddy rice agriculture in southern China using multi-temporal MODIS images. *Remote Sens. Environ.* **2005**, *95*, 480–492.
8. Bazzi, H.; Baghdadi, N.; El Hajj, M.; Zribi, M.; Minh, D.H.T.; Ndikumana, E.; Courault, D.; Belhouchette, H. Mapping paddy rice using Sentinel-1 SAR time series in Camargue, France. *Remote Sens.* **2019**, *11*, 887. [CrossRef]
9. Onojeghuo, A.O.; Blackburn, G.A.; Wang, Q.; Atkinson, P.M.; Kindred, D.; Miao, Y. Mapping paddy rice fields by applying machine learning algorithms to multi-temporal Sentinel-1A and Landsat data. *Int. J. Remote Sens.* **2018**, *39*, 1042–1067. [CrossRef]
10. Xiao, X.; Boles, S.; Frolking, S.; Li, C.; Babu, J.Y.; Salas, W.; Moore, B., III. Mapping paddy rice agriculture in South and Southeast Asia using multi-temporal MODIS images. *Remote Sens. Environ.* **2006**, *100*, 95–113. [CrossRef]
11. Yin, Q.; Liu, M.; Cheng, J.; Ke, Y.; Chen, X. Mapping Paddy Rice Planting Area in Northeastern China Using Spatiotemporal Data Fusion and Phenology-Based Method. *Remote Sens.* **2019**, *11*, 1699. [CrossRef]
12. Torbick, N.; Salas, W.A.; Hagen, S.; Xiao, X. Monitoring rice agriculture in the Sacramento Valley, USA with multitemporal PALSAR and MODIS imagery. *IEEE J. Sel. Top. Appl. Earth Obs. Remote Sens.* **2010**, *4*, 451–457. [CrossRef]

13. Zhang, Y.; Wang, C.; Wu, J.; Qi, J.; Salas, W.A. Mapping paddy rice with multitemporal ALOS/PALSAR imagery in southeast China. *Int. J. Remote Sens.* **2009**, *30*, 6301–6315. [CrossRef]
14. Aschbacher, J.; Pongsrihadulchai, A.; Karnchanasutham, S.; Rodprom, C.; Paudyal, D.; Le Toan, T. Assessment of ERS-1 SAR data for rice crop mapping and monitoring. In Proceedings of the 1995 International Geoscience and Remote Sensing Symposium, IGARSS'95, Florence, Italy, 10–14 July 1995; pp. 2183–2185.
15. Wu, F.; Wang, C.; Zhang, H.; Zhang, B.; Tang, Y. Rice crop monitoring in South China with RADARSAT-2 quad-polarization SAR data. *IEEE Geosci. Remote Sens. Lett.* **2010**, *8*, 196–200. [CrossRef]
16. Clauss, K.; Ottinger, M.; Künzer, C. Mapping rice areas with Sentinel-1 time series and superpixel segmentation. *Int. J. Remote Sens.* **2018**, *39*, 1399–1420. [CrossRef]
17. Nguyen, D.B.; Gruber, A.; Wagner, W. Mapping rice extent and cropping scheme in the Mekong Delta using Sentinel-1A data. *Remote Sens. Lett.* **2016**, *7*, 1209–1218. [CrossRef]
18. Le Toan, T.; Ribbes, F.; Wang, L.-F.; Floury, N.; Ding, K.-H.; Kong, J.A.; Fujita, M.; Kurosu, T. Rice crop mapping and monitoring using ERS-1 data based on experiment and modeling results. *IEEE Trans. Geosci. Remote Sens.* **1997**, *35*, 41–56. [CrossRef]
19. Guan, X.; Huang, C.; Liu, G.; Meng, X.; Liu, Q. Mapping rice cropping systems in Vietnam using an NDVI-based time-series similarity measurement based on DTW distance. *Remote Sens.* **2016**, *8*, 19. [CrossRef]
20. Belgiu, M.; Csillik, O. Sentinel-2 cropland mapping using pixel-based and object-based time-weighted dynamic time warping analysis. *Remote Sens. Environ.* **2018**, *204*, 509–523. [CrossRef]
21. Guan, X.; Liu, G.; Huang, C.; Meng, X.; Liu, Q.; Wu, C.; Ablat, X.; Chen, Z.; Wang, Q. An Open-Boundary Locally Weighted Dynamic Time Warping Method for Cropland Mapping. *Isprs Int. J. Geo-Inf.* **2018**, *7*, 75. [CrossRef]
22. Li, M.; Bijker, W. Vegetable classification in Indonesia using Dynamic Time Warping of Sentinel-1A dual polarization SAR time series. *Int. J. Appl. Earth Obs. Geoinf.* **2019**, *78*, 268–280. [CrossRef]
23. Chen, C.-F.; Son, N.-T.; Chen, C.-R.; Chang, L.-Y. Wavelet filtering of time-series moderate resolution imaging spectroradiometer data for rice crop mapping using support vector machines and maximum likelihood classifier. *J. Appl. Remote Sens.* **2011**, *5*, 053525. [CrossRef]
24. McCloy, K.; Smith, F.; Robinson, M. Monitoring rice areas using Landsat MSS data. *Int. J. Remote Sens.* **1987**, *8*, 741–749. [CrossRef]
25. Clauss, K.; Yan, H.; Kuenzer, C. Mapping paddy rice in China in 2002, 2005, 2010 and 2014 with MODIS time series. *Remote Sens.* **2016**, *8*, 434. [CrossRef]
26. Kücük, C.; Taskın, G.; Erten, E. Paddy-rice phenology classification based on machine-learning methods using multitemporal co-polar X-band SAR images. *IEEE J. Sel. Top. Appl. Earth Obs. Remote Sens.* **2016**, *9*, 2509–2519. [CrossRef]
27. Park, S.; Im, J.; Park, S.; Yoo, C.; Han, H.; Rhee, J. Classification and mapping of paddy rice by combining Landsat and SAR time series data. *Remote Sens.* **2018**, *10*, 447. [CrossRef]
28. Teluguntla, P.; Thenkabail, P.S.; Oliphant, A.; Xiong, J.; Gumma, M.K.; Congalton, R.G.; Yadav, K.; Huete, A. A 30-m Landsat-derived cropland extent product of Australia and China using random forest machine learning algorithm on Google Earth Engine cloud computing platform. *ISPRS J. Photogramm. Remote Sens.* **2018**, *144*, 325–340. [CrossRef]
29. Gumma, M.K.; Nelson, A.; Thenkabail, P.S.; Singh, A.N. Mapping rice areas of South Asia using MODIS multitemporal data. *J. Appl. Remote Sens.* **2011**, *5*, 053547. [CrossRef]
30. Wang, J.; Yang, Y.; Mao, J.; Huang, Z.; Huang, C.; Xu, W. CNN-RNN: A unified framework for multi-label image classification. In Proceedings of the IEEE Conference on Computer Vision and Pattern Recognition, Las Vegas, NV, USA, 26–30 June 2016; pp. 2285–2294.
31. Kussul, N.; Lavreniuk, M.; Skakun, S.; Shelestov, A. Deep learning classification of land cover and crop types using remote sensing data. *IEEE Geosci. Remote Sens. Lett.* **2017**, *14*, 778–782. [CrossRef]
32. Zhang, M.; Lin, H.; Wang, G.; Sun, H.; Fu, J. Mapping paddy rice using a convolutional neural network (CNN) with Landsat 8 datasets in the Dongting Lake Area, China. *Remote Sens.* **2018**, *10*, 1840. [CrossRef]
33. Sun, Z.; Di, L.; Fang, H. Using long short-term memory recurrent neural network in land cover classification on Landsat and Cropland data layer time series. *Int. J. Remote Sens.* **2019**, *40*, 593–614. [CrossRef]
34. Clement, M.; Kilsby, C.; Moore, P. Multi-temporal synthetic aperture radar flood mapping using change detection. *J. Flood Risk Manag.* **2018**, *11*, 152–168. [CrossRef]

35. Cai, Y.; Guan, K.; Peng, J.; Wang, S.; Seifert, C.; Wardlow, B.; Li, Z. A high-performance and in-season classification system of field-level crop types using time-series Landsat data and a machine learning approach. *Remote Sens. Environ.* **2018**, *210*, 35–47. [CrossRef]

36. Sakoe, H.; Chiba, S. Dynamic programming algorithm optimization for spoken word recognition. *IEEE Trans. Acoust. SpeechSignal Process.* **1978**, *26*, 43–49. [CrossRef]

37. Maus, V.; Câmara, G.; Cartaxo, R.; Sanchez, A.; Ramos, F.M.; De Queiroz, G.R. A time-weighted dynamic time warping method for land-use and land-cover mapping. *IEEE J. Sel. Top. Appl. Earth Obs. Remote Sens.* **2016**, *9*, 3729–3739. [CrossRef]

38. Petitjean, F.; Inglada, J.; Gancarski, P. Satellite image time series analysis under time warping. *IEEE Trans. Geosci. Remote Sens.* **2012**, *50*, 3081–3095. [CrossRef]

39. Berndt, D.J.; Clifford, J. Using dynamic time warping to find patterns in time series. In Proceedings of the KDD Workshop, Seattle, WA, USA, 31 July 1994; pp. 359–370.

40. Hochreiter, S.; Schmidhuber, J. Long short-term memory. *Neural Comput.* **1997**, *9*, 1735–1780. [CrossRef] [PubMed]

41. Understanding LSTM Networks. Available online: https://colah.github.io/posts/2015-08-Understanding-LSTMs (accessed on 16 June 2020).

42. Goodfellow, I.; Bengio, Y.; Courville, A.; Bengio, Y. Deep Learning. MIT press Cambridge: Cambridge, MA, USA, 2016; pp. 226–227.

43. Zhong, L.; Hu, L.; Zhou, H. Deep learning based multi-temporal crop classification. *Remote Sens. Environ.* **2019**, *221*, 430–443. [CrossRef]

44. Qiu, B.; Li, W.; Tang, Z.; Chen, C.; Qi, W. Mapping paddy rice areas based on vegetation phenology and surface moisture conditions. *Ecol. Indic.* **2015**, *56*, 79–86. [CrossRef]

45. Zhang, G.; Xiao, X.; Dong, J.; Kou, W.; Jin, C.; Qin, Y.; Zhou, Y.; Wang, J.; Menarguez, M.A.; Biradar, C. Mapping paddy rice planting areas through time series analysis of MODIS land surface temperature and vegetation index data. *Isprs J. Photogramm. Remote Sens.* **2015**, *106*, 157–171. [CrossRef]

46. Xiao, X.; Boles, S.; Frolking, S.; Salas, W.; Moore, B., III; Li, C.; He, L.; Zhao, R. Observation of flooding and rice transplanting of paddy rice fields at the site to landscape scales in China using VEGETATION sensor data. *Int. J. Remote Sens.* **2002**, *23*, 3009–3022. [CrossRef]

47. Chen, C.; McNairn, H. A neural network integrated approach for rice crop monitoring. *Int. J. Remote Sens.* **2006**, *27*, 1367–1393. [CrossRef]

Publisher's Note: MDPI stays neutral with regard to jurisdictional claims in published maps and institutional affiliations.

Article

Time Series Prediction with Artificial Neural Networks: An Analysis Using Brazilian Soybean Production

Emerson Rodolfo Abraham [1,*], João Gilberto Mendes dos Reis [1,2,3,*], Oduvaldo Vendrametto [1], Pedro Luiz de Oliveira Costa Neto [1], Rodrigo Carlo Toloi [1,4], Aguinaldo Eduardo de Souza [1,5,6] and Marcos de Oliveira Morais [1,7,8]

[1] Postgraduate Program in Production Engineering, Universidade Paulista-UNIP, Dr. Bacelar Street 1212, São Paulo 04026-002, Brazil; oduvaldo.vendrametto@docente.unip.br (O.V.); pedro.neto@docente.unip.br (P.L.d.O.C.N.); rodrigo.toloi@roo.ifmt.edu.br (R.C.T.); souza.eduaguinaldo@gmail.com (A.E.d.S.); marcostecnologia2001@gmail.com (M.d.O.M.)

[2] Postgraduate Program in Business Administration, Universidade Paulista-UNIP, Dr. Bacelar Street 1212, São Paulo 04026-002, Brazil

[3] Postgraduate Program in Agribusiness, Universidade Federal da Grande Dourados, Dourados 79804-970, Brazil

[4] Instituto Federal de Mato Grosso, Campus Rondonópolis, Rondonópolis 78721-520, Brazil

[5] Faculdade de Tecnologia de São Sebastião, Centro Paula Souza, São Sebastião 11600-970, Brazil

[6] Faculdade de São Vicente-UNIBR, São Vicente 11310-200, Brazil

[7] Universidade Santo Amaro-UNISA, Isabel Schmidt Street 349, São Paulo 04743-030, Brazil

[8] Centro Universitário Estácio São Paulo, Eng. Armando de Arruda Pereira Avenue, São Paulo 04309-010, Brazil

* Correspondence: emerson.abraham@stricto.unip.br (E.R.A.); joao.reis@docente.unip.br (J.G.M.d.R.)

Received: 3 September 2020; Accepted: 4 October 2020; Published: 15 October 2020

Abstract: Food production to meet human demand has been a challenge to society. Nowadays, one of the main sources of feeding is soybean. Considering agriculture food crops, soybean is sixth by production volume and the fourth by both production area and economic value. The grain can be used directly to human consumption, but it is highly used as a source of protein for animal production that corresponds 75% of the total, or as oil and derived food products. Brazil and the US are the most important players responsible for more than 70% of world production. Therefore, a reliable forecasting is essential for decision-makers to plan adequate policies to this important commodity and to establish the necessary logistical resources. In this sense, this study aims to predict soybean harvest area, yield, and production using Artificial Neural Networks (ANN) and compare with classical methods of Time Series Analysis. To this end, we collected data from a time series (1961–2016) regarding soybean production in Brazil. The results reveal that ANN is the best approach to predict soybean harvest area and production while classical linear function remains more effective to predict soybean yield. Moreover, ANN presents as a reliable model to predict time series and can help the stakeholders to anticipate the world soybean offer.

Keywords: artificial neural networks; time series forecasting; soybean; food production

---

1. Introduction

World's population is projected to reach 9.8 billion in 2050 [1] and food production needs to increase by 60% to meet the demand [2,3]. One reason for that is the developing countries—that have been growing much more rapidly than the industrial countries—are creating implications for world food demand mainly in products such as animal-based, fruits, and vegetables [4]. However, declining rates of growth in crop yields, slowing investment in agricultural research, and rising commodity prices have raised concerns of a general slowdown in global agricultural harvest area, yield, and production [5].

The rapid per capita income growth in countries like China and India (40% world population) pressure food supply chains shifting towards animal-based products that require disproportionately more agricultural resources in production [4,6] such as land, water, and vegetable protein [7]. Moreover, there is a concern revolving around big agriculture growers such as Brazil and the US using their agriculture areas to produce biofuels [6].

It is not only in the economy that this relationship between food demand and income are finding shelter. It is possible to verify in the literature a connection about technology and agricultural production. Crop yield and production, for instance, have been studied in the light of artificial intelligence. Khan et al. [8] predicted fruit production using deep neural networks. García-Martiínez et al. [9] estimated corn grain yield with a neural network using multispectral and RGB images acquired with unmanned aerial vehicles. Maimaitijiang et al. [10] predicted soybean yield using multimodal data fusion and deep learning. These applications are a clear attempt to improve knowledge about food production and provide decision-makers with valuable information to face the challenges of food demand.

Another possible solution discussed is the use of areas in Latin America and the Caribbean to expand agriculture production [11]. Brazil, for instance, has more than 8 million km^2 of area and uses only 15% of its arable land—approximately 60 of 400 million hectares [12]. The country is an important global food supplier, and it is estimated that one out of four agribusiness products in circulation around the world came from Brazil [13]. Despite the concern of biofuel production, sugarcane occupies only 8.9 million hectares of arable land [14], and the majority is used for sugar production rather than ethanol.

Brazil has more than 300 different crops and exports 350 types of products to 180 countries. The main export products are sugar, coffee, maize, orange juice, cotton, and soybean. Among these products, soybean is the main global source of protein, and the country is the major exporter that corresponded for approximately 29.9% of agribusiness external sales in 2016—USD 25.4 billion [13]. According to the Department of Agriculture of United States—USDA [15], Brazilian exports of the soybean complex are 81% grains, 15.7% meal, and 3.3% refined oil.

Soybean production has overspread inside the country from south, through the center-western to the northeast area. These movements are motivated by low land cost, and investments in agriculture inputs, mechanization, and infrastructure [16–18]. Other factors contributing to soybean growth in Brazil include the genetic improvement of seeds, increasingly productive planting systems [19], favourable climatic conditions, predictable precipitation patterns, and public financing policies for soybean plantations [20].

The soybean production is evaluated considering three categories: harvest area, yield, and production. The two main players are Brazil and the US, the former planted a harvest area of 36.9 million of hectares that produced 120.9 million of tones with a yield of 3.3 tones per hectare [21] and the latter planted a harvest area around 30.8 million of hectares with a production of 96.8 million tons and a yield of 3.1 tones per hectare [22]. These values are constantly predicted using classical methods and presented to stakeholders by government agencies. However, the respective literature is sparse and relates to agronomy aspects of soybean yield [10,23,24]. In this paper, we focus on the prediction of these soybean indicators based on the previous crop data.

Therefore, our study aims to estimate Brazilian soybean harvest area, yield, and production adopting Artificial Neural Networks (ANN) and comparing with classical methods of Time Series Analysis. To this end, we collected the values of harvest area, production, and yield over a period of 56 years (1961–2016). We established the trend lines for five functions: Linear, Exponential, Logarithmic, Polynomial, and Power, and compared these results with an ANN model with 10 neurons and six delays computed using a Nonlinear Autoregressive Network with External Input-NARX with Levenberg—Marquardt backpropagation for training the network.

The results show that the ANN model is the most efficient method to predict soybean harvest area and production. The novelty of this paper is to obtain a reliable prediction for soybean production measures using an ANN model and dealing with a short data period time series (50 years) [25]. The period of 1961–1966 was used only for ANN model delay.

This paper is divided into sections: Section 1 presents this introduction and literature review, Section 2 shows the methodological procedures, Section 3 deal with results and discussion, and Section 4 presents the conclusions of the study.

1.1. Artificial Neural Networks

Artificial Neural Networks, as the name proposed, use artificial neurons connected in layers to simulate human synapse (Figure 1). A mathematical model mimics the neural structure to learn and to acquire knowledge via experiences (Equations (1) and (2)). This technology is effective to solve problems—dynamic and nonlinear—such as pattern recognition and prediction [25–30].

$$ne = \sum_{i=1}^{n} x_i w_i + b_i \tag{1}$$

$$u = f(ne) \tag{2}$$

where $x_1 \ldots x_n$ are the input values (data set), $w_1 \ldots w_n$ are the weights, and b is the activation threshold (bias) in the neuron potential ne [25,26,31].

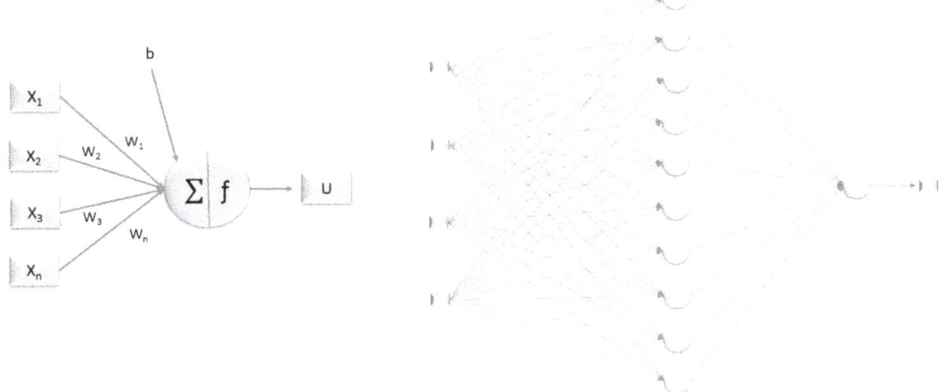

Figure 1. Artificial neuron (**left**) and ANN multilayer (**right**).

Among several types of neuron activation functions, the most common are: hyperbolic tangent (Equation (3)), hidden layer, and linear. The last one always assumes values identical to the activation potential n [25,26,31]:

$$f(ne) = \frac{1 - e^{-\beta u}}{1 + e^{-\beta u}} \tag{3}$$

where β is the constant associated with the slope of the hyperbolic tangent function and the output values assume numbers between -1 and 1.

ANN uses previous data for training the network and minimizes errors between the insertion and the estimation. This process adjusts the weights and possible bias for each neuron interaction. The training usually stops when finding out the optimal learning rate [25–30].

There are various ANN techniques such as General Regression Neural Network (GRNN), Backpropagation Neural Network (BNN), Radial Base Function Neural Network (RBFNN), and Adaptive Neuro-Fuzzy Inference System (ANFIS) [32]. Backpropagation (BP) is a learning algorithm widely used in forecasting problems with ANN, and the networks [30]. The weights between the different layers may be updated using the BP algorithm, with momentum and learning rate. Moreover, the weights between the different layers may be updated where the error is then propagated backward from the output to the input layer [33].

Some studies have been using ANN to study the agricultural environment. Garg et al. [30] compare the performance between different training methods using an ANN model to forecast wheat production in India. The data contain 95 years of wheat production (1919–2013), and the results revealed that the algorithms most effective in training methods are Bayesian regularization and Levenberg–Marquardt.

Almomani [34] adopted artificial neural networks to predict the biofuel production from agricultural wastes and cow manure at high accuracy. The training and testing of the ANN used to predict the cumulative methane production was assessed by using the root mean square method. The study confirms the capacity of the ANN model to predict the behavior of biofuel production and to identify the optimum conditions in a short time.

Sankhadeep et al. [35] use an ANN model for soil moisture quantity prediction for sustainable agricultural applications. They study soil moisture prediction in terms of soil temperature, air temperature, and relative humidity. The nonlinear relation between soil moisture and the features is realized using a hybrid modified flower pollination algorithm supported by the ANN model. They conclude that for sustainable agricultural application the model is highly suitable.

Khan et al. [8] use deep neural networks to fruit production prediction. They considered different types of fruit production such as apples, bananas, citrus, pears, and grapes with data from the National Bureau of Statistics of Pakistan. They adopted Levenberg–Marquardt optimization, backpropagation, and Bayesian regularization backpropagation. The results reveal that the government of Pakistan needs to further increase fruit production and create better policies for farmers to improve their production.

Wang and Xiao [36] studied recycle agriculture in West China to make a prediction on the comprehensive development status applying a neural network model with the application of backpropagation through the MATLAB program. They conclude that China needs to take measures to promote resources' decrement input and resource reuse efficiency, protect the forest resources, and reinforce harnessing of water loss and soil erosion.

Liu et al. [37] create an artificial neural network model for crop yield responding to soil parameters. The model was established by training a backpropagation neural network with 58 samples and tested with other 14 samples. They conclude that the model can precisely describe crop yield responding to soil parameters.

Fegade and Pawar [38] describe that, in India, farmers have difficulties to select proper crop for farming due to factors such as rainfall, temperature humidity, soil, and so on. Therefore, they used support vector machine and artificial neural networks to predict crop with 86.80% of accuracy.

Regarding grains, Maimaitijiang et al. [10] evaluate the power of an unmanned aerial vehicle (UAV) to estimate soybean grain yield within the framework of deep neural networks (DNN). Thermal images were collected using a low-cost multi-sensory UAV. The results propose that multimodal data fusion improves the yield prediction accuracy and is more adaptable to spatial variations; DNN-based models improve yield prediction model accuracy and were less prone to saturation effects.

Zhang et al. [39] establish a model for forecasting soybean price in China using quantile regression models to describe the distribution of the soybean price range, and using regression-radial basis function neural networks to approximate the nonlinear component of the soybean price. They collected the monthly domestic soybean price in China, and the results of the model indicate that the proposed model is effective.

García-Martínez et al. [9] analyze different multispectral and red-green-blue vegetation indices, canopy cover, and plant density in order to estimate corn grain yield using a neural network model. The neural network model provided a high correlation coefficient between the estimated and the observed corn grain yield with acceptable errors in the yield estimation.

Abraham et al. [40] propose to design, train, and simulate an ANN on to forecast the demand of soybean production in Mato Grosso state, Brazil that is exported by the port of Santos. A nonlinear autoregressive solution was adopted considering 80% of data for training, 5% to validation, and 15% for testing the network—a value of 9.0 million tons for 2017 as an increase of about 26.5% compared with the 2016.

Eventually, Abraham et al. [41] also analyze the relationship between soybean supply (production) and soybean demand (export) using artificial intelligence in a hybrid model neuro-fuzzy. Data from 20 years of soybean production and exportation were used, and the results indicate that the supply tends to be low when the demands of the ports are overloaded.

Specifically, in the present article, we raised two questions regarding ANN in soybean production:

- Can soybean harvest area, yield, and production be predicted efficiently using Artificial Neural Networks?
- If so, are Artificial Neural Networks more effective than classical methods of Time Series Analysis to predict soybean production measures?

To answer these two questions, we develop an ANN model using NARX with the Levenberg–Marquardt algorithm for backpropagation and data of Brazilian soybean production.

1.2. Time Series and Classical Methods

Time series analysis studies the past behavior of historical series using different methods (Table 1). It verifies trends, seasonality, and randomness in a dataset in two ways: stationary, when observations oscillate around a central horizontal axis; and non-stationary when oscillates around changing values [42,43]. The most appropriate model for a specific dataset is the coefficient of determination (R), the mean absolute error (MAE), and the mean squared error (MSE) [42,43].

Table 1. Classical methods, equations, and characteristics.

Method	Formulas	Features
Linear function	$y = ax \pm b$	Linear is defined as a curve of the first degree or a simple straight line—where y is the trend, x represents the period of time, a is a slope, and b is the intercept. The intercept will determine how far from the x-axis the trend begins. The slope will determine the direction and the steepness.
Exponential function	$y = ae^{bx}$	Exponential is defined as a transcendental curve, where e represents the basis for natural logarithms, and its constant value is 2.7813. It grows exponentially, but they never reach the attracting value.
Logarithmic function	$y = aln(x) \pm b$	The inverse of the exponential function is a logarithmic function.
Polynomial function	$y = ax^2 \pm bx \pm c$	The second-degree polynomial curve is a parabola. The polynomial model can go up to the sixth degree. A larger magnitude corresponds to a greater adjustment than that in the original data; however, this does not mean that it is best for forecasting. The best method is the one that can perform well with minimum parameters.
Power function	$y = ax^b$	The graph of a power curve is a hyperbola.

Source: adapted from [42,43].

The coefficient of determination (Equation (4)) measures the linear regression adjustment, which aims to explain the relationship of the variables. The closer this number is to one, the more fitted is the model. However, a measure higher than 0.7 is satisfactory [25,42,43]:

$$R = \frac{\sum (\hat{y} - y)^2}{\sum (y - \bar{y})^2}. \tag{4}$$

The coefficient of determination is calculated based on the ratio between the explained and the total variance where y represents the real value of the series, $\hat{y}$ is the expected value (value of the regression line approaching the actual value), and $\bar{y}$ is the average value of the series.

Note that the variance is the difference between the expected value and the mean, and the total variance is the difference between the original and mean value [25,42,43]. The MAE and MSE are calculated according to Equations (2) and (3), where n is the number of elements in the series.

$$MAE = \frac{\sum |y - \hat{y}|}{n}, \tag{5}$$

$$MSE = \frac{\sum (y - \hat{y})^2}{n}. \tag{6}$$

Finally, functions with error values close to 0 are the most effective in predicting future values. These time series applications are described in the Results and Discussion section.

2. Materials and Methods

2.1. Dataset

To perform this study, we collected data from the Food and Agriculture Organization of the United Nations (FAO) [44] regarding harvest area (million hectares), yield (tons per hectare), and production (million tons) between 1961–2016. The dataset was imported from MS© Excel 2016 spreadsheet to Matlab© R2017b arrays. However, the period from 1961–1966 was used only for delay configuration, and it was not plotted on the time series [29,45].

Firstly, we conducted Time Series Analysis. The historical series was extracted and processed in MS© Excel 2016 spreadsheet format generating graphs with trend lines. Tables 2–4 present the formulas.

Table 2. Harvest area.

Model	Trend Formulas
Linear function	$y = 0.5523x - 1.1485$
Exponential function	$y = 2.1503e^{0.0582x}$
Logarithmic function	$y = 7.8508ln(x) - 10.379$
Polynomial function	$y = 0.009x^2 + 0.0937x + 2.8259$
Power function	$y = 0.4302x^{1.0419}$

x = timestep(year). y = million hectares.

Table 3. Yield.

Model	Trend Formulas
Linear function	$y = 0.0388x + 1.0523$
Exponential function	$y = 1.1748e^{0.0199x}$
Logarithmic function	$y = 0.5735ln(x) + 0.3397$
Polynomial function	$y = 0.0001x^2 + 0.0321x + 1.111$
Power function	$y = 0.7747x^{0.3113}$

x = timestep(year). y = tons per hectare.

Table 4. Production.

Model	Trend Formulas
Linear function	$y = 1.6935x - 12.323$
Exponential function	$y = 2.5259e^{0.0782x}$
Logarithmic function	$y = 22.599ln(x) - 36.247$
Polynomial function	$y = 0.045x^2 - 0.6007x + 7.5607$
Power function	$y = 0.3332x^{1.3534}$

x = timestep(year). y = million tons.

Secondly, we used neural networks toolbox of the Matlab©R2017b software to create, train, and validate the ANN model—we tested with 10 neurons and six delays (Figure 2).

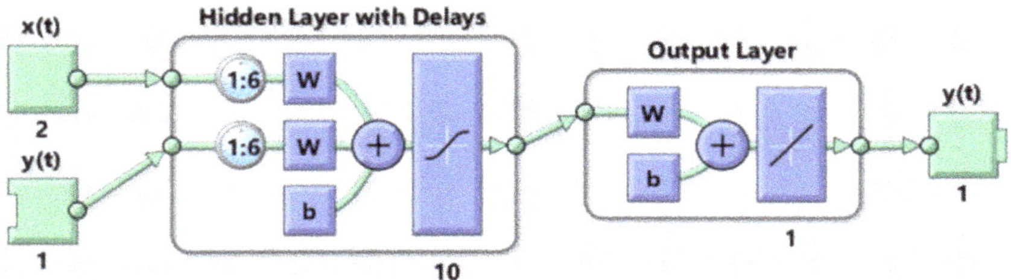

Figure 2. Neural network created in Matlab 2017b.

We adopted the Nonlinear Autoregressive Network with External Input-NARX type because it has proven to be the most effective and accurate solution for multivariable data series [27,46–48]. The NARX network applies historical input data with time delay operators [9]. We used 70% of data for training, 15% for validation, and 15% for testing. We defined the percentage based on k-cross validation that utilizes efficiently the learning abilities of the ANN model [49], and data are distributed randomly by NARX [46]. Moreover, we adopted the Levenberg–Marquardt algorithm for backpropagation due to being the fastest supervised algorithm for training and widely used for time series prediction in the ANN model [25,30,46].

For harvested area (target), we used yield and production as input variables; for yield (target), we used harvested area and production as input variables; for production (target), harvested area and yield were used as input variables.

After that, Matlab© R2017b provided algorithms for closed-loop form simulation (named multistep prediction). This type of simulation is important to verify the ability of the networks to make predictions (calculation of errors) [25]. Figure 3 shows the overall flowchart of the ANN model.

2.2. Model Classification

The differences between the original and predicted values were computed using MAE and MSE, even for ANN. We compared classical models and neural networks where the errors of each model were sorted from lowest to highest. However, regression measures were sorted from highest to lowest. Depending on the use of two measures of error the weighted average was used (Equation (7)):

$$\text{Rank} = \frac{(\text{MAE} \times 0.5) + (\text{MSE} \times 0.5) + (\text{R} \times 1)}{2} \tag{7}$$

where MAE is mean absolute error, MSE is mean squared error, and R is the coefficient of determination.

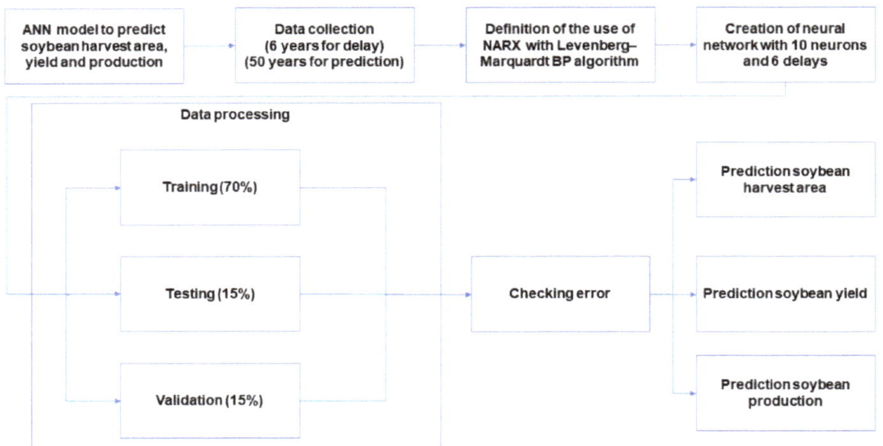

Figure 3. Flowchart of the ANN model.

3. Results and Discussion

3.1. Time Series Analysis Using Classical Predictive Methods (Functions)

3.1.1. Harvested Area

The first application of classical methods for prediction uses the time series for harvest area (target). Figure 4 illustrates the 1967–2016 timespan in million hectares.

The harvested area raised continuously, mainly after the harvest of 1997 (Timestep 31), and reached around 33 million hectares in 2016 (Timestep 50). Looking back over 1967 year (Timestep 1), the planted soybean area was 2% (around 0.6 million hectares) of the area planted in 2016. There has been a more than a 50-fold increase while the US, the main Brazilian competitor, had in 1967 54% of the current planted area [22].

Regarding the fit of functions, polynomial and power were more effective in predicting harvest area considering R, MAE, and MSE (Table 5).

Table 5. Effective functions for forecasting harvested area.

Rank	Model	R	MAE	MSE
1°	Polynomial function	0.944	1.813	3.915
2°	Power function	0.949	1.927	6.901
3°	Linear function	0.904	1.996	6.716
4°	Exponential function	0.797	2.481	8.859
5°	Logarithmic function	0.680	3.825	22.482

R is the coefficient of determination (value between 0 and 1), MAE is the mean absolute error (value in millions of hectares), and MSE is the mean squared error (value in millions of hectares).

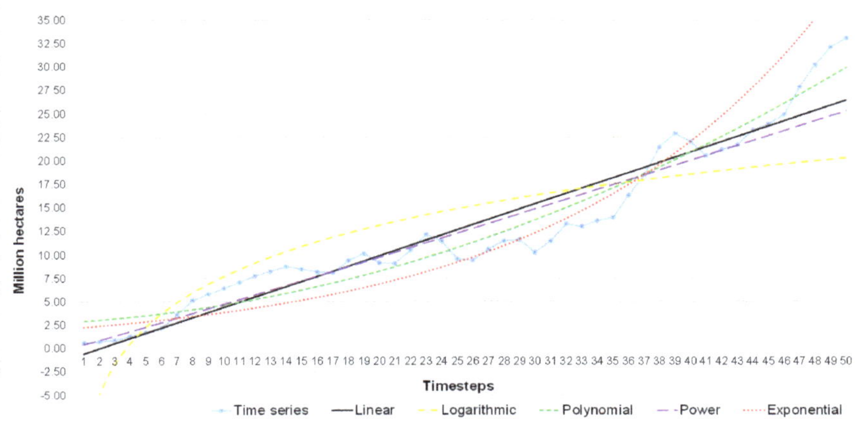

Figure 4. Original time series for harvested area.

3.1.2. Yield

The second application of classical methods of predication verifies soybean yield (target). Figure 5 depicts the time series analysis results considering data from 1967–2016 in hectares.

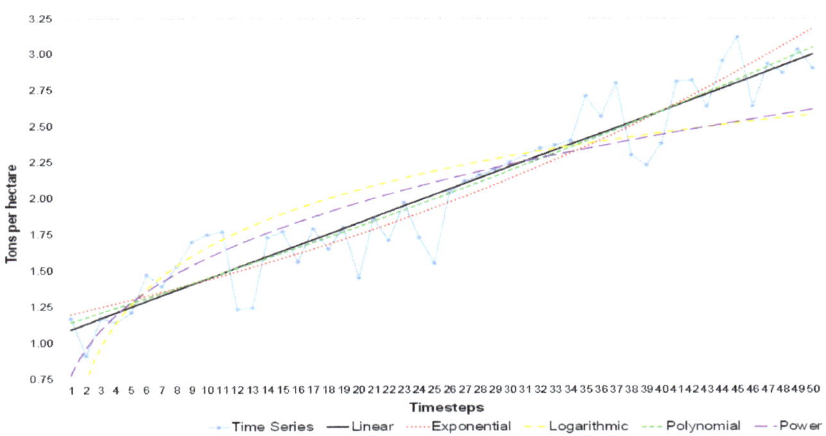

Figure 5. Original time series for yield.

The average yield over the 50-year period was approximately two tons per hectare. The lowest value was 0.9 hectares in 1968 (Timestep 2), and the highest value 3.1 tons per hectare in 2011 (Timestep 45). During the 1967 (Timestep 1) crop season, the national yield was approximately 1.2 ton per hectare, which corresponds to less than half of the yield in 2016 (Timestep 50). In addition, compared with the US in the same period, the lowest value was 1.6 ton per hectare in 1974 and the highest value 3.5 in 2016 [22]. This means that the Brazilian soybean yield varies 158% against 119% of the US.

The yield increase in soybean production was affected by changes in production processes and the use of new technologies. Moreover, genetic improvements of soybean created a variety of grain with better adaptation to the climate that affected productivity [12]. According to Pereira [50], the last 30 years have demonstrated innovative solutions in food production, such as new crop varieties and new irrigation techniques. However, Dani [12] argue that the evolution has generally been technological without improving governance processes causing logistics issues throughout the supply chain. Logistics has a huge impact on soybean production and directly affect the trade [51].

Related to the fit of the functions, linear and polynomial predict more precisely soybean yield given that R, MAE, and MSE (Table 6).

Table 6. Effective functions for forecasting yield.

Rank	Model	R	MAE	MSE
1°	**Linear function**	0.898	0.148	0.036
2°	**Polynomial function**	0.899	0.158	0.037
3°	**Exponential function**	0.874	0.170	0.038
4°	**Power function**	0.794	0.202	0.068
5°	**Logarithmic function**	0.728	0.239	0.095

R is the coefficient of determination (value between 0 and 1), MAE is the mean absolute error (value in tons per hectare), and MSE is the mean squared error (value in tons per hectare).

3.1.3. Production

Finally, we use classic methods to predict production. Figure 6 shows the time series analysis for soybean production from 1967–2016 in millions of tons.

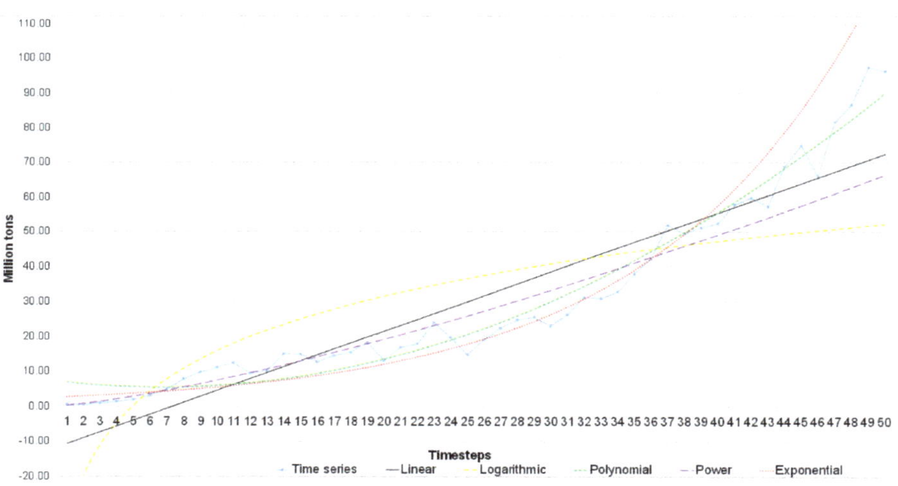

Figure 6. Original time series for production.

Brazilian soybean production raised 130-fold, moving from 700 thousand tons in 1967 (Timestep 1) to 96.3 million tons in 2016 (Timestep 50). In the same period, the US soybean production moves from 26.5 million tons to 116.9 million tons [22]. In 2019, the Brazilian soybean production was 25% higher [21] than 2016.

Furthermore, we identify that in 1970 (Timestep 4) the production was 1.5 million tons, but, in 1977 (Timestep 11), it reached 12.5 million tons. This great expansion of soybean cultivation occurs due to the expansion of international demand and the national soybean oil industry [52].

Regarding soybean production, polynomial and power were more effective considering R, MAE, and MSE (Table 7).

Table 7. Effective functions for forecasting production.

Rank	Model	R	MAE	MSE
1°	**Polynomial function**	0.968	3.990	21.755
2°	**Power function**	0.952	6.058	88.784
3°	**Exponential function**	0.853	5.847	77.116
4°	**Linear function**	0.867	7.658	91.879
5°	**Logarithmic function**	0.574	13.843	293.441

R is the coefficient of determination (value between 0 and 1), MAE is the mean absolute error (value in millions of tons), and MSE is the mean squared error (value in millions of tons).

3.2. ANN Model

3.2.1. Training, Validation, and Testing of Neural Network

Harvested area training reached an optimal value for the regression and correlation among variables after nine interactions (Figure 7). The training procedure stops when the performance on the test data does not improve following a fixed number of training iterations [39]. The main purpose of the training phase is to find the optimal set of weights for the ANN model where the error is minimized [35].

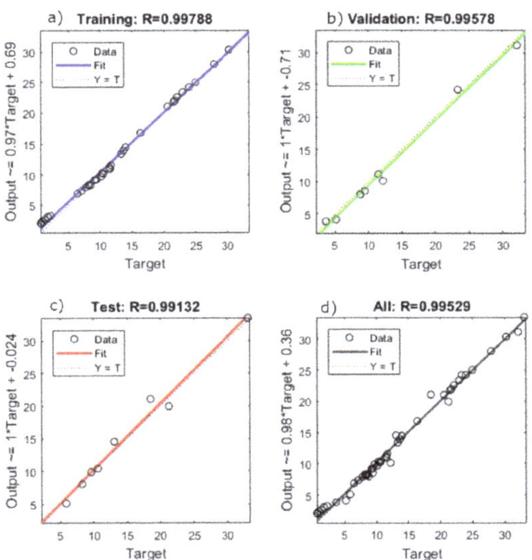

Figure 7. Regression and correlation of the harvested area using Matlab R2017b.

The training, validation, and testing indicate that the network learned from the data (R > 0.99). Moreover, the fit was well-aligned, which means the model has a good capacity for generalization and prediction.

Yield training reached an optimal value for the regression and correlation among variables after 12 iterations and pose an R higher than 0.9 (Figure 8).

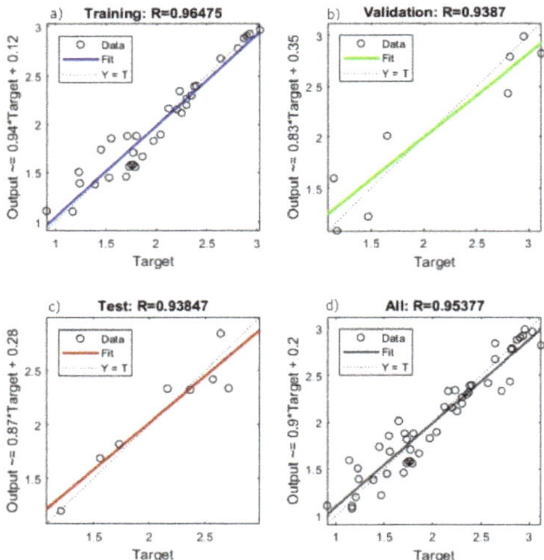

Figure 8. Regression and correlation for yield using Matlab R2017b

Yield presents correlation and regression results similar to the other two networks. However, fit shows a reasonable alignment representing a capacity of the network generalize and predict.

Finally, the production network was trained and after nine interactions reached an optimal value for the regression and correlation among variables (Figure 9).

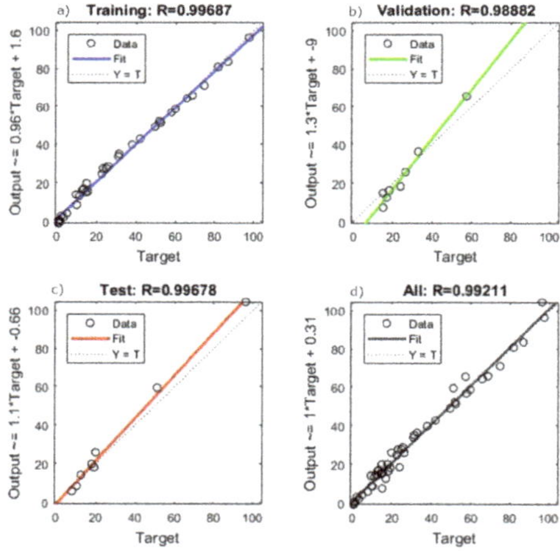

Figure 9. Regression and correlation for production using Matlab R2017b.

The network shows an excellent rate of learning, with reasonable values of alignment. However, the validation and test pose deviations on fit. The overall results present proper alignment confirming its ability to generalize and make predictions.

Given that there are three networks, the harvested area indicated the best results, followed by production and yield.

3.2.2. Time Series Results with an Artificial Neural Network

Figures 10–12 depict the results of the time series generated by the neural networks in closed-loop form (multistep prediction). The blue line (target) represents the original data, and the red line (prediction) represents the obtained values for each period.

Neural network prediction shows better adjustment to the original data than time series analysis. In other words, these predictions show a smoother follow-up. The trend lines of the classical models follow the randomness of the series increasing the error between the original and predicted values. The base graphic Figure 10 presents the error of the prediction in millions of hectares, in Figure 11 in tons per hectare and Figure 12 in millions of tons.

The classical models are based on elements dependent on the analysis of their predecessors. On the flipside, ANN is a generalization of the classical models, where an element to be predicted also depends on the previous elements of other related time series [27].

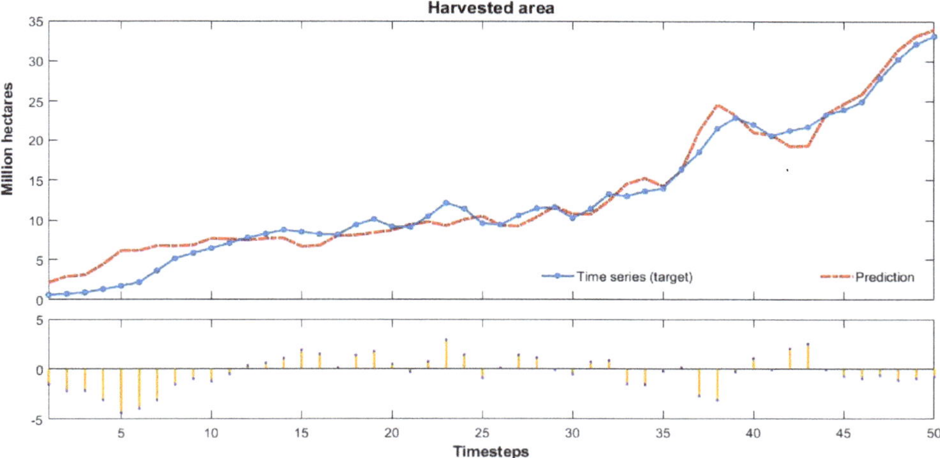

Figure 10. Multistep prediction of harvested area using Matlab R2017b.

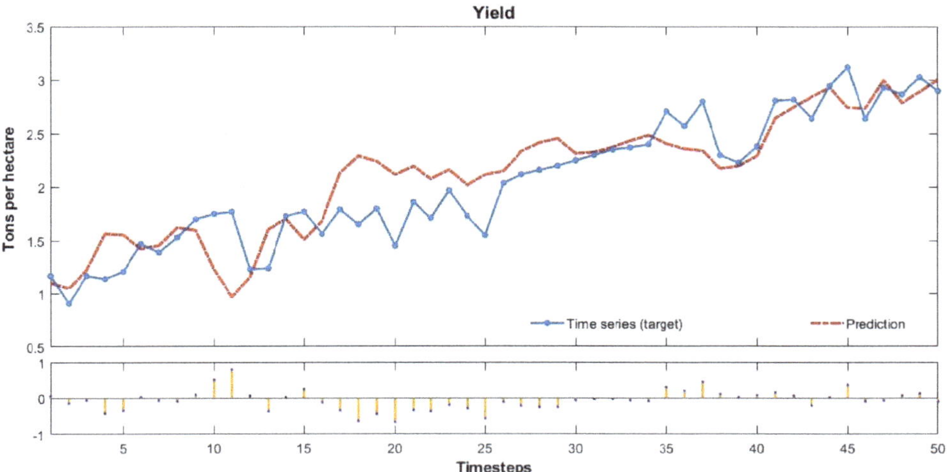

Figure 11. Multistep prediction of yield using Matlab R2017b.

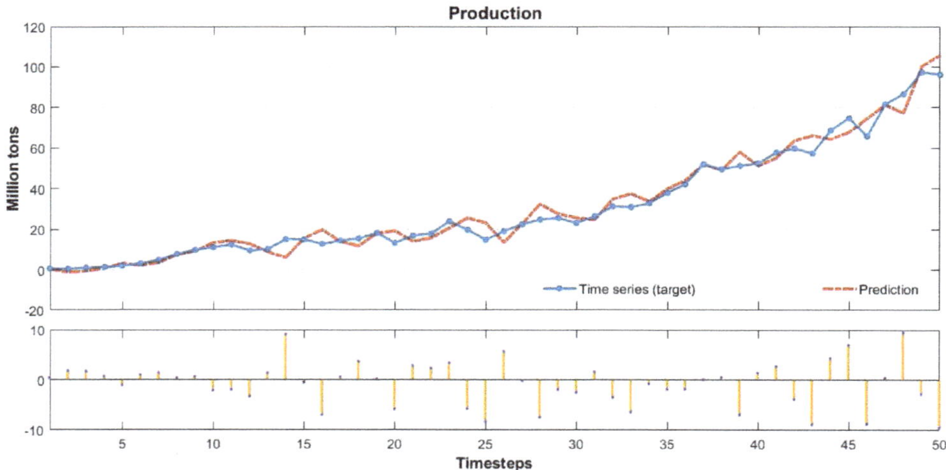

Figure 12. Multistep prediction of production using Matlab R2017b.

3.2.3. Comparison between Artificial Neural Networks and Time Series Classical Models

Considering R, MAE, and MSE, Tables 8–10 present the ranking of ANN model versus classical functions for forecast harvested area, yield, and production, respectively.

Considering the results, Artificial Neural Networks ranked first for predicting harvested area and production and third place to yield. The polynomial model ranked second in all three series showing the reliability of the model to estimate future values. The logarithmic model is the least fitted and should be discarded for these series.

Table 8. Effective functions versus ANN for forecasting harvested area.

Rank	Model	R	MAE	MSE
1°	**ANN**	0.995	1.309	2.763
2°	**Polynomial function**	0.944	1.813	3.915
3°	**Power function**	0.949	1.927	6.901
4°	**Linear function**	0.904	1.996	6.716
5°	**Exponential function**	0.797	2.481	8.859
6°	**Logarithmic function**	0.680	3.825	22.482

R is the coefficient of determination (value between 0 and 1), MAE is the mean absolute error (value in millions of hectares), and MSE is the mean squared error (value in millions of hectares).

Table 9. Effective functions versus ANN for forecasting yield.

Rank	Model	R	MAE	MSE
1°	**Linear function**	0.898	0.148	0.036
2°	**Polynomial function**	0.899	0.158	0.037
3°	**ANN**	0.954	0.220	0.084
4°	**Exponential function**	0.874	0.170	0.038
5°	**Power function**	0.794	0.202	0.068
6°	**Logarithmic function**	0.728	0.239	0.095

R is the coefficient of determination (value between 0 and 1), MAE is the mean absolute error (value in tons per hectare), and MSE is the mean squared error (value in tons per hectare).

Table 10. Effective functions versus ANN for forecasting production.

Rank	Model	R	MAE	MSE
1°	**ANN**	0.992	3.362	19.713
2°	**Polynomial function**	0.968	3.990	21.755
3°	**Power function**	0.952	6.058	88.784
4°	**Exponential function**	0.853	5.847	77.116
5°	**Linear function**	0.867	7.658	91.879
6°	**Logarithmic function**	0.574	13.843	293.441

R is the coefficient of determination (value between 0 and 1), MAE is the mean absolute error (values in millions of tons), and MSE is the mean squared error (values in millions of tons).

Based on these results, it is possible to infer that predictive capabilities of the developed ANN model are efficient to soybean prediction with short data time series. This fact confirms the superior performance of the ANN model against classical methods. Similar results are obtained for Nedic et al. [53] when compared to an ANN model with classical statistical models to predict traffic noise.

ANN has been recognized as a valuable predictive tool due to its ability to learn, adapt, and generalize the results of a sample of noise data and are more effective and flexible than conventional statistics for dealing with nonlinearity [54]. Moreover, there is a tendency towards the adopting of artificial intelligence in decision models.

4. Conclusions

This study compares classical methods of time series prediction with Artificial Neural Networks using Brazilian soybean harvest area, yield and production from 1961–2016. The results indicate that ANN is the best approach to predict soybean harvest area and production while classical linear function remains more effective to predict soybean yield. However, ANN is a reliable model to predict using time series and can help farmers, government, and trading companies anticipate the soybean world offer to organize efficiently logistics resources and public policies.

Our results confirm the important role of neural networks in dealing with agriculture issues as showed in previous studies in the literature [8,10,35,39]. The R value above 0.9 confirms the high performance of the model. Nevertheless, regarding the agriculture concerns about low availability for planting areas, yield, and production [4–6], your results demonstrated that, at least in case of the soybean, this is not a concern.

Furthermore, we can conclude that the ANN model can be effective even using a short time series—that, in our case, was 50 years. This fact reveals a robustness of the model. However, despite the advantages of the ANN model, classical methods also can produce very good models. A comparison in other agriculture commodities can be made to confirm or refuse the behavior presented in a soybean case.

Finally, we also suggest for further studies to combine neural networks in hybrid systems using, for example, ANN and Fuzzy Logic, similar to that proposed by [41]. Literature has shown that hybrid systems are more efficient. The goal is to achieve a synergy between hybrid systems to compensate for the disadvantage of one by the advantage of another [55–57].

Author Contributions: Conceptualization, E.R.A., J.G.M.d.R., O.V., P.L.d.O.C.N., and R.C.T.; methodology, E.R.A., J.G.M.d.R., O.V., and P.L.d.O.C.N.; software, E.R.A.; validation, J.G.M.d.R., O.V., P.L.d.O.C.N., R.C.T., and A.E.d.S.; formal analysis, J.G.M.d.R., O.V., P.L.d.O.C.N., M.d.O.M., and R.C.T.; investigation, E.R.A., J.G.M.d.R., and R.C.T.; resources, E.R.A. and J.G.M.d.R. Data curation, E.R.A. and J.G.M.d.R.; writing—original draft preparation, E.R.A. and J.G.M.d.R.; writing—review and editing, J.G.M.d.R., O.V., P.L.d.O.C.N., M.d.O.M., A.E.d.S., and R.C.T.; visualization, E.R.A. J.G.M.d.R., O.V., P.L.d.O.C.N., M.d.O.M., A.E.d.S., and R.C.T.; supervision, J.G.M.d.R., O.V., and P.L.d.O.C.N.; project administration, E.R.A.; funding acquisition, M.d.O.M., A.E.d.S. and R.C.T. All authors have read and agreed to the published version of the manuscript.

Funding: This research was funded by the Coordenação de Aperfeiçoamento de Pessoal de Nível Superior Grant No. 0001.

Acknowledgments: The authors would like to thank Sunsetti Treinamentos e Serviços and Universidade Paulista UNIP for the financial incentives.

Conflicts of Interest: The authors declare no conflict of interest.

References

1. United Nations. World Population Prospects. In *The 2017 Revision. Key Findings and Advance Tables*; Technical Report; United Nations Department of Economic and Social Affairs: New York, NY, USA, 2017.
2. Alexandratos, N.; Bruinsma, J. *World Agriculture towards 2030/2050: The 2012 Revision*; Technical Report; Food and Agriculture Organization of the United Nations: Rome, Italy, 2012.
3. ONUBR. FAO: Se o Atual Ritmo de Consumo Continuar, em 2050 Mundo Precisará de 60% Mais Alimentos e 40% Mais água. 2015. Available online: https://brasil.un.org/pt-br/68525-fao-se-o-atual-ritmo-de-consumo-continuar-em-2050-mundo-precisara-de-60-mais-alimentos-e-40 (accessed on 19 November 2019).
4. Fukase, E.; Martin, W. Economic growth, convergence, and world food demand and supply. *World Dev.* **2020**, *132*, 104954. [CrossRef]
5. Fuglie, K.O. Is agricultural productivity slowing? *Glob. Food Secur.* **2018**, *17*, 73–83. [CrossRef]
6. Rask, K.J.; Rask, N. Economic development and food production–Consumption balance: A growing global challenge. *Food Policy* **2011**, *36*, 186–196. [CrossRef]
7. Fraanje, W.; Garnett, T. *Soy: Food, Feed, and Land Use Change (Foodsource: Building Blocks)*; Technical Report; Food Climate Research Network, University of Oxford: Oxford, UK, 2020.
8. Khan, T.; Qiu, J.; Ali Qureshi, M.A.; Iqbal, M.S.; Mehmood, R.; Hussain, W. Agricultural Fruit Prediction Using Deep Neural Networks. *Procedia Comput. Sci.* **2020**, *174*, 72–78. [CrossRef]
9. García-Martínez, H.; Flores-Magdaleno, H.; Ascencio-Hernández, R.; Khalil-Gardezi, A.; Tijerina-Chávez, L.; Mancilla-Villa, O.R.; Vázquez-Peña, M.A. Corn Grain Yield Estimation from Vegetation Indices, Canopy Cover, Plant Density, and a Neural Network Using Multispectral and RGB Images Acquired with Unmanned Aerial Vehicles. *Agriculture* **2020**, *10*, 277. [CrossRef]
10. Maimaitijiang, M.; Sagan, V.; Sidike, P.; Hartling, S.; Esposito, F.; Fritschi, F.B. Soybean yield prediction from UAV using multimodal data fusion and deep learning. *Remote Sens. Environ.* **2020**, *237*, 111599. [CrossRef]
11. The World Bank. Future Looks Bright for Food Production in Latin America and Caribbean. 2013. Available online: http://www.worldbank.org/en/news/feature/2013/10/16/food-production-trade-latin-america-caribbean-future (accessed on 11 December 2019).
12. Dani, S. *Food Supply Chain Management and Logistics: From Farm to Fork*, 1st ed.; Kogan Page: London, UK; Philadelphia, PA, USA, 2015.
13. EMBRAPA. *Embrapa em Números*; Technical Report; Empresa Brazileira de Pesquisa Agropecuária-EMBRAPA. Ministério da Agricultura, Pecuária e Abastecimento: Brasília, Brazil, 2017.
14. Defante, L.R.; Vilpoux, O.F.; Sauer, L. Rapid expansion of sugarcane crop for biofuels and influence on food production in the first producing region of Brazil. *Food Policy* **2018**, *79*, 121–131. [CrossRef]
15. USDA. *World Agricultural Production: Circular Series November 2019*; Technical Report 11-19; USDA: Washington, DC, USA, 2019.
16. Horvat, R.; Watanabe, M.; Yamaguchi, C.K. Fertilizer consumption in the region Matopiba and their reflections on Brazilian soybean production. *Int. J. Agric. For.* **2015**, *5*, 52–59.
17. Sauer, S.; Pereira Leite, S. Agrarian structure, foreign investment in land, and land prices in Brazil. *J. Peasant Stud.* **2012**, *39*, 873–898. [CrossRef]
18. Kumagai, E.; Sameshima, R. Genotypic differences in soybean yield responses to increasing temperature in a cool climate are related to maturity group. *Agric. For. Meteorol.* **2014**, *198–199*, 265–272. [CrossRef]
19. Castanheira, É.G.; Freire, F. Greenhouse gas assessment of soybean production: Implications of land use change and different cultivation systems. *J. Clean. Prod.* **2013**, *54*, 49–60. [CrossRef]
20. Gil, J.; Garrett, R.; Berger, T. Determinants of crop-livestock integration in Brazil: Evidence from the household and regional levels. *Land Use Policy* **2016**, *59*, 557–568. [CrossRef]
21. EMBRAPA. Soja em Números. Available online: embrapa.br/web/portal/soja/cultivos/soja1/dados-economicos (accessed on 10 August 2020).

22. United States Department of Agriculture-Economic Research Service. Overview. Available online: https://www.ers.usda.gov/data-products/oil-crops-yearbook/oil-crops-yearbook/#So%20and%20Soybean%20Products (accessed on 10 August 2020).

23. Kaul, M.; Hill, R.L.; Walthall, C. Artificial neural networks for corn and soybean yield prediction. *Agric. Syst.* **2005**, *85*, 1–18. [CrossRef]

24. Ma, B.L.; Dwyer, L.M.; Costa, C.; Cober, E.R.; Morrison, M.J. Early Prediction of Soybean Yield from Canopy Reflectance Measurements. *Agron. J.* **2001**, *93*, 1227–1234. [CrossRef]

25. Demuth, H.; Beale, M.; Hagan, M. *Neural Network Toolbox User's Guide*; The MathWorks, Inc.: Natick, MA, USA, 2017.

26. Russell, S.; Norvig, P. *Artificial Intelligence: A Modern Approach*, 3rd ed.; Pearson Education India: Bengaluru, India, 2015.

27. Aizenberg, I.; Sheremetov, L.; Villa-Vargas, L.; Martinez-Muñoz, J. Multilayer Neural Network with Multi-Valued Neurons in Time Series Forecasting of Oil Production. *Neurocomputing* **2016**, *175*, 980–989. [CrossRef]

28. Gomes, L.F.A.M.; Machado, M.A.S.; Caldeira, A.M.; Santos, D.J.; Nascimento, W.J.D.d. Time Series Forecasting with Neural Networks and Choquet Integral. *Procedia Comput. Sci.* **2016**, *91*, 1119–1129. [CrossRef]

29. Wang, J.; Tsapakis, I.; Zhong, C. A space–Time delay neural network model for travel time prediction. *Eng. Appl. Artif. Intell.* **2016**, *52*, 145–160. [CrossRef]

30. Garg, B.; Kirar, N.; Menon, S.; Sah, T. A performance comparison of different back propagation neural networks methods for forecasting wheat production. *CSI Trans. ICT* **2016**, *4*, 305–311. [CrossRef]

31. Silva, I.N.D. *Redes Neurais Artificiais Para Engenharia e Ciencias Aplicadas: Fundamentos Teoricos e Aspectos Praticos*; ARTLIBER: São Paulo, Brazil, 2016.

32. Golnaraghi, S.; Zangenehmadar, Z.; Moselhi, O.; Alkass, S. Application of Artificial Neural Network(s) in Predicting Formwork Labour Productivity. *Adv. Civ. Eng.* **2019**, *2019*, 1–11. [CrossRef]

33. Mohamed, Z.E. Using the artificial neural networks for prediction and validating solar radiation. *J. Egypt. Math. Soc.* **2019**, *27*, 47. [CrossRef]

34. Almomani, F. Prediction of biogas production from chemically treated co-digested agricultural waste using artificial neural network. *Fuel* **2020**, *280*, 118573. [CrossRef]

35. Chatterjee, S.; Dey, N.; Sen, S. Soil moisture quantity prediction using optimized neural supported model for sustainable agricultural applications. *Sustain. Comput. Inform. Syst.* **2018**, 100279. [CrossRef]

36. Wang, F.; Xiao, H. Prediction on Development Status of Recycle Agriculture in West China Based on Artificial Neural Network Model. In *Information Computing and Applications*; Zhu, R., Zhang, Y., Liu, B., Liu, C., Eds.; Communications in Computer and Information Science; Springer: Berlin/Heidelberg, Germany, 2010; Volume 105, pp. 423–429. [CrossRef]

37. Liu, G.; Yang, X.; Li, M. An Artificial Neural Network Model for Crop Yield Responding to Soil Parameters. In *Advances in Neural Networks—ISNN 2005*; Hutchison, D., Kanade, T., Kittler, J., Kleinberg, J.M., Mattern, F., Mitchell, J.C., Naor, M., Nierstrasz, O., Pandu Rangan, C., Steffen, B., et al., Eds.; Lecture Notes in Computer Science; Springer: Berlin/Heidelberg, Germany, 2005; Volume 3498, pp. 1017–1021. [CrossRef]

38. Fegade, T.K.; Pawar, B.V. Crop Prediction Using Artificial Neural Network and Support Vector Machine. In *Data Management, Analytics and Innovation*; Sharma, N., Chakrabarti, A., Balas, V.E., Eds.; Advances in Intelligent Systems and Computing; Springer: Singapore, 2020; Volume 1016, pp. 311–324. [CrossRef]

39. Zhang, D.; Zang, G.; Li, J.; Ma, K.; Liu, H. Prediction of soybean price in China using QR-RBF neural network model. *Comput. Electron. Agric.* **2018**, *154*, 10–17. [CrossRef]

40. Abraham, E.R.; dos Reis, J.G.M.; Colossetti, A.P.; de Souza, A.E.; Toloi, R.C. Neural Network System to Forecast the Soybean Exportation on Brazilian Port of Santos. In *Advances in Production Management Systems. The Path to Intelligent, Collaborative and Sustainable Manufacturing*; Lödding, H., Riedel, R., Thoben, K.D., von Cieminski, G., Kiritsis, D., Eds.; Springer: Cham, Switzerland, 2017; pp. 83–90.

41. Abraham, E.R.; dos Reis, J.G.M.; de Souza, A.E.; Colossetti, A.P. Neuro-Fuzzy System for the Evaluation of Soya Production and Demand in Brazilian Ports. In *Advances in Production Management Systems. Production Management for the Factory of the Future*; Ameri, F., Stecke, K.E., von Cieminski, G., Kiritsis, D., Eds.; IFIP Advances in Information and Communication Technology; Springer: Cham, Switzerland, 2019; Volume 566, pp. 87–94. [CrossRef]

42. Escolano, N.R.; Espin, J.J.L. *Econometría: Series Temporales y Modelos de Ecuaciones Simultáneas*; Limencop: Alicante, Spain, 2012.
43. Pecar, B.; Davis, G. *Time Series Based Predictive Analytics Modelling: Using MS Excel*, 3rd ed.; Amazon Kindle: Seattle, WA, USA, 2018.
44. FAO. FAOSTAT. Available online: http://www.fao.org/faostat/en/#data/QC (accessed on 15 December 2019).
45. Shao, Y.E.; Lin, S.C. Using a Time Delay Neural Network Approach to Diagnose the Out-of-Control Signals for a Multivariate Normal Process with Variance Shifts. *Mathematics* **2019**, *7*, 959. [CrossRef]
46. Di Nunno, F.; Granata, F. Groundwater level prediction in Apulia region (Southern Italy) using NARX neural network. *Environ. Res.* **2020**, *190*, 110062. [CrossRef] [PubMed]
47. Zhang, J.; Zhang, X.; Niu, J.; Hu, B.X.; Soltanian, M.R.; Qiu, H.; Yang, L. Prediction of groundwater level in seashore reclaimed land using wavelet and artificial neural network-based hybrid model. *J. Hydrol.* **2019**, *577*, 123948. [CrossRef]
48. Guzman, S.M.; Paz, J.O.; Tagert, M.L.M. The Use of NARX Neural Networks to Forecast Daily Groundwater Levels. *Water Resour. Manag.* **2017**, *31*, 1591–1603. [CrossRef]
49. Javed, S.; Zakirulla, M.; Baig, R.U.; Asif, S.M.; Meer, A.B. Development of artificial neural network model for prediction of post-streptococcus mutans in dental caries. *Comput. Methods Programs Biomed.* **2020**, *186*, 105198. [CrossRef]
50. Pereira, L.S. Water, Agriculture and Food: Challenges and Issues. *Water Resour Manag.* **2017**, *31*, 2985–2999. [CrossRef]
51. Mendes dos Reis, J.G.; Sanches Amorim, P.; Sarsfield Pereira Cabral, J.A.; Toloi, R.C. The Impact of Logistics Performance on Argentina, Brazil, and the US Soybean Exports from 2012 to 2018: A Gravity Model Approach. *Agriculture* **2020**, *10*, 338. [CrossRef]
52. Schnepf, R.D.; Dohlman, E.; Bolling, C. *Agriculture in Brazil and Argentina: Developments and Prospects for Major Field Crops*; Technical Report; International Agriculture and Trade Outlook No. (WRS-013); United States Department of Agriculture Economic Research Service: Washington DC, USA, 2001.
53. Nedic, V.; Despotovic, D.; Cvetanovic, S.; Despotovic, M.; Babic, S. Comparison of classical statistical methods and artificial neural network in traffic noise prediction. *Environ. Impact Assess. Rev.* **2014**, *49*, 24–30. [CrossRef]
54. Ko, M.; Tiwari, A.; Mehnen, J. A review of soft computing applications in supply chain management. *Appl. Soft Comput.* **2010**, *10*, 661–674. [CrossRef]
55. Rajasekaran, S.; Pai, G.A.V. *Neural Networks, Fuzzy Logic and Genetic Algorithm: Synthesis and Applications*; PHI Learning Pvt. Ltd.: Delhi, India, 2012.
56. Roy, S.; Chakraborty, U. *Soft Computing: Neuro-Fuzzy and Genetic Algorithms*, 1st ed.; Pearson: London, UK, 2013.
57. Barrios Rolanía, D.; Delgado Martínez, G.; Manrique, D. Multilayered neural architectures evolution for computing sequences of orthogonal polynomials. *Ann. Math. Artif. Intell.* **2018**, *84*, 161–184. [CrossRef]

Publisher's Note: MDPI stays neutral with regard to jurisdictional claims in published maps and institutional affiliations.

 *agriculture*

Review

Machine Learning for Plant Breeding and Biotechnology

Mohsen Niazian [1,*] and Gniewko Niedbała [2,*]

[1] Field and Horticultural Crops Research Department, Kurdistan Agricultural and Natural Resources Research and Education Center, Agricultural Research, Education and Extension Organization (AREEO), Jam-e Jam cross way, P.O. Box 741, 66169-36311 Sanandaj, Iran
[2] Department of Biosystems Engineering, Faculty of Environmental Engineering and Mechanical Engineering, Poznań University of Life Sciences, Wojska Polskiego 50, 60-627 Poznań, Poland
[*] Correspondence: mniazian@ut.ac.ir (M.N.); gniewko@up.poznan.pl (G.N.)

Received: 18 August 2020; Accepted: 25 September 2020; Published: 27 September 2020

Abstract: Classical univariate and multivariate statistics are the most common methods used for data analysis in plant breeding and biotechnology studies. Evaluation of genetic diversity, classification of plant genotypes, analysis of yield components, yield stability analysis, assessment of biotic and abiotic stresses, prediction of parental combinations in hybrid breeding programs, and analysis of in vitro-based biotechnological experiments are mainly performed by classical statistical methods. Despite successful applications, these classical statistical methods have low efficiency in analyzing data obtained from plant studies, as the genotype, environment, and their interaction (G × E) result in nondeterministic and nonlinear nature of plant characteristics. Large-scale data flow, including phenomics, metabolomics, genomics, and big data, must be analyzed for efficient interpretation of results affected by G × E. Nonlinear nonparametric machine learning techniques are more efficient than classical statistical models in handling large amounts of complex and nondeterministic information with "multiple-independent variables versus multiple-dependent variables" nature. Neural networks, partial least square regression, random forest, and support vector machines are some of the most fascinating machine learning models that have been widely applied to analyze nonlinear and complex data in both classical plant breeding and in vitro-based biotechnological studies. High interpretive power of machine learning algorithms has made them popular in the analysis of plant complex multifactorial characteristics. The classification of different plant genotypes with morphological and molecular markers, modeling and predicting important quantitative characteristics of plants, the interpretation of complex and nonlinear relationships of plant characteristics, and predicting and optimizing of in vitro breeding methods are the examples of applications of machine learning in conventional plant breeding and in vitro-based biotechnological studies. Precision agriculture is possible through accurate measurement of plant characteristics using imaging techniques and then efficient analysis of reliable extracted data using machine learning algorithms. Perfect interpretation of high-throughput phenotyping data is applicable through coupled machine learning-image processing. Some applied and potentially applicable capabilities of machine learning techniques in conventional and in vitro-based plant breeding studies have been discussed in this overview. Discussions are of great value for future studies and could inspire researchers to apply machine learning in new layers of plant breeding.

Keywords: artificial neural networks; big data; classification; high-throughput phenotyping; modeling; predicting

---

1. Introduction

Due to climate change (global warming), increasing food requirements and depletion of resources in consequence of increasing global population, it is necessary to use modern technologies in agriculture and food sciences [1]. Plant breeding is a dynamic branch of agricultural science. It started with simple selection of impressive plants with superior characteristics. Later, genetics and statistics were involved in classical plant breeding, mainly after the discoveries of Gregor Mendel and Sir Ronald Aylmer Fisher. Next, modern plant breeding emerged with the advancements in genetic and biotechnology approaches. Classical plant breeding methods mainly included assessment and classification of genetic diversity, yield components analysis (indirect selection of superior genotypes with impressive economic characteristics), yield stability analysis (genotype × environment interaction), enhanced tolerance to biotic and abiotic stresses, and hybrid breeding programs. In vitro-based biotechnological breeding methods mainly included in vitro micropropagation, doubled haploid production, artificial polyploidy induction, and *Agrobacterium*-mediated gene transformation. In in vitro micropropagation studies, researchers want to investigate the effects of influential factors (inputs), such as combination of culture medium components, combination and concentrations of plant growth regulators (PGRs), and interactions of plant genotype × culture medium × PGRs × explant type × explant age × elicitor additives × type and concentration of carbohydrate source × etc., on regeneration efficiency (outputs) of their desired plants. Classical statistical techniques have been employed to analyze and interpret the results of both classical and in vitro-based plant breeding studies. These analytical techniques are mainly based on variance and linear regression models to assess the relationship of variables and predict the effect of independent variables on dependent variables. One regression model is required to assess the effect of a group of independent variables (X1, X2, X3, ... , Xn) on one dependent variable (Y), according to the multiple linear relationships [2]. However, nonlinear and nondeterministic properties are inextricably linked with plant biological systems [3]. Therefore, despite of successful applications, the classical linear regression-based models are unable to interpret highly nonlinear and complex relationships between dependent and independent variables. Most of these plant breeding approaches are "multiple-independent variables versus multiple-dependent variables." Under these conditions, one regression model is required for each output [4]. Powerful data mining tools are employed in plant breeding studies to predict and explain complex data.

Machine learning—the science of programming computers so they can learn from data—has been widely applied in both classical and in vitro-based plant breeding studies to interpret the flow of information about plants from the DNA sequence to the observed phenotypes. There are three ways to classify machine learning methods, including supervised and supervised models, linear and nonlinear algorithms, and shallow and deep learning models (Figure 1). Artificial neural networks (ANNs), deep neural networks (DNNs), convolutional neural networks (CNNs), random forest (RF), and support vector machines (SVMs) are examples of nonlinear nonparametric machine learning algorithms, applied for processing nonlinear data in plant studies [5]. These data-driven models are able to parse and interpret non-normal, nonlinear, and nondeterministic unpredictable data sets, through the full use of all spectral data and avoid irrelevant spectral bands and multicollinearity [6,7]. Among different learning algorithms, including supervised, unsupervised, reinforcement, sparse dictionary, and rule-based, supervised learning is more suitable and efficient for life science problems [8]. Supervised learning can be used for classification (predicting non-numeric answers) and regression (predicting numeric answers) [9]. Formless datasets such as data obtained by photo imaging or sequencing can be interpreted through machine learning algorithms [10]. Genome sequencing data can be used in machine learning models for the identification and classification of transposable elements [11]. By using machine learning algorithms, breeders are able to predict multiple outputs (multiple-dependent variables) through different combinations of multiple inputs in one model and reduce required analyses.

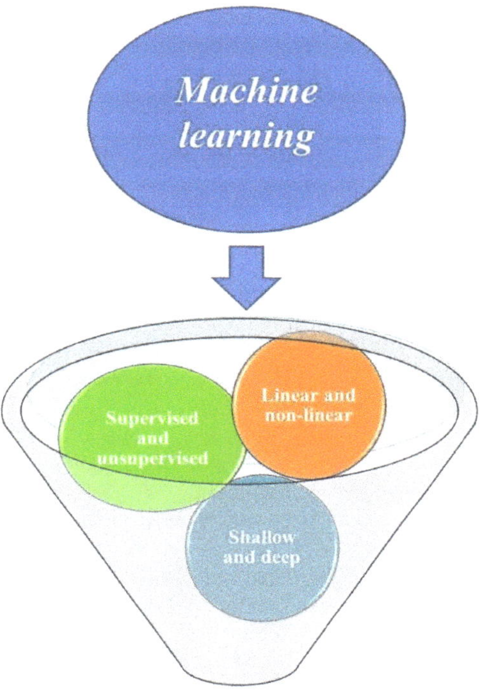

Figure 1. Different categories of machine learning algorithms.

Artificial neural networks, consist of an input, an output, and several hidden layers, are nonlinear nonparametric models which do not require a prior structure for data and detailed information about the physical processes to be modeled and to tolerate data loss [12,13]. Because of their more hidden layers, DNNs have greater predictive power than ANNs. Convolutional neural networks, as state-of-the-art deep learning architecture, are inspired by the natural visual perception mechanism of the living creatures and consist of convolutional, pooling, fully-connected layers, and an output layer [14]. CNNs are suitable for classification studies because of automatic feature extraction [9]. Image classification, object detection, object tracking, pose estimation, text detection and recognition, visual saliency detection, action recognition, scene labeling, speech, and natural language processing are some of the typical applications of CNNs [14]. Neural networks have low interpretability of the features (lack the interpretation capability), especially CNN in which the features extracted are hidden. More advanced machine learning technique of SVMs, which uses a supervised learning algorithm to find both linear and nonlinear relationships in data, can be used for clustering, classification, and regression analysis of data sets. In comparison with multilayer perceptron (MLP) of ANN, SVM uses a large number of hidden units and has better performance in the formulation of the learning problem, subsequently quadratic optimization task [15]. Random forest regression is a regression tree-based machine learning that uses multiple decision trees to classify data and needs setting the number of trees, the number of random features, and the stop criteria for training. RF is more suitable for spectral data analysis and overfitting can be controlled through combining different independent predictors [16,17]. In semantic segmentation methods, such as automated phenotyping and plant disease detection, deep learning CNN can be more effective than shallow learning models of SVMs and RF and problem of required large manually crafted features can be solved by using image augmentation and small manually annotated empirical dataset for fine-tuning a synthetically bootstrapped CNN [18]. Through the integrating image feature extraction with classification in a single pipeline, deep convolutional neural networks have been considered as mainstream in biotic and abiotic stress diagnosis and classification [19]. A nine-layer deep CNN model was trained for identification of plant leaf diseases using data set with 39 different classes of plant leaves and background images and 96.46% classification accuracy was reported, which is greater than traditional machine learning approaches of SVM, decision tree, logistic regression, and K-NN [20]. CNNs are also applicable in remote sensing for object detection and pattern recognition. High accuracy (84%) for fine-grained mapping of vegetation species and communities using deep CNN-based segmentation, trained by data directly derived from visual interpretation of unmanned aerial vehicles (UAV)-based high-resolution Red-Green-Blue (RGB) imagery, has been reported [21].

A lot of training data is required in ANN for the optimization of sigmoid functions belonging to the hidden layer's neurons, as overfitting and local minima may happen by small number of training data. Therefore, the optimization process cannot be properly carried using back-propagation algorithms, when the number of training samples is small [8]. Through the short review on studies that used SVM and ANN techniques for identifying disease in plants, it was concluded that the ANN-based methods are better than SVM-based methods, as few samples and features are used in SVM-based methods to identify the disease-affected plants [22]. Conversely, in modeling in vitro culture of *Chrysanthemum* (*Dendranthema* × *grandiflorum*), better performance accuracy of SVR ($R^2 > 0.92$) than MLP ($R^2 > 0.82$) has been reported [15]. Applying different algorithm and comparing their performance is an appropriate solution to find the best algorithm in a particular data set. In tea plant (*Camellia sinensis* L.), partial least squares discriminative analysis (PLS-DA) and least squares-support vector machines (LS-SVM) were used for the classification of different nitrogen nutrition status under field condition and better performance with correct classification of LS-SVM than PLS-DA was reported [23]. The pros and cons of different nonlinear machine learning methods under similar scenarios are presented in Table 1.

Table 1. Pros and cons of nonlinear machine learning algorithms applied in classical and in vitro-based plant breeding studies.

Leaning Algorithm	Advantages	Disadvantages
ANNs	• Good learning capabilities	• Lack the interpretation capability • Overfitting and local minima in small number of training data • Implementing very small number of hidden neurons
CNNs	• Ability of automatic feature extraction	• Lack the interpretation capability • Require large amounts of data for training • Require considerable skill and experience to select suitable hyperparameters
SVMs	• Uses a large number of hidden units • Quadratic optimization task in the formulation of the learning problem	• Shallow architecture
RF	• Ability to handle noise • Prevent overfitting • Ability to manage a large number of features	• Shallow architecture

ANNs—artificial neural networks; CNN—convolutional neural networks; RF—random forest; SVMs—support vector machines.

Different application areas for nonlinear machine learning technologies in classical and in vitro-based plant breeding studies are shown in Figure 2. The following sections of the article provide a comprehensive review of the applications of these nonlinear machine learning techniques in classical and in vitro-based plant breeding studies.

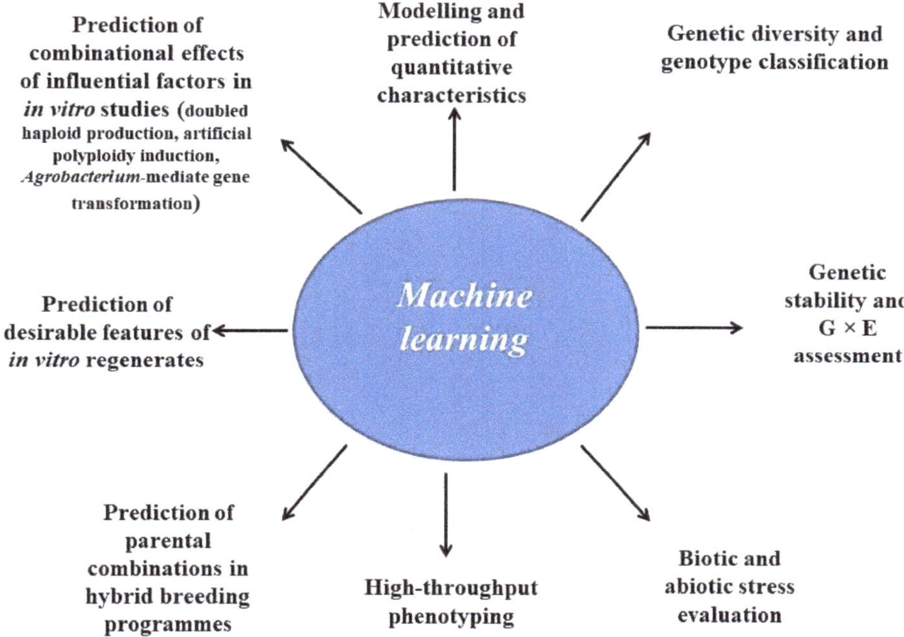

Figure 2. Potential applications of machine learning techniques in classical and modern plant breeding.

Some recently applied nonlinear machine learning models in both classical and in vitro-based plant breeding studies are listed in Table 2.

Table 2. Examples of recently applied nonlinear machine learning models in classical and modern plant breeding studies.

Plant Species	Type of Machine Learning	Techniques	Purpose(s)	Reference
Ajowan (*Trachyspermum ammi* L.)	ANN	MLR	Modeling and predicting of seed yield	[2]
	ANN	MLR	Modeling and predicting of essential oil content	[24]
	ANN	MLR, IP	Predicting physical properties of embryogenic callus and number of somatic embryos	[25]
Arabidopsis thaliana	DT, SVMs, NB	Gaussian kernel	Predict the plant abiotic stresses response through the miRNAs' concentration	[8]
Carrot (*Daucus carota*)	RF	-	Precision agriculture-yield mapping	[26]
Chrysanthemum	ANN	GA	Modeling and optimizing of in vitro sterilization	[27]
	ANFIS	GA	Modeling and optimizing of somatic embryogenesis	[3]
	ANN, SVMs	MLP	Modeling effect of plant growth regulators on somatic embryogenesis	[15]
Cucumber (*Cucumis sativus*)	CNN	IP	Segmentation and quantification of powdery mildew disease	[28]
Garnem (G × N15) *Prunus* rootstock	ANN	GA	Prediction and optimization of mineral salts of in vitro culture medium	[29]
	ANN	GA	Modeling and optimizing of in vitro hormonal combination	[30]
	ANN	GA	Modeling and optimizing of new in vitro culture medium	[31]
Grapevine rootstock	ANN	Principal coordinate analysis, UPGMA	Genetic diversity assessment through molecular markers (RAPD-SSR) dataset	[32]
Maize (*Zea mays* L.)	CNN	IP	Identification of haploid and diploid maize seeds	[33]
	CNN	IP	Classification model to identify the infected and healthy leaves	[34]
	CNN	IP	Plant diseases recognition	[35]
	CNN	IP	Identification and classification of drought stress	[19]
Okra (*Abelmoschus esculentus* L.)	DNN	IP	High-throughput salt-stress phenotyping	[36]
Pearl millet (*Pennisetum glaucum*)	DNN	IP	Identification of mildew disease	[37]
Potato (*Solanum tuberosum*)	ANN	IP	Identification and discrimination of potato varieties	[38]
	RF		Classification of *Phytophthora infestans* infected cultivars	[17]
Rapeseed (*Brassica napus*)	ANN	MLP	Seed yield modeling	[39]
	CNN	IP	Stand count estimation	[40]
	ANN	MLP	Multicriteria yield prediction based on meteorological data and mineral fertilization data	[41]
	ANN	MLP	Early prediction and simulation of seed yield based on meteorological and mineral fertilization data	[42]
Rice (*Oryza sativa*)	CNN		Plant diseases and pest recognition	[43,44]
Safflower (*Carthamus tinctorius* L.)	ANN	MLR	Seed yield modeling	[45]
Sesame (*Sesamum indicum* L.)	ANN	MLR	Oil content modeling	[46]
	ANN, SVMs	RBF, ERBF, GRNN, M5-Rule, M5-Tree, MLR	Estimation of oil and protein content	[47]
Soybean (*Glycine max*)	CNN	IP	Estimation of seeds per pod	[48]
	DNN	IP	Evaluation of stomatal density diversity	[49]
Tomato (*Lycopersicon esculentum* L.)	ANN	MLR, IP	Modeling of callus induction and regeneration in anther culture	[50]
	CNN	IP	Evaluation of disease severity	[51]
	ANN	MLP	Estimation of salinity tolerance	[52]
	ANN	MLP	Prediction of seed yield based on meteorological data and information on mineral fertilization	[53]
	ANN	MLP	Prediction and simulation of seed yield with qualitative and quantitative data sets	[54]
Wheat (*Triticum aestivum* L.)	CNN	IP	Quantification of spikes	[55]
	DNN	LSTM	Production forecasting	[56]
	CNN	-	Genomic selection	[57]
	ANN, GRNN	MLP	Modeling in vitro shoot regeneration	[58]
	ANN	MLP	Analysis of concentration of ferulic acid, deoxynivalenol, and nivalenol	[59]
White ginger (*Hedychium coronarium*)	ANN	MLP	Prediction and optimization of coronarin D content	[60]

ANFIS—adaptive neuro-fuzzy inference system; ANN—artificial neural networks; CNN—convolutional neural networks; DNN—deep neural network; DT—decision tree; ERBF—extended radial bases function; GA—genetic algorithm; GRNN—generalized regression neural network; IP—image processing; LSTM—long short term memory; MLR—multiple linear regressions; MLP—multilayer perceptron; NB—Naïve Bayes; RBF—radial bases function; RF—random forest; SVMs—support vector machines; UPGMA—unweighted pair group method with arithmetic mean.

2. Traditional Plant Breeding

2.1. Assessment and Classification of Genetic Diversity

One of the most important prerequisites of plant breeding programs is genetic diversity, which enables selection of important accessions and their use in future breeding programs [61]. Morphological, biochemical, and physiological markers have been analyzed to investigate the genetic diversity of different plants. Morphological features are the simplest to measure and do not require special tools or techniques. The statistical analysis of these markers can prove the existence of genetic diversity among studied genotypes. Niazian et al. [62] used analysis of variance (ANOVA) and estimated the coefficient of variation (CV) of different agro-morphological traits (plant height, number of branches, number of umbels, number of umbellets in an inflorescence, biological yield, and single plant yield) of eight ecotypes of ajowan medicinal plant (*Carum copticum* L.) and observed significant genetic diversity.

Molecular/genetic markers are another group of markers which enable assessment of genetic diversity and discrimination of the genotype. Amplified fragment-length polymorphism (AFLP), restriction-fragment length (RFLP), randomly amplified polymorphic DNA (RAPD), simple sequence repeat (SSR), intersimple sequence repeats (ISSR), and single nucleotide polymorphism (SNP) are the most commonly used molecular markers to study genetic diversity and species identification in different target plants [63]. These genetic markers estimate phylogenetic relationships and identify varieties more reliably and effectively than morphological markers [64]. Although molecular markers are more effective than morphological markers in the assessment of genetic diversity and discrimination and identification of various plant genotypes, there are some technical and/or economic limitations [65].

Classical multivariate analyses such as cluster analysis, discriminant function analysis, and principal component analysis (PCA) have been used for the classification and grouping of different genotypes in various plant species by means of morphological, biochemical, physiological, and molecular markers [61,64,66–68]. Object detection through deep learning algorithms could be used for efficient genetic diversity assessment and classification of plant genotypes. The use of CNN to classify morphological parameters is an appropriate alternative to conventional classification methods, such as k-nearest neighbor, probabilistic neural network, support vector machine, genetic algorithm, and PCA, all of which are time consuming and require feature extraction [65,69]. In soybeans (*Glycine max* (L.) Merr.), the genetic diversity of 90 accessions was detected through high-throughput evaluation of stomatal density [49]. In *Cinnamomum osmophloeum* Kanehira (Lauraceae), deep CNN was applied for differentiating between morphologically similar species, and accuracy of CNN classifiers was better than SVMs classifiers (96.7% vs. 74.6%) [70]. Sant'Anna et al. [71] compared the performance of ANN with Fisher's classical multivariate statistical technique and Anderson's discriminant functions to assess the genetic diversity and classify 10 plant populations. They observed that ANN-classified populations with high and low differentiation were better than classical methods, as there were fewer wrongly classified individuals. Linear discriminant analysis and nonlinear artificial neural network methods were applied to identify and discriminate 10 potato varieties with morphological data obtained through image processing. The correctness of classification of the ANN method was 100% [38].

As was mentioned above, machine learning can also be used for classification through molecular markers data. DNA/RNA sequences can be used for training CNNs and applications in plant molecular biology and classification of genotypes through molecular markers [9]. Different machine learning models were used to identify true single nucleotide polymorphisms (SNPs) in allopolyploid peanuts (*Arachis hypogaea* L.). The selection of true SNPs by means of real peanut RNA sequencing (RNA-seq) and whole-genome shotgun (WGS) resequencing data resulted in above 80% accuracy [72]. Costa et al. [32] applied a neural network algorithm to infer the genetic diversity and group allelic frequencies obtained by RAPD and SSR molecular markers in grapevine rootstock varieties and found three genetically diversified clusters among 64 grapevine rootstocks analyzed. Deep learning techniques enable prediction of plant phenotypes from their genome data [9].

Artificial neural networks have also been applied for genomic prediction and genomic selection in different plant species [73–75]. The phenotypes of 2000 Iranian bread wheat landrace accessions were predicted from genomic dataset collected from 33,709 DArT markers using a deep convolutional neural network. Authors reported that the Pearson's correlation coefficients between observed and predicted phenotypic values (grain length, grain width grain hardness, thousand-kernel weight, test weight, sodium dodecyl sulfate sedimentation, grain protein, and plant height) in deep CNN were more than other genomic selection methods [57].

2.2. Yield Component Analysis and Indirect Selection (Prediction)

An increase in the economic yield (seed yield, oil yield, sugar yield, essential oil yield, biomass yield, straw yield, lint percentage, etc.) is always the target of most breeding programs. However, yield is a highly complex quantitative trait, which is usually controlled by several genes, and it is strongly influenced by the environment. Therefore, yield traits have low heritability and direct selection does not improve such complex traits. Instead, plant breeders usually prefer to use simpler highly correlated traits to have greater influence on yield. Selected yield component(s) will be used as "selection criteria" in future studies, i.e., indirect selection [2,76]. Classical single variable and multivariate linear methods, such as correlation coefficient analysis, PCA, path analysis, and multiple regression analyses (stepwise, forward, and reverse), have been used in classical plant breeding to interpret relationships between plant traits and improve important quantitative properties like yield and tolerance to biotic and abiotic stresses. The correlation coefficient analysis and path analysis have been used to evaluate a simple relationship between two traits and identify cause/effect relationships between correlated variables, respectively [24]. Regression-based methods are the most effective multivariate statistical methods for indirect selection purposes. They are based on a linear relationship of a dependent variable (Y) as a function of multiple independent variables. These multiple variables create a complicated condition for interpretation. However, some reduction techniques, such as PCA and factor analysis, are able to concentrate the original multiple variables in a few complex variables [77].

Stepwise, forward, and reverse regression analyses have been used to determine the effects of yield components on different economic quantitative characteristics in various crops. Backward stepwise regression was used to find the relationship between changes in grain yield and yield components of rice (*Oryza sativa* L.) in terms of the relative response ratio to elevated CO_2 [78]. Stepwise regression was used to determine the components of sugar beet (*Beta vulgaris* L.) yield affecting the yield of sugar under water deficit regimes and foliar application of jasmonic acid [76]. Zou et al. [77] applied stepwise regression analysis to identify the yield components involved in drought resistance of cotton seedlings (*Gossypium hirsutum* L.). Despite all the advantages, there is one major drawback to regression-based models in classical plant breeding studies—it is impossible to analyze nonlinear relationships of dependent and independent variables [2,79]. The application of nonlinear machine learning algorithms in yield component analysis and indirect selection studies enables the interpretation of nonlinear relationships between dependent and independent variables, the contribution of yield components to yield and prediction of economic quantitative characteristics. ANN was more efficient than multiple linear regressions (MLR) in the prediction of seed yield [2] and essential oil content [24] of ajowan (*Trachyspermum ammi* L.). Emamgholizadeh et al. [79] found that ANN predicted the yield of sesame seeds (*Sesamum indicum* L.) better than MLR. The ANN model was characterized by lower root mean square error (RMSE) and higher determination coefficient (R^2). The analysis of the sensitivity of the ANN model showed that the number of capsules per plant and the flowering time were, respectively, the most and the least significant variables to the yield of sesame seeds. Artificial neural networks have successfully predicted the yield of apples, pears, chives, and onions, allowing for data on crop diseases, time until harvest (based on the date), current temperature, humidity and precipitation (amount of snowfall) in the area, amount of sunshine, ground temperature, atmospheric pressure, and moisture evaporation in the ground [80]. ANNs have also been used to predict the yield of winter rapeseed and winter wheat on the basis of meteorological data (air temperature and precipitation) and information on the use of mineral fertilizer [41,42,53,54]. The superiority of DNN (Long Short-Term Memory) over the auto regressive integrated moving average (ARIMA) model in predicting wheat production has been reported [56]. Deep CNN classification has been applied for image-based ear counting of wheat with high level of robustness, without considering of variables, such as growth stage and weather conditions [55].

Neural networks have also been used to estimate and predict the qualitative characteristics of different plants. The ANN model predicted the oil content in sesame more accurately and efficiently than the MLR model [46]. Parsaeian et al. [47] applied a multilayer perceptron (MLP)-ANN to estimate the oil and protein content in sesame on the basis of 138 morphological features measured in 125 sesame seed genotypes. The morphological characteristics of seeds were measured through image processing. The qualitative parameters of oil and the protein content in sesame seeds estimated by means of R^2 and RMSE statistics revealed the superiority of MLP over the radial basis function (RBF), extended RBF (ERBF), GRNN, M5-Rule, M5-Tree, support vector machine regression, and linear regression models [47]. Niedbała et al. [59] developed a multilayer perceptron ANN to assess the influence of the cultivar and weather conditions on the concentration of ferulic acid and correlate its content with the concentration of deoxynivalenol and nivalenol in 23 winter wheat genotypes with different Fusarium resistance. Independent variables consisted of 14 features, including 12 quantitative data and 2 qualitative data. The sensitivity analysis of neural networks showed that the experiment variant and winter wheat cultivar were the most important determinants of the concentration of ferulic acid, deoxynivalenol, and nivalenol in winter wheat seeds [59]. Ray et al. [60] applied an MLP-ANN model to assess the effects of topographic, soil, and environmental factors (18 input parameters, including soil nutrients and climate factors) on the content of active constituent of coronarin D in white ginger lily (*Hedychium coronarium*). The sensitivity analysis of the ANN showed that altitude, manganese, and zinc were the most important variables predicting the coronarin D content.

2.3. Yield Stability and Genotype × Environment Interaction

The environment (climate and soil), agricultural operations (sowing-cultivation-harvesting), and plant genotype are the factors that affect the yield and productivity of crops. The relationships (direct and/or indirect) and interactions between these factors create a complex situation determining the potential yield of plants [39]. Environmental variations and the genotype × environment interaction (GEI) are the factors that cause year to year variations in the yield and phenotypic trait of a specific genotype. The choice of a genotype for a target trait is a complex and difficult task because of the GEI, as different genotypes respond differently to varied environmental conditions. The estimation of relative performance of genotypes over the environments, through stability analysis is a perfect solution to these yearly variations [81]. Finlay and Wilkenson's regression analysis and coefficient [82]; Eberhart and Russel's coefficient of regression (S^2_{di}) [83]; Wricke's ecovalence (Wi) [84], Shukla's procedure of stability variance [85], coefficient of variance (CV) [86], and Lin and Binns cultivar performance measure [87] are classical univariate approaches used for the assessment of the GEI. Linear regression analysis and variance components are the main aspects of these methods [88]. Apart from the aforementioned statistic methods, the sustainable yield index (SYI) [89] is used to evaluate the effects of agricultural practices on crop yield sustainability [90]. All these methods are parametric, and therefore, the assumptions of the distribution of data and the homogeneity of variance should be considered before they are applied [91]. There are nonparametric univariate methods to evaluate the GEI, including S_i^1, S_i^2, S_i^3, and S_i^6 stability parameters [92,93], Kang parameter [94], Ketrank and Ketyield plots [95], Fox-rank [96], and Star [91]. These nonparametric stability statistics are analytical clustering procedures that determine the stability of genotypes on the basis of ranks rather than data and free from modeling assumptions. A genotype is considered stable if its ranking is relatively constant across environments [91].

Principal component analysis, cluster analysis, additive main effects and multiplicative interactions (AMMI), and genotype plus genotype × environment interaction biplot (GGE) are multivariate procedures enabling examination of multidirectional aspects of the GEI by imaging the response of a genotype in an *E*-dimensional space [91]. Multivariate stability analyses are more powerful and precise than univariate approaches. However, these are complex methods that do not provide a simple measure of yield stability for a reliable ranking of genotypes. Limited access to software is another bottleneck of these methods [91]. In a recent study, both linear and nonlinear regression models were applied to estimate the influence of climate variables (precipitation, sunshine duration, average relative humidity, maximum temperature, minimum temperature, and average temperature) on the growth and yield-related characteristics of cotton (the cotton height at the flowering stage, stalk weight, yield of cotton seeds, and lint percentage). The authors found that the interpretation of linear regression equations was generally lower than the interpretation of nonlinear equations [97]. There was a linear relationship or a relatively complex nonlinear relationship between the cotton growth indicators and climate variables in one site of their study, but they did not find the best equations for the cotton growth indices and the influence of climate variables on the cotton growth indices at several sites. In addition, the authors developed one regression model for each condition [97]. When several independent variables and several dependent variables are of interest, i.e., multiple-independent variables versus multiple-dependent variables, ANN can reduce the required analyses and result in higher accuracy [25]. It is clear that an ANN model can find the best equations in all studied environments in a faster and more precise manner by considering other factors such as soil and cotton properties. Plant growth indices and climate variables could be entered into an ANN model as dependent and independent variables, respectively. Then, linear and nonlinear relationships between the variables can be considered through powerful ANN models. There are well-recognized statistical and biological limitations to regression approaches. ANN modeling would enable breeders to evaluate the GEI and genetic stability of a large number of genotypes faster and more precisely. Coupled artificial intelligence (ANN) with deep phenotyping is a valuable tool for understanding plant–environment interactions [98].

2.4. Biotic and Abiotic Stress Assessment

Plants are exposed to various biotic and abiotic stresses. Different approaches have been applied to assess the tolerance and resistance of plant genotypes to these stresses and to identify superior genotypes.

There have been numerous breeding attempts to combat drought stress. Plants' tolerance to drought has been studied through some statistical indices, such as tolerance (TOL), mean productivity (MP), stress susceptibility index (SSI), geometric mean productivity (GMP), harmonic mean (HARM), relative drought index (RDI), stress tolerance index (STI), yield index (YI), yield stability index (YSI), and modified stress tolerance index (K1STI and K2STI) [99–106]. These classical approaches are based on morphological data, mainly yield generated under nonstress (Y_p) and stress (Y_s) conditions. However, apart from morphological attributes, there are many physiological and biochemical pathways involved in plants' response to environmental stresses. Secondary metabolites, cellular antioxidants, plant growth regulators, compatible solutes, and polyamines are all involved in plants' response to biotic and abiotic stresses [107,108]. Combining phenomic data with metabolomic and genomic data is an efficient strategy to assess plants' responses to biotic and abiotic stresses [109]. Classical multivariate statistical methods are not efficient enough to manage such a large volume of data (multiple independent variables versus multiple dependent variables). Linear regression is the most common technique used for the detection of nutrition deficiencies through the RGB image technique. However, features extracted from digital images with nonlinear relationship with nutrient content cannot be explained through linear regression model [110]. Machine learning techniques, along with digital images, could be used to model and predict genotypes' responses to stressful conditions and find the ones that are more resistant to stress and nonstress environments by analyzing all phenomic and omics (metabolomic and genomic) data. Big data—imaging and remote-sensing data—can be interpreted through machine learning for high-throughput stress phenotyping [111]. Ravari et al. [52] applied an MLP-ANN and the TOL, MP, GMP, HM, SSI, STI, YI, and YSI indices to predict the salinity tolerance of 41 Iranian wheat cultivars (*Triticum aestivum* L.). They found that the YSI, MP, GMP, and STI were the best predictors of salinity-tolerant cultivars. In Arabidopsis (*Arabidopsis thaliana*), miRNA expressions were used as input features to predict plant responses to abiotic stresses of drought, salinity, cold, and heat using machine learning models of decision tree (DT), SVM, least-square support vector machines (LSSVM), and Naïve Bayes (NB). It was concluded that miRNA-169, miRNA-159, miRNA-396, and miRNA-393 had the highest contributions to plant response towards abiotic stresses and the SVM with Gaussian kernel had better performance than other machine learning methods in prediction of plant stress response (R^2 = 0.96) [8]. Deep CNN along with traditional machine learning method was used for identification and classification of maize drought stress through the field-obtained data under optimum moisture, light drought, and moderate drought stress. Authors reported identification accuracy of 98.14%, which was more than Gradient Boosting Decision Tree (GBDT) method [19].

Deep CNNs have been widely used to classify and detect various plant diseases—biotic stress [112–115]. Image recognition and classification of maize leaf diseases, including northern corn leaf blight (*Exserohilum*), common rust (*Puccinia sorghi*), and gray leaf spot (*Cercospora*) diseases, have been conducted using deep CNN with an accuracy of 93.35% [116]. In cucumber (*Cucumis sativus*), a semantic segmentation model based on CNN was developed to segment the powdery mildew disease on leaf images at pixel level, and pixel accuracy of CNN model (96.08%) was more than segmentation methods of K-means, Random forest, and GBDT [28]. In pearl millet (*Pennisetum glaucum*), DNNs has been applied for identification of mildew disease, and accuracy of 95.00% was reported for the developed model [37].

2.5. Classical Mating Designs and Hybrid Breeding Programs

The integration of statistics into genetics led to some classical mating designs such as mean generation analysis [117], diallel crosses analysis [118–120], line × tester analysis [121], North Carolina designs [122], and triple test cross [123,124]. These methods have been used for genetic analysis of crops in order to find the nature of gene actions (additive, dominance, and epistasis) involved in controlling important morphological, phenological, and yield component characteristics, to calculate broad and narrow sense heritability and predict the outcomes of cross-breeding programs.

The prediction of parental combinations is critical to the choice of superior combinational homozygous parental lines in F1-hybrid breeding programs [125]. However, it is a challenging task with a large number of cross combinations when there are many inbred parental lines. Therefore, the prediction of the yield performance of cross combinations of parental lines may significantly reduce the required time and budget of F1-hybrid breeding programs [126]. ANN could be used to predict parental combinations and calculate the correct values of general and specific combining abilities (GCA and SCA) in mating designs, such as topcross, line × tester, and diallel cross. Khaki et al. [126] applied matrix factorization and a neural network to predict the yield performance of cross combinations of inbreds and testers of unsown maize on the basis of historical yield data collected from the crossing of other inbreds and testers. The proposed model was significantly better than other models such as deep factorization machines (DeepFM), generalized matrix factorization (GMF), LASSO, RF, and neural networks.

3. Applications of Machine Learning in In Vitro-Based Plant Biotechnology

Biotechnology-based breeding methods (BBBMs) complement classical breeding methods in rapid plant improvement. In vitro regeneration, as the main core of many in-vitro-based breeding methods, has numerous plant breeding applications. In situ and ex situ conservation and micropropagation (proliferation) are direct applications of in vitro regeneration [127]. In endangered rare plant species, like medicinal plants, in vitro culture is an effective strategy for mass propagation, germplasm conservation, and production of bioactive compounds [128]. Several factors determine the fate of cultured cells in in vitro regeneration of plants. These are the plant genotype, plant growth regulators (PGRs), culture medium components, explant type, explant age, enhancer additives-elicitors, etc. [127]. These factors can be divided into three main categories: initial triggers of regeneration (environmental signal inputs and physical stimuli), epigenetic and transcriptional cellular responses to the initial triggers, and molecules that manage the formation and development of the new stem cell niche [129]. The combination and interactions between these factors lead to multifactorial nature of the in vitro plant regeneration process. Basal culture medium components, plant genotype, PGRs, explant type, and explant age are all multilevel factors with different applicable combinations. The inclusion of other factors results in a very complex situation for interpretation. Plant cells and tissues have nondeterministic and nonlinear developmental patterns in a stressful in vitro environment [130]. The analysis of variance of factorial experiments and simple means comparison analysis with classical methods such as LSD, Tukey's HSD, and Duncan's test, are the main statistical methods used to interpret the effects of interaction between effective factors in most in vitro regeneration studies [128,131,132].

Murashige and Skoog (MS), modified MS (MMS), Gamborg's B5 medium Woody Plant Medium (WPM), and Driver and Kuniyuki Woody Plant Medium (DKW) are the most commonly used basal culture media in in vitro regeneration studies. Basal medium manipulation is a promoting strategy that has been applied to increase the output of in vitro studies [133]. However, due to the large number of micro- and macroelements in the culture medium, it is difficult to manipulate their concentrations. In this situation, prediction of the effect of culture media components on the target characteristics of in vitro regenerants is the right solution. Artificial neural networks have been applied in these experiments to predict the best culture media components for efficient propagation of different plant species [29,31,134].

Different combinations of auxin and cytokinin PGRs can determine the developmental fate of cultured cells and tissues toward organogenesis and/or somatic embryogenesis. The cytokinin/auxin ratio is also very important in in vitro studies [135]. Niazian et al. [131] found that 2,4-dichlorophenoxyacetic acid (2,4-D) combined with kinetin resulted in indirect somatic embryogenesis of cultured hypocotyl segments of ajowan medicinal plants, whereas a combination of 3-methoxy(-6-benzylamino-9-tetrahydropyran-2-yl) purine and naphthalene acetic acid led cultivated explants toward an indirect shoot regeneration pathway. Arab et al. [30] combined artificial neural networks and genetic algorithms to predict and optimize the effect of cytokinin–auxin plant hormone (BAP, KIN, TDZ, IBA, and NAA) combinations and concentrations on the number of microshoots per explant, the length of microshoots, developed callus weight, and the quality index of plantlets in in vitro proliferation of Garnem (G × N15) rootstock. The ANN model predicted the number and length of microshoots with high accuracy. The highest values of the variable sensitivity ratio for the proliferation rate were related to the BAP (19.3), KIN (9.64), and IBA (2.63) inputs. An MLP-ANN was developed to predict the physical properties of embryogenic callus and the number of somatic embryos in in vitro regeneration of ajowan under the effect of different combinations of the explant age, concentrations of 2,4-D, kinetin, and sucrose inputs [25]. The ANN model predicted the physical properties of embryogenic callus (area, perimeter, Feret diameter, roundness, and true density) and the number of somatic embryos better than the multiple linear regressions. Fifteen-day-old hypocotyl explants × 1.5 mg/L 2,4-D × 0.5 mg/L Kin × 2.5% (*w/v*) sucrose was the best combination of inputs with the highest measured and predicted number of somatic embryos [25].

Apart from culture medium components and PGRs combination, ANN has been applied to model the sterilization step of in vitro regeneration. Hesami et al. [27] applied an MLP-ANN along with a genetic algorithm to model and optimize the contamination frequency and explant viability under the influence of seven input variables, i.e., $HgCl_2$, $Ca(ClO)_2$, nanosilver, H_2O_2, NaOCl, $AgNO_3$, and immersion times, in an in vitro culture of chrysanthemum. The lowest contamination frequency (0%) and the highest explant viability (99.98%) resulted from 1.62% NaOCl at 13.96 min immersion time. The sensitivity analysis of the ANN showed that the immersion time was the most important variable affecting the contamination frequency and explant viability [27]. ANNs are also used to simulate in vitro growth of plant tissue cultures, distinguish embryos from nonembryos, predict the formation of plantlets from embryos, estimate the biomass of cell cultures, simulate the distribution of temperature in a culture vessel, identify and estimate the in vitro induced shoot length, and cluster in vitro regenerated plantlets [130].

Other in vitro-based breeding methods, such as artificial polyploidy induction, doubled haploid production, plant gene transformation, and genome editing methods also have multifactorial nature and require multivariate statistical methods to interpret the results. Different chemical enhancers can be used in in vitro doubled haploid production methods (induced parthenogenesis and androgenesis) to improve the haploid induction efficiency, e.g., PGRs, osmoprotectants, cellular antioxidants, reactive oxygen species scavengers, polyamins, stress hormones, chlormequat chloride, compatible solutes, DNA demethylating agents, histone deacetylase inhibitors, cell wall remodeling agents, ethylene inhibitors, and other applicable additives. They enhance tolerance to inductive stresses and improve the final efficiency of doubled haploid production [108]. ANN models may improve the efficiency of in vitro doubled haploid production and solve the problem of recalcitrant species/genotypes by predicting the best combination(s) of these additives in interaction with other influencing factors, such as the plant genotype, the surrounding environment of donor plants, physical treatments (inductive stresses) of cultured gametophytic cells, the developmental stage of initial gametophytic cells, and culture medium components. The ANN predicted the callus induction percentage in androgenesis (anther culture) of tomato (*Lycopersicon esculentum* L.) under the influence of plant genotype, the concentrations of 2,4-D and kinetin PGRs, and the concentration of gum Arabic better than the MLR model [50].

Plants' vigor and performance are commonly enhanced by mitotic-induced polyploidy. It consists in in vivo and in vitro application of mitotic spindle poisons [136]. In vitro-induced polyploidy is a multifactorial procedure. The efficiency of in vitro-induced polyploidy may be affected not only by in vitro regeneration parameters (basal culture medium components, combination of PGRs, additives, etc.) but also by the plant genotype, the developmental stage of initial explants as well as the type, dosage, and duration (exposure time) of the application of the antimitotic agent. Due to the genotype dependency, different genotypes of plant species exhibit different responses to concentrations of the antimitotic agent applied [137]. This results in significant interaction of the plant genotype and antimitotic agent in artificial polyploidy induction. Although there have been no reports on the application of ANN to model and predict the results of in vitro-induced artificial polyploidy, it might increase the efficiency by predicting and finding the best combination and interaction of all influential factors.

Agrobacterium-mediated gene transformation is a well-known method of plant gene transformation and genetic engineering. However, various parameters must be optimized for an efficient gene delivery, including the *Agrobacterium* strain cell density, the time of inoculation, the type and concentration of antibiotics to kill *Agrobacterium*, the type and concentration of selectable antibiotics, and the concentration of acetosyringone [138]. These influencing factors along with in vitro regeneration factors result in a multi-variable nature of *Agrobacterium*-mediated gene transformation [127]. It is obvious that machine learning algorithms could be used to predict and optimize *Agrobacterium*-mediated gene transformation, especially in important *Agrobacterium*-recalcitrant plant species.

4. Coupled Machine Learning-Image Processing for High-Throughput Phenotyping and Precision Agriculture

Classical measurement of plants' physical features by visual assessment is a laborious, time-consuming, costly, and error-prone process in both conventional and in vitro-based plant breeding studies. This step can be accelerated and facilitated by the machine vision method, which is more accurate and precise than visual assessment. Nondestructive measurement of physical features, both outdoors and in vitro, is another important advantage of image processing [25]. Automated non-invasive fast scoring of several plant traits through high-throughput phenotyping platforms can speed up and facilitate the phenotyping of plant populations and selection of superior varieties [139]. The integration of precise measured image-based characteristics with omics data could help to identify the key traits involved in the mechanisms of stress tolerance and acclimation [109]. On the other hand, the ability of deep learning in the identification of plants' features provides a great opportunity for further advances in image analysis [98]. Combined image processing (for feature extraction) and machine learning (for data analysis) is a powerful strategy required for faster and precise image-based plant phenotyping [140]. The use of deep learning techniques in computer vision can accelerate plant breeding programs such as plant phenotyping and classification of genotypes [141]. Coupled image processing-ANN has been used to measure phenotypic characteristics and assess genetic diversity and classify different plant species [38,65,142,143]. Deep learning, especially CNN, has become a powerful tool for image analysis in recent years [49]. Uzal et al. [48] applied a computer vision method for feature extraction along with developed convolutional neural networks to estimate the number of seeds in soybean pods and then to classify the obtained data. In most cases, the convolutional neural networks learnt to detect each seed in the pod, which indicates their high classification efficiency. There are other advanced imaging techniques, which are more efficient than simple visualization techniques and can be used to analyze in-field images instead of indoor methods. Recently, an R-based pipeline has been developed, which enables analysis of orthomosaic images from agricultural field trials and calculation of the number of plants per plot, canopy cover percentage, vegetation indices, and plant height [139]. A deep neural network model trained with such in-field images could very effectively classify and estimate desired characteristics from in-field images [48]. Coupled image processing-artificial neural network has been used in BBBMs for in vitro modeling of somatic embryogenesis in ajowan [25] and androgenesis-based haploid induction in tomato [50]. Plant phenotyping and precision agriculture could be significantly different in terms of the spatial and temporal resolutions, although both generate big data sets in a format of image. These are information- and technology-based domains with specific demand and challenges. Precision agriculture is an agricultural management system based on spatial and temporal variability in crop and soil factors within a field (with environmental parameters). However, in phenotyping systems, the crop field parameters are homogenous and datasets in molecular, cellular, and whole plant levels are considered for plant phenotyping. Precision agriculture examines spatial heterogeneities within crop stands, whereas the appearance and performance of a genotype under distinct environmental conditions are examined in plant phenotyping [144,145]. High-throughput salt-stress phenotyping has been reported in okra (*Abelmoschus esculentus* L.) through a trained DNN using physiological and biochemical traits, such as fresh weight, SPAD, elemental contents, and photosynthesis-related parameters, measured from 13 genotypes under salt stress treatment [36]. Establishment of high-throughput phenotyping platforms (HTPPs) to phenotype physiomorphological traits under highly heterogeneous field environment, in a precise, labor-, and cost-effective manner, is essential to bridge the gap between genomics and phenomics [146]. Machine learning algorithms can be used for image-based plant stress phenotyping in a wide scale from leaf and canopy to filed range. Identification, classification, quantification, and prediction of big data, obtained from higher-throughput phenotyping systems such as unmanned aerial system (UAS) technology and ground robots, can be conducted through deep learning algorithms [147]. In carrot (*Daucus carota*), a precision agriculture approach was conducted through on-farm punctual carrot sampling data incorporated into the satellite imagery data using a random forest regression algorithm. Accuracy of developed model to predict carrot yield using database composed of spectral bands was acceptable ($R^2 = 0.82$; RMSE = 2.64 mg ha^{-1}; MAE = 1.74 mg ha^{-1}) [26].

5. A Proposed Idea for Plant Ploidy Level Determination through Image Processing-Machine Learning

In chromosome engineering studies (polyploidy and haploid induction), one important step is taken to verify the ploidy level. It can be confirmed through direct (chromosome counting) or indirect methods (morphological and anatomical indicators and flow cytometry). Although the direct method of chromosome counting is reliable and unambiguous [148], it is laborious, time consuming, and complicated and requires highly skilled operators [149]. Indirect verification of the ploidy level through classical markers, such as stomatal morphometric data (stomatal density per unit area, the number and size of stomata), the density of chloroplasts per stomatal guard cells, size of guard cells and pollen size, is a rapid and simple method [150], but not completely reliable. Flow cytometry is a reliable method based on direct correlation between the nuclear DNA content and ploidy level. However, according to a recent study, the comparison of the DNA content in standardized leaf punch samples is not a reliable method to recognize putative doubled haploids, as there is a DNA content equivalence between haploid and diploid samples [151].

Machine learning algorithms can be used for ploidy level identification of plants. Recently, a deep learning-based object detection algorithm has been developed for evaluating the stomatal density and elucidating the variation in the stomatal density among various soybean accessions [49]. This DNN could also be useful for ploidy level prediction. There have been two reports on the use of other methods to identify the ploidy level in plants. Altuntaş et al. [33] used convolutional neural networks to recognize haploid and diploid maize seeds through *R1-nj* anthocyanin color marker data of 1230 haploid and 1770 diploid maize seed images. The accuracy and sensitivity of the model amounted to 94.22% and 94.58%, respectively [33]. Remote sensing has also been applied to determine the ploidy level of quaking aspen (*Populus tremuloides* Michx.) [152].

Here, we offer another idea to identify the ploidy level of plants through coupled image processing-supervised deep neural network using visual data of cellular patterning of the epidermal layer. Haploids have smaller and more densely packed epidermal and mesophyll cells (more cells per same unit area) than doubled haploids. This results in an equivalent DNA content per unit leaf area for haploids and their counterpart diploids [151]. Cellular patterning in the epidermis and mesophyll can be specific to each ploidy group. Therefore, epidermal cell patterning (size shape and number) could be used as ploidy level recognition and classification criteria [151]. The use of imaging techniques for precise feature extraction of leaf punch samples (the cellular pattern, including the cell size and number) and the subsequent modeling of captured images (classification modeling) through deep learning approaches, particularly CNN, results in an image-based model, which can be used to estimate the ploidy level in chromosome engineering studies of different plant species (Figure 3). It is a more precise, fast, and cost-effective method of ploidy level distinction, which could also be used in other branches of plant science, e.g., in genetic diversity, evolutionary, and species invasiveness studies.

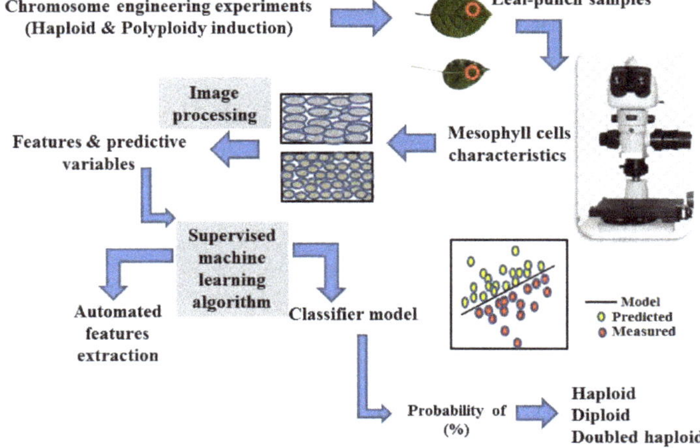

Figure 3. The proposed coupled image processing-supervised machine learning to determine the ploidy level in different plant species through cellular patterning in leaf epidermis and mesophyll layers.

6. Conclusions

Most classical statistical methods use only simple statistics and few influential factors to assess the biological features of plants. For example, Y_p and Y_s are the only indices used to identify drought-tolerant plant genotypes in yield-based drought tolerance assessment methods. However, there are other influential factors, such as cellular, physiological, and phytochemical pathways, which are involved in plants' responses to environmental stress. The tolerance of different plant species to biotic and abiotic stresses, as complex biological processes, can be efficiently enhanced through large-scale analysis of phenomic, metabolomic, and genomic data. Machine learning models are capable of processing large amounts of data (imaging and remote-sensing data) for high-throughput stress phenotyping. The analysis of different omics and phenomic data may result in more precise interpretation of GEI and yield stability. Plants' qualitative and quantitative characteristics can be predicted more precisely by analysis of climate data (temperature, humidity, sunshine, precipitation, etc.), soil factors, agricultural operations data (harvest date, information on diseases, crop status, ground temperature, etc.), topographic, and meteorological data. Big data analysis enables more efficient classification of plants' phenotypes and genotypes. Machine learning techniques are able to manage large amounts of data in various areas of plant breeding, which can lead to more accurate results and better interoperation than classical statistical methods. Artificial neural networks can be used for pattern recognition, nonlinear regression, and classification purposes in plant tissue culture studies because they can handle binary, continuous, categorical, and fuzzy datasets. The present review can give an overview of applications of machine learning to plant breeders. It would be helpful to adopt the correct method of data analysis in future studies, which in turn can increase the output of studies.

Author Contributions: M.N. and G.N. contributed equally to this work. All authors have read and agreed to the published version of the manuscript.

Funding: This research received no external funding.

Acknowledgments: The publication was co-financed within the framework of the Ministry of Science and Higher Education program titled "Regional Initiative Excellence" in 2019–2022, project no. 005/RID/2018/19.

Conflicts of Interest: The authors declare no conflict of interest.

References

1. Ali, S.; Shafique, O.; Mahmood, T.; Hanif, M.A.; Ahmed, I.; Khan, B.A. A Review about Perspectives of Nanotechnology in Agriculture. *Pakistan J. Agric. Res.* **2018**, *31*, 116–121. [CrossRef]
2. Niazian, M.; Sadat-Noori, S.A.; Abdipour, M. Modeling the seed yield of Ajowan (*Trachyspermum ammi* L.) using artificial neural network and multiple linear regression models. *Ind. Crops Prod.* **2018**, *117*, 224–234. [CrossRef]
3. Hesami, M.; Naderi, R.; Tohidfar, M.; Yoosefzadeh-Najafabadi, M. Application of Adaptive Neuro-Fuzzy Inference System-Non-dominated Sorting Genetic Algorithm-II (ANFIS-NSGAII) for Modeling and Optimizing Somatic Embryogenesis of Chrysanthemum. *Front. Plant Sci.* **2019**, *10*. [CrossRef] [PubMed]
4. Chegini, G.R.; Khazaei, J.; Ghobadian, B.; Goudarzi, A.M. Prediction of process and product parameters in an orange juice spray dryer using artificial neural networks. *J. Food Eng.* **2008**, *84*, 534–543. [CrossRef]
5. Zheng, H.; Li, W.; Jiang, J.; Liu, Y.; Cheng, T.; Tian, Y.; Zhu, Y.; Cao, W.; Zhang, Y.; Yao, X.A. Comparative Assessment of Different Modeling Algorithms for Estimating Leaf Nitrogen Content in Winter Wheat Using Multispectral Images from an Unmanned Aerial Vehicle. *Remote Sens.* **2018**, *10*, 2026. [CrossRef]
6. Hesami, M.; Naderi, R.; Yoosefzadeh-Najafabadi, M.; Rahmati, M. Data-driven modeling in plant tissue culture. *J. Appl. Environ. Biol. Sci* **2017**, *7*, 37–44.
7. Salehi, M.; Farhadi, S.; Moieni, A.; Safaie, N.; Ahmadi, H. Mathematical Modeling of Growth and Paclitaxel Biosynthesis in Corylus avellana Cell Culture Responding to Fungal Elicitors Using Multilayer Perceptron-Genetic Algorithm. *Front. Plant Sci.* **2020**, *11*, 1148. [CrossRef]
8. Asefpour Vakilian, K. Machine learning improves our knowledge about miRNA functions towards plant abiotic stresses. *Sci. Rep.* **2020**, *10*, 3041. [CrossRef]

9. Wang, H.; Cimen, E.; Singh, N.; Buckler, E. Deep learning for plant genomics and crop improvement. *Curr. Opin. Plant Biol.* **2020**, *54*, 34–41. [CrossRef]

10. Hu, H.; Scheben, A.; Edwards, D. Advances in Integrating Genomics and Bioinformatics in the Plant Breeding Pipeline. *Agriculture* **2018**, *8*, 75. [CrossRef]

11. Orozco-Arias, S.; Isaza, G.; Guyot, R. Retrotransposons in Plant Genomes: Structure, Identification, and Classification through Bioinformatics and Machine Learning. *Int. J. Mol. Sci.* **2019**, *20*, 3837. [CrossRef]

12. Alvarez, R. Predicting average regional yield and production of wheat in the Argentine Pampas by an artificial neural network approach. *Eur. J. Agron.* **2009**, *30*, 70–77. [CrossRef]

13. Azevedo, A.M.; Andrade Júnior, V.C.D.; Pedrosa, C.E.; Oliveira, C.M.D.; Dornas, M.F.S.; Cruz, C.D.; Valadares, N.R. Application of artificial neural networks in indirect selection: A case study on the breeding of lettuce. *Bragantia* **2015**, *74*, 387–393. [CrossRef]

14. Gu, J.; Wang, Z.; Kuen, J.; Ma, L.; Shahroudy, A.; Shuai, B.; Liu, T.; Wang, X.; Wang, G.; Cai, J.; et al. Recent advances in convolutional neural networks. *Pattern Recognit.* **2018**, *77*, 354–377. [CrossRef]

15. Hesami, M.; Naderi, R.; Tohidfar, M.; Yoosefzadeh-Najafabadi, M. Development of support vector machine-based model and comparative analysis with artificial neural network for modeling the plant tissue culture procedures: Effect of plant growth regulators on somatic embryogenesis of chrysanthemum, as a case study. *Plant Methods* **2020**, *16*, 112. [CrossRef] [PubMed]

16. Ali, A.M.; Darvishzadeh, R.; Skidmore, A.; Gara, T.W.; Heurich, M. Machine learning methods' performance in radiative transfer model inversion to retrieve plant traits from Sentinel-2 data of a mixed mountain forest. *Int. J. Digit. Earth* **2020**, 1–15. [CrossRef]

17. Gold, K.M.; Townsend, P.A.; Herrmann, I.; Gevens, A.J. Investigating potato late blight physiological differences across potato cultivars with spectroscopy and machine learning. *Plant Sci.* **2020**, *295*, 110316. [CrossRef]

18. Barth, R.; IJsselmuiden, J.; Hemming, J.; Van Henten, E.J. Synthetic bootstrapping of convolutional neural networks for semantic plant part segmentation. *Comput. Electron. Agric.* **2019**, *161*, 291–304. [CrossRef]

19. An, J.; Li, W.; Li, M.; Cui, S.; Yue, H. Identification and Classification of Maize Drought Stress Using Deep Convolutional Neural Network. *Symmetry* **2019**, *11*, 256. [CrossRef]

20. Geetharamani, G.; Arun Pandian, J. Identification of plant leaf diseases using a nine-layer deep convolutional neural network. *Comput. Electr. Eng.* **2019**, *76*, 323–338.

21. Kattenborn, T.; Eichel, J.; Wiser, S.; Burrows, L.; Fassnacht, F.E.; Schmidtlein, S. Convolutional Neural Networks accurately predict cover fractions of plant species and communities in Unmanned Aerial Vehicle imagery. *Remote Sens. Ecol. Conserv.* **2020**. [CrossRef]

22. Iniyan, S.; Jebakumar, R.; Mangalraj, P.; Mohit, M.; Nanda, A. Plant Disease Identification and Detection Using Support Vector Machines and Artificial Neural Networks. In *Artificial Intelligence and Evolutionary Computations in Engineering Systems*; Advances in Intelligent Systems and Computing; Dash, S., Lakshmi, C., Das, S., Panigrahi, B., Eds.; Springer: Singapore, 2020; pp. 15–27.

23. Wang, Y.; Li, T.; Jin, G.; Wei, Y.; Li, L.; Kalkhajeh, Y.K.; Ning, J.; Zhang, Z. Qualitative and quantitative diagnosis of nitrogen nutrition of tea plants under field condition using hyperspectral imaging coupled with chemometrics. *J. Sci. Food Agric.* **2020**, *100*, 161–167. [CrossRef]

24. Niazian, M.; Sadat-Noori, S.A.; Abdipour, M. Artificial neural network and multiple regression analysis models to predict essential oil content of ajowan (*Carum copticum* L.). *J. Appl. Res. Med. Aromat. Plants* **2018**, *9*, 124–131. [CrossRef]

25. Niazian, M.; Sadat-Noori, S.A.; Abdipour, M.; Tohidfar, M.; Mortazavian, S.M.M. Image Processing and Artificial Neural Network-Based Models to Measure and Predict Physical Properties of Embryogenic Callus and Number of Somatic Embryos in Ajowan (*Trachyspermum ammi* (L.) Sprague). *Vitr. Cell. Dev. Biol. Plant* **2018**, *54*, 54–68. [CrossRef]

26. Wei, M.C.F.; Maldaner, L.F.; Ottoni, P.M.N.; Molin, J.P. Carrot Yield Mapping: A Precision Agriculture Approach Based on Machine Learning. *AI* **2020**, *1*, 229–241.

27. Hesami, M.; Naderi, R.; Tohidfar, M. Modeling and Optimizing in vitro Sterilization of Chrysanthemum via Multilayer Perceptron-Non-dominated Sorting Genetic Algorithm-II (MLP-NSGAII). *Front. Plant Sci.* **2019**, *10*. [CrossRef] [PubMed]

28. Lin, K.; Gong, L.; Huang, Y.; Liu, C.; Pan, J. Deep Learning-Based Segmentation and Quantification of Cucumber Powdery Mildew Using Convolutional Neural Network. *Front. Plant Sci.* **2019**, *10*. [CrossRef]

29. Arab, M.M.; Yadollahi, A.; Shojaeiyan, A.; Ahmadi, H. Artificial Neural Network Genetic Algorithm as Powerful Tool to Predict and Optimize In vitro Proliferation Mineral Medium for G × N15 Rootstock. *Front. Plant Sci.* **2016**, *7*. [CrossRef]

30. Arab, M.M.; Yadollahi, A.; Ahmadi, H.; Eftekhari, M.; Maleki, M. Mathematical Modeling and Optimizing of in Vitro Hormonal Combination for G × N15 Vegetative Rootstock Proliferation Using Artificial Neural Network-Genetic Algorithm (ANN-GA). *Front. Plant Sci.* **2017**, *8*. [CrossRef]

31. Arab, M.M.; Yadollahi, A.; Eftekhari, M.; Ahmadi, H.; Akbari, M.; Khorami, S.S. Modeling and Optimizing a New Culture Medium for In Vitro Rooting of G×N15 Prunus Rootstock using Artificial Neural Network-Genetic Algorithm. *Sci. Rep.* **2018**, *8*, 9977. [CrossRef]

32. Costa, M.O.; Capel, L.S.; Maldonado, C.; Mora, F.; Mangolin, C.A.; Machado, M. High genetic differentiation of grapevine rootstock varieties determined by molecular markers and artificial neural networks. *Acta Sci. Agron.* **2019**, *42*, e43475. [CrossRef]

33. Altuntaş, Y.; Cömert, Z.; Kocamaz, A.F. Identification of haploid and diploid maize seeds using convolutional neural networks and a transfer learning approach. *Comput. Electron. Agric.* **2019**, *163*, 104874. [CrossRef]

34. Darwish, A.; Ezzat, D.; Hassanien, A.E. An optimized model based on convolutional neural networks and orthogonal learning particle swarm optimization algorithm for plant diseases diagnosis. *Swarm Evol. Comput.* **2020**, *52*, 100616. [CrossRef]

35. Mishra, S.; Sachan, R.; Rajpal, D. Deep Convolutional Neural Network based Detection System for Real-time Corn Plant Disease Recognition. *Procedia Comput. Sci.* **2020**, *167*, 2003–2010. [CrossRef]

36. Feng, X.; Zhan, Y.; Wang, Q.; Yang, X.; Yu, C.; Wang, H.; Tang, Z.; Jiang, D.; Peng, C.; He, Y. Hyperspectral imaging combined with machine learning as a tool to obtain high-throughput plant salt-stress phenotyping. *Plant J.* **2020**, *101*, 1448–1461. [CrossRef]

37. Coulibaly, S.; Kamsu-Foguem, B.; Kamissoko, D.; Traore, D. Deep neural networks with transfer learning in millet crop images. *Comput. Ind.* **2019**, *108*, 115–120. [CrossRef]

38. Azizi, A.; Abbaspour-Gilandeh, Y.; Nooshyar, M.; Afkari-Sayah, A. Identifying Potato Varieties Using Machine Vision and Artificial Neural Networks. *Int. J. Food Prop.* **2016**, *19*, 618–635. [CrossRef]

39. Niedbała, G.; Piekutowska, M.; Weres, J.; Korzeniewicz, R.; Witaszek, K.; Adamski, M.; Pilarski, K.; Czechowska-Kosacka, A.; Krysztofiak-Kaniewska, A. Application of Artificial Neural Networks for Yield Modeling of Winter Rapeseed Based on Combined Quantitative and Qualitative Data. *Agronomy* **2019**, *9*, 781. [CrossRef]

40. Zhang, J.; Zhao, B.; Yang, C.; Shi, Y.; Liao, Q.; Zhou, G.; Wang, C.; Xie, T.; Jiang, Z.; Zhang, D.; et al. Rapeseed Stand Count Estimation at Leaf Development Stages With UAV Imagery and Convolutional Neural Networks. *Front. Plant Sci.* **2020**, *11*. [CrossRef]

41. Niedbała, G. Application of artificial neural networks for multi-criteria yield prediction of winter rapeseed. *Sustainability* **2019**, *11*, 533.

42. Niedbała, G. Simple model based on artificial neural network for early prediction and simulation winter rapeseed yield. *J. Integr. Agric.* **2019**, *18*, 54–61. [CrossRef]

43. Li, D.; Wang, R.; Xie, C.; Liu, L.; Zhang, J.; Li, R.; Wang, F.; Zhou, M.; Liu, W. A Recognition Method for Rice Plant Diseases and Pests Video Detection Based on Deep Convolutional Neural Network. *Sensors* **2020**, *20*, 578. [CrossRef] [PubMed]

44. Rahman, C.R.; Arko, P.S.; Ali, M.E.; Iqbal Khan, M.A.; Apon, S.H.; Nowrin, F.; Wasif, A. Identification and recognition of rice diseases and pests using convolutional neural networks. *Biosyst. Eng.* **2020**, *194*, 112–120. [CrossRef]

45. Abdipour, M.; Younessi-Hmazekhanlu, M.; Ramazani, S.H.R.; Omidi, A. Hassan Artificial neural networks and multiple linear regression as potential methods for modeling seed yield of safflower (*Carthamus tinctorius* L.). *Ind. Crops Prod.* **2019**, *127*, 185–194. [CrossRef]

46. Abdipour, M.; Ramazani, S.H.R.; Younessi-Hmazekhanlu, M.; Niazian, M. Modeling Oil Content of Sesame (*Sesamum indicum* L.) Using Artificial Neural Network and Multiple Linear Regression Approaches. *J. Am. Oil Chem. Soc.* **2018**, *95*, 283–297. [CrossRef]

47. Parsaeian, M.; Shahabi, M.; Hassanpour, H. Estimating Oil and Protein Content of Sesame Seeds Using Image Processing and Artificial Neural Network. *J. Am. Oil Chem. Soc.* **2020**, *97*, 691–702. [CrossRef]

48. Uzal, L.C.; Grinblat, G.L.; Namías, R.; Larese, M.G.; Bianchi, J.S.; Morandi, E.N.; Granitto, P.M. Seed-per-pod estimation for plant breeding using deep learning. *Comput. Electron. Agric.* **2018**, *150*, 196–204. [CrossRef]

49. Sakoda, K.; Watanabe, T.; Sukemura, S.; Kobayashi, S.; Nagasaki, Y.; Tanaka, Y.; Shiraiwa, T. Genetic Diversity in Stomatal Density among Soybeans Elucidated Using High-throughput Technique Based on an Algorithm for Object Detection. *Sci. Rep.* **2019**, *9*, 7610. [CrossRef]

50. Niazian, M.; Shariatpanahi, M.E.; Abdipour, M.; Oroojloo, M. Modeling callus induction and regeneration in an anther culture of tomato (*Lycopersicon esculentum* L.) using image processing and artificial neural network method. *Protoplasma* **2019**, *256*, 1317–1332. [CrossRef]

51. Verma, S.; Chug, A.; Singh, A.P. Application of convolutional neural networks for evaluation of disease severity in tomato plant. *J. Discret. Math. Sci. Cryptogr.* **2020**, *23*, 273–282. [CrossRef]

52. Ravari, S.Z.; Dehghani, H.; Naghavi, H. Assessment of salinity indices to identify Iranian wheat varieties using an artificial neural network. *Ann. Appl. Biol.* **2016**, *168*, 185–194. [CrossRef]

53. Niedbała, G.; Kozłowski, R.J. Application of Artificial Neural Networks for Multi-Criteria Yield Prediction of Winter Wheat. *J. Agric. Sci. Technol.* **2019**, *21*, 51–61.

54. Niedbała, G.; Nowakowski, K.; Rudowicz-Nawrocka, J.; Piekutowska, M.; Weres, J.; Tomczak, R.J.; Tyksiński, T.; Pinto, A.Á. Multicriteria prediction and simulation of winter wheat yield using extended qualitative and quantitative data based on artificial neural networks. *Appl. Sci.* **2019**, *9*, 2773. [CrossRef]

55. Sadeghi-Tehran, P.; Virlet, N.; Ampe, E.M.; Reyns, P.; Hawkesford, M.J. DeepCount: In-Field Automatic Quantification of Wheat Spikes Using Simple Linear Iterative Clustering and Deep Convolutional Neural Networks. *Front. Plant Sci.* **2019**, *10*. [CrossRef] [PubMed]

56. Haider, S.A.; Naqvi, S.R.; Akram, T.; Umar, G.A.; Shahzad, A.; Sial, M.R.; Khaliq, S.; Kamran, M. LSTM Neural Network Based Forecasting Model for Wheat Production in Pakistan. *Agronomy* **2019**, *9*, 72. [CrossRef]

57. Ma, W.; Qiu, Z.; Song, J.; Li, J.; Cheng, Q.; Zhai, J.; Ma, C. A deep convolutional neural network approach for predicting phenotypes from genotypes. *Planta* **2018**, *248*, 1307–1318. [CrossRef]

58. Hesami, M.; Condori-Apfata, J.A.; Valderrama Valencia, M.; Mohammadi, M. Application of Artificial Neural Network for Modeling and Studying In Vitro Genotype-Independent Shoot Regeneration in Wheat. *Appl. Sci.* **2020**, *10*, 5370. [CrossRef]

59. Niedbała, G.; Kurasiak-Popowska, D.; Stuper-Szablewska, K.; Nawracała, J. Application of Artificial Neural Networks to Analyze the Concentration of Ferulic Acid, Deoxynivalenol, and Nivalenol in Winter Wheat Grain. *Agriculture* **2020**, *10*, 127. [CrossRef]

60. Ray, A.; Halder, T.; Jena, S.; Sahoo, A.; Ghosh, B.; Mohanty, S.; Mahapatra, N.; Nayak, S. Application of artificial neural network (ANN) model for prediction and optimization of coronarin D content in Hedychium coronarium. *Ind. Crops Prod.* **2020**, *146*, 112186. [CrossRef]

61. Srivastava, A.; Gupta, S.; Shanker, K.; Gupta, N.; Gupta, A.K.; Lal, R.K. Genetic diversity in Indian poppy (*P. somniferum* L.) germplasm using multivariate and SCoT marker analyses. *Ind. Crops Prod.* **2020**, *144*, 112050. [CrossRef]

62. Niazian, M.; Sadat Noori, S.A.; Tohidfar, M.; Mortazavian, S.M.M. Essential Oil Yield and Agro-morphological Traits in Some Iranian Ecotypes of Ajowan (*Carum copticum* L.). *J. Essent. Oil Bear. Plants* **2017**, *20*, 1151–1156. [CrossRef]

63. Schulman, A.H. Molecular markers to assess genetic diversity. *Euphytica* **2007**, *158*, 313–321. [CrossRef]

64. Boonsrangsom, T. Genetic diversity of 'Wan Chak Motluk' (Curcuma comosa Roxb.) in Thailand using morphological characteristics and random amplification of polymorphic DNA (RAPD) markers. *South Afr. J. Bot.* **2020**, *130*, 224–230. [CrossRef]

65. Pandolfi, C.; Mugnai, S.; Azzarello, E.; Bergamasco, S.; Masi, E.; Mancuso, S. Artificial neural networks as a tool for plant identification: A case study on Vietnamese tea accessions. *Euphytica* **2009**, *166*, 411–421. [CrossRef]

66. Raza, A.; Mehmood, S.S.; Ashraf, F.; Khan, R.S.A. Genetic Diversity Analysis of Brassica Species Using PCR-Based SSR Markers. *Gesunde Pflanz.* **2019**, *71*, 1–7. [CrossRef]

67. Bird, C.; Schweizer, M.; Roberts, A.; Austin, W.E.N.; Knudsen, K.L.; Evans, K.M.; Filipsson, H.L.; Sayer, M.D.J.; Geslin, E.; Darling, K.F. The genetic diversity, morphology, biogeography, and taxonomic designations of Ammonia (Foraminifera) in the Northeast Atlantic. *Mar. Micropaleontol.* **2020**, *155*, 101726. [CrossRef]

68. Poletto, T.; Poletto, I.; Moraes Silva, L.M.; Brião Muniz, M.F.; Silveira Reiniger, L.R.; Richards, N.; Stefenon, V.M. Morphological, chemical and genetic analysis of southern Brazilian pecan (*Carya illinoinensis*) accessions. *Sci. Hortic. Amst.* **2020**, *261*, 108863. [CrossRef]

69. Saini, G.; Khamparia, A.; Luhach, A.K. Classification of Plants Using Convolutional Neural Network. In *Advances in Intelligent Systems and Computing, Proceedings of the First International Conference on Sustainable Technologies for Computational Intelligence*; Luhach, A., Kosa, J., Poonia, R., Gao, X., Singh, D., Eds.; Springer: Singapore, 2020; Volume 1045, pp. 551–561.

70. Yang, H.-W.; Hsu, H.-C.; Yang, C.-K.; Tsai, M.-J.; Kuo, Y.-F. Differentiating between morphologically similar species in genus Cinnamomum (Lauraceae) using deep convolutional neural networks. *Comput. Electron. Agric.* **2019**, *162*, 739–748. [CrossRef]

71. Sant'Anna, I.C.; Tomaz, R.S.; Silva, G.N.; Nascimento, M.; Bhering, L.L.; Cruz, C.D. Superiority of artificial neural networks for a genetic classification procedure. *Genet. Mol. Res.* **2015**, *14*, 9898–9906. [CrossRef]

72. Korani, W.; Clevenger, J.P.; Chu, Y.; Ozias-Akins, P. Machine Learning as an Effective Method for Identifying True Single Nucleotide Polymorphisms in Polyploid Plants. *Plant Genome* **2019**, *12*, 1–10. [CrossRef]

73. González-Camacho, J.M.; de los Campos, G.; Pérez, P.; Gianola, D.; Cairns, J.E.; Mahuku, G.; Babu, R.; Crossa, J. Genome-enabled prediction of genetic values using radial basis function neural networks. *Theor. Appl. Genet.* **2012**, *125*, 759–771. [CrossRef] [PubMed]

74. Peixoto, L.A.; Bhering, L.L.; Cruz, C.D. Artificial neural networks reveal efficiency in genetic value prediction. *Genet. Mol. Res.* **2015**, *14*, 6796–6807. [CrossRef]

75. Zingaretti, L.M.; Gezan, S.A.; Ferrão, L.F.V.; Osorio, L.F.; Monfort, A.; Muñoz, P.R.; Whitaker, V.M.; Pérez-Enciso, M. Exploring Deep Learning for Complex Trait Genomic Prediction in Polyploid Outcrossing Species. *Front. Plant Sci.* **2020**, *11*. [CrossRef]

76. Ghaffari, H.; Tadayon, M.R.; Razmjoo, J.; Bahador, M.; Soureshjani, H.K.; Yuan, T. Impact of Jasmonic Acid on Sugar Yield and Physiological Traits of Sugar Beet in Response to Water Deficit Regimes: Using Stepwise Regression Approach. *Russ. J. Plant Physiol.* **2020**, *67*, 482–493. [CrossRef]

77. Zou, J.; Hu, W.; Li, Y.; He, J.; Zhu, H.; Zhou, Z. Screening of drought resistance indices and evaluation of drought resistance in cotton (*Gossypium hirsutum* L.). *J. Integr. Agric.* **2020**, *19*, 495–508. [CrossRef]

78. Lv, C.; Huang, Y.; Sun, W.; Yu, L.; Zhu, J. Response of rice yield and yield components to elevated [CO_2]: A synthesis of updated data from FACE experiments. *Eur. J. Agron.* **2020**, *112*, 125961. [CrossRef]

79. Emamgholizadeh, S.; Parsaeian, M.; Baradaran, M. Seed yield prediction of sesame using artificial neural network. *Eur. J. Agron.* **2015**, *68*, 89–96. [CrossRef]

80. Lee, S.; Jeong, Y.; Son, S.; Lee, B. A Self-Predictable Crop Yield Platform (SCYP) Based on Crop Diseases Using Deep Learning. *Sustainability* **2019**, *11*, 3637. [CrossRef]

81. Ajay, B.C.; Bera, S.K.; Singh, A.L.; Kumar, N.; Gangadhar, K.; Kona, P. Evaluation of Genotype × Environment Interaction and Yield Stability Analysis in Peanut Under Phosphorus Stress Condition Using Stability Parameters of AMMI Model. *Agric. Res.* **2020**, 1–10. [CrossRef]

82. Finlay, K.; Wilkinson, G. The analysis of adaptation in a plant-breeding programme. *Aust. J. Agric. Res.* **1963**, *14*, 742. [CrossRef]

83. Eberhart, S.A.; Russell, W.A. Stability Parameters for Comparing Varieties 1. *Crop Sci.* **1966**, *6*, 36–40. [CrossRef]

84. Wricke, G. Über eine Methode zur Erfassung der ökologischen Streubreite in Feldversuchen. *Z. Pflanzenzuchtg* **1962**, *47*, 92–96.

85. Shukla, G.K. Some statistical aspects of partitioning genotype-environmental components of variability. *Heredity Edinb.* **1972**, *29*, 237–245. [CrossRef] [PubMed]

86. Francis, C.A.; Prager, M.; Laing, D.R.; Flor, C.A. Genotype × Environment Interactions in Bush Bean Cultivars in Monoculture and Associated with Maize. *Crop Sci.* **1978**, *18*, 237–242. [CrossRef]

87. Lin, C.S.; Binns, M.R. A superiority measure of cultivar performance for cultivar × location data. *Can. J. Plant Sci.* **1988**, *68*, 193–198. [CrossRef]

88. Karimizadeh, R.; Mohammadi, M.; Sabaghni, N.; Mahmoodi, A.A.; Roustami, B.; Seyyedi, F.; Akbari, F. GGE Biplot Analysis of Yield Stability in Multi-environment Trials of Lentil Genotypes under Rainfed Condition. *Not. Sci. Biol.* **2013**, *5*, 256–262. [CrossRef]

89. Singh, R.P.; Das, S.K.; Bhaskarrao, U.M.; Reddy, M.N. *Sustainability Index Under Different Management*; Annual Report; CRIDA: Hyderabad, India, 1990.

90. Han, X.; Hu, C.; Chen, Y.; Qiao, Y.; Liu, D.; Fan, J.; Li, S.; Zhang, Z. Crop yield stability and sustainability in a rice-wheat cropping system based on 34-year field experiment. *Eur. J. Agron.* **2020**, *113*, 125965. [CrossRef]

91. Flores, F.; Moreno, M.; Cubero, J. A comparison of univariate and multivariate methods to analyze G×E interaction. *Field Crop. Res.* **1998**, *56*, 271–286. [CrossRef]

92. Hühn, M. Beiträge zur Erfassung der phänotypischen Stabilität. I. Vorschlag einiger auf Ranginformationen beruhender Stabilitätsparameter. *EDV Medizin Biol.* **1979**, *10*, 112–117.

93. Nassar, R.; Hühn, M. Studies on Estimation of Phenotypic Stability: Tests of Significance for Nonparametric Measures of Phenotypic Stability. *Biometrics* **1987**, *43*, 45. [CrossRef]

94. Kang, M.S. A rank-sum method for selecting high-yielding, stable corn genotypes. *Cereal Res. Commun.* **1988**, *16*, 113–115.

95. Ketata, H.; Yan, S.K.; Nachit, M. Relative consistency performance across environments. In Proceedings of the International Symposium on Physiology and Breeding of Winter Cereals for Stressed Mediterranean Environments, Montpellier, France, 3–6 July 1989.

96. Fox, P.N.; Skovmand, B.; Thompson, B.K.; Braun, H.-J.; Cormier, R. Yield and adaptation of hexaploid spring triticale. *Euphytica* **1990**, *47*, 57–64. [CrossRef]

97. Li, N.; Lin, H.; Wang, T.; Li, Y.; Liu, Y.; Chen, X.; Hu, X. Impact of climate change on cotton growth and yields in Xinjiang, China. *Field Crop. Res.* **2020**, *247*, 107590. [CrossRef]

98. Harfouche, A.L.; Jacobson, D.A.; Kainer, D.; Romero, J.C.; Harfouche, A.H.; Scarascia Mugnozza, G.; Moshelion, M.; Tuskan, G.A.; Keurentjes, J.J.B.; Altman, A. Accelerating Climate Resilient Plant Breeding by Applying Next-Generation Artificial Intelligence. *Trends Biotechnol.* **2019**, *37*, 1217–1235. [CrossRef]

99. Fischer, R.; Maurer, R. Drought resistance in spring wheat cultivars. I. Grain yield responses. *Aust. J. Agric. Res.* **1978**, *29*, 897. [CrossRef]

100. Fischer, R.; Wood, J. Drought resistance in spring wheat cultivars. III.* Yield associations with morpho-physiological traits. *Aust. J. Agric. Res.* **1979**, *30*, 1001. [CrossRef]

101. Rosielle, A.A.; Hamblin, J. Theoretical Aspects of Selection for Yield in Stress and Non-Stress Environment. *Crop Sci.* **1981**, *21*, 943–946. [CrossRef]

102. Bouslama, M.; Schapaugh, W.T. Stress Tolerance in Soybeans. I. Evaluation of Three Screening Techniques for Heat and Drought Tolerance. *Crop Sci.* **1984**, *24*, 933–937. [CrossRef]

103. Fernandez, G.C.J. Effective selection criteria for assessing stress tolerance. In Proceedings of the International Symposium on Adaptation of Vegetables and Other Food Crops in Temperature and Water Stress, Tainan, Taiwan, 13–16 August 1992; Kuo, C.G., Ed.; AVRDC Publication: Tainan, Taiwan, 1992; pp. 257–270.

104. Gavuzzi, P.; Rizza, F.; Palumbo, M.; Campanile, R.G.; Ricciardi, G.L.; Borghi, B. Evaluation of field and laboratory predictors of drought and heat tolerance in winter cereals. *Can. J. Plant Sci.* **1997**, *77*, 523–531. [CrossRef]

105. Schneider, K.A.; Rosales-Serna, R.; Ibarra-Perez, F.; Cazares-Enriquez, B.; Acosta-Gallegos, J.A.; Ramirez-Vallejo, P.; Wassimi, N.; Kelly, J.D. Improving Common Bean Performance under Drought Stress. *Crop Sci.* **1997**, *37*, 43–50. [CrossRef]

106. Farshadfar, E.; Sutka, J. Screening drought tolerance criteria in maize. *Acta Agron. Hungarica* **2002**, *50*, 411–416. [CrossRef]

107. Niazian, M.; Sadat-Noori, S.A.; Tohidfar, M.; Galuszka, P.; Mortazavian, S.M.M. Agrobacterium-mediated genetic transformation of ajowan (*Trachyspermum ammi* (L.) Sprague): An important industrial medicinal plant. *Ind. Crops Prod.* **2019**, *132*, 29–40. [CrossRef]

108. Niazian, M.; Shariatpanahi, M.E. In vitro-based doubled haploid production: Recent improvements. *Euphytica* **2020**, *216*, 69. [CrossRef]

109. Marchetti, C.F.; Ugena, L.; Humplík, J.F.; Polák, M.; Ćavar Zeljković, S.; Podlešáková, K.; Fürst, T.; De Diego, N.; Spíchal, L. A Novel Image-Based Screening Method to Study Water-Deficit Response and Recovery of Barley Populations Using Canopy Dynamics Phenotyping and Simple Metabolite Profiling. *Front. Plant Sci.* **2019**, *10*. [CrossRef]

110. Barbedo, J.G.A. Detection of nutrition deficiencies in plants using proximal images and machine learning: A review. *Comput. Electron. Agric.* **2019**, *162*, 482–492. [CrossRef]

111. Singh, A.; Ganapathysubramanian, B.; Singh, A.K.; Sarkar, S. Machine Learning for High-Throughput Stress Phenotyping in Plants. *Trends Plant Sci.* **2016**, *21*, 110–124. [CrossRef]

112. Picon, A.; Alvarez-Gila, A.; Seitz, M.; Ortiz-Barredo, A.; Echazarra, J.; Johannes, A. Deep convolutional neural networks for mobile capture device-based crop disease classification in the wild. *Comput. Electron. Agric.* **2019**, *161*, 280–290. [CrossRef]

113. Agarwal, M.; Sinha, A.; Gupta, S.K.; Mishra, D.; Mishra, R. Potato crop disease classification using convolutional neural network. In *Smart Systems and IoT: Innovations in Computing*; Somani, A.K., Shekhawat, R.S., Mundra, A., Srivastava, S., Verma, V.K., Eds.; Smart Innovation, Systems and Technologies; Springer: Singapore, 2020; Volume 141, ISBN 978-981-13-8405-9.

114. Anagnostis, A.; Asiminari, G.; Papageorgiou, E.; Bochtis, D. A Convolutional Neural Networks Based Method for Anthracnose Infected Walnut Tree Leaves Identification. *Appl. Sci.* **2020**, *10*, 469. [CrossRef]

115. Khamparia, A.; Saini, G.; Gupta, D.; Khanna, A.; Tiwari, S.; de Albuquerque, V.H.C. Seasonal Crops Disease Prediction and Classification Using Deep Convolutional Encoder Network. *Circuits Syst. Signal Process.* **2020**, *39*, 818–836. [CrossRef]

116. Sibiya, M.; Sumbwanyambe, M. A Computational Procedure for the Recognition and Classification of Maize Leaf Diseases Out of Healthy Leaves Using Convolutional Neural Networks. *AgriEngineering* **2019**, *1*, 9. [CrossRef]

117. Kearsey, M.; Pooni, H. *The Genetical Analysis of Quantitative Traits*; Stanley Thornes Ltd: Cheltenham, UK, 1998.

118. Griffing, B. Concept of General and Specific Combining Ability in Relation to Diallel Crossing Systems. *Aust. J. Biol. Sci.* **1956**, *9*, 463. [CrossRef]

119. Hayman, B.I. The Theory and Analysis of Diallel Crosses. II. *Genetics* **1958**, *43*, 63–85. [PubMed]

120. Jinks, J.L. The Analysis of Continuous Variation in a Diallel Cross of Nicotiana Rustica Varieties. *Genetics* **1954**, *39*, 767–788. [PubMed]

121. Kempthorne, O. *An Introduction to Genetic Statistics*; John Wiley & Sons Inc.: New York, NY, USA, 1957.

122. Comstock, R.E.; Robinson, H.F. The Components of Genetic Variance in Populations of Biparental Progenies and Their Use in Estimating the Average Degree of Dominance. *Biometrics* **1948**, *4*, 254. [CrossRef] [PubMed]

123. Opsahl, B. The Discrimination of Interactions and Linkage in Continuous Variation. *Biometrics* **1956**, *12*, 415. [CrossRef]

124. Kearsey, M.J.; Jinks, J.L. A general method of detecting additive, dominance and epistatic variation for metrical traits I. Theory. *Heredity Edinb.* **1968**, *23*, 403–409. [CrossRef]

125. Dezfouli, P.M.; Sedghi, M.; Shariatpanahi, M.E.; Niazian, M.; Alizadeh, B. Assessment of general and specific combining abilities in doubled haploid lines of rapeseed (*Brassica napus* L.). *Ind. Crop. Prod.* **2019**, *141*, 111754. [CrossRef]

126. Khaki, S.; Khalilzadeh, Z.; Wang, L. Predicting yield performance of parents in plant breeding: A neural collaborative filtering approach. *PLoS ONE* **2020**, *15*, e0233382. [CrossRef]

127. Niazian, M. Application of genetics and biotechnology for improving medicinal plants. *Planta* **2019**, *249*, 953–973. [CrossRef]

128. Ayuso, M.; García-Pérez, P.; Ramil-Rego, P.; Gallego, P.P.; Barreal, M.E. In vitro culture of the endangered plant Eryngium viviparum as dual strategy for its ex situ conservation and source of bioactive compounds. *Plant Cell Tissue Organ Cult.* **2019**, *138*, 427–435. [CrossRef]

129. Sugimoto, K.; Temman, H.; Kadokura, S.; Matsunaga, S. To regenerate or not to regenerate: Factors that drive plant regeneration. *Curr. Opin. Plant Biol.* **2019**, *47*, 138–150. [CrossRef] [PubMed]

130. Prasad, V.S.S.; Gupta, S.D. Applications and potentials of artificial neural networks in plant tissue culture. In *Plant Tissue Culture Engineering*; Gupta, S.D., Ibaraki, Y., Eds.; Focus on Biotechnology; Springer: Dordrecht, The Netherlands, 2008; Volume 6, ISBN 978-1-4020-3594-4.

131. Niazian, M.; Noori, S.A.S.; Galuszka, P.; Tohidfar, M.; Mortazavian, S.M.M. Genetic stability of regenerated plants via indirect somatic embryogenesis and indirect shoot regeneration of *Carum copticum* L. *Ind. Crops Prod.* **2017**, *97*, 330–337. [CrossRef]

132. Orłowska, R.; Bednarek, P.T. Precise evaluation of tissue culture-induced variation during optimisation of in vitro regeneration regime in barley. *Plant Mol. Biol.* **2020**, *103*, 33–50. [CrossRef] [PubMed]

133. Phillips, G.C.; Garda, M. Plant tissue culture media and practices: An overview. *Vitr. Cell. Dev. Biol. Plant* **2019**, *55*, 242–257. [CrossRef]

134. Alanagh, E.N.; Garoosi, G.; Haddad, R.; Maleki, S.; Landín, M.; Gallego, P.P. Design of tissue culture media for efficient Prunus rootstock micropropagation using artificial intelligence models. *Plant Cell Tissue Organ Cult.* **2014**, *117*, 349–359. [CrossRef]

135. Hesami, M.; Naderi, R.; Tohidfar, M. Modeling and Optimizing Medium Composition for Shoot Regeneration of Chrysanthemum via Radial Basis Function-Non-dominated Sorting Genetic Algorithm-II (RBF-NSGAII). *Sci. Rep.* **2019**, *9*, 18237. [CrossRef]

136. Sadat Noori, S.A.; Norouzi, M.; Karimzadeh, G.; Shirkool, K.; Niazian, M. Effect of colchicine-induced polyploidy on morphological characteristics and essential oil composition of ajowan (*Trachyspermum ammi* L.). *Plant Cell Tissue Organ Cult.* **2017**, *130*, 543–551. [CrossRef]

137. Castillo, A.M.; Cistué, L.; Vallés, M.P.; Soriano, M. Chromosome Doubling in Monocots. In *Advances in Haploid Production in Higher Plants*; Springer: Dordrecht, The Netherlands, 2009; pp. 329–338.

138. Niazian, M.; Sadat Noori, S.A.; Galuszka, P.; Mortazavian, S.M.M. Tissue culture-based Agrobacterium-mediated and in planta transformation methods. *Czech J. Genet. Plant Breed.* **2017**, *53*, 133–143. [CrossRef]

139. Matias, F.I.; Caraza-Harter, M.V.; Endelman, J.B. FIELDimageR: An R package to analyze orthomosaic images from agricultural field trials. *Plant Phenome J.* **2020**, *3*, e20005. [CrossRef]

140. Tsaftaris, S.A.; Minervini, M.; Scharr, H. Machine Learning for Plant Phenotyping Needs Image Processing. *Trends Plant Sci.* **2016**, *21*, 989–991. [CrossRef]

141. Ubbens, J.R.; Stavness, I. Deep Plant Phenomics: A Deep Learning Platform for Complex Plant Phenotyping Tasks. *Front. Plant Sci.* **2017**, *8*, 1190. [CrossRef] [PubMed]

142. Li, X.; He, Y. Discriminating varieties of tea plant based on Vis/NIR spectral characteristics and using artificial neural networks. *Biosyst. Eng.* **2008**, *99*, 313–321. [CrossRef]

143. Khoshroo, A.; Arefi, A.; Masoumiasl, A.; Jowkar, G.H. Classification of wheat cultivars using image processing and artificial neural networks. *Agric. Commun.* **2014**, *2*, 17–22.

144. Mahlein, A.-K. Plant Disease Detection by Imaging Sensors—Parallels and Specific Demands for Precision Agriculture and Plant Phenotyping. *Plant Dis.* **2016**, *100*, 241–251. [CrossRef] [PubMed]

145. Maes, W.H.; Steppe, K. Perspectives for Remote Sensing with Unmanned Aerial Vehicles in Precision Agriculture. *Trends Plant Sci.* **2019**, *24*, 152–164. [CrossRef] [PubMed]

146. Khadka, K.; Earl, H.J.; Raizada, M.N.; Navabi, A. A Physio-Morphological Trait-Based Approach for Breeding Drought Tolerant Wheat. *Front. Plant Sci.* **2020**, *11*, 715. [CrossRef]

147. Singh, A.; Jones, S.; Ganapathysubramanian, B.; Sarkar, S.; Mueller, D.; Sandhu, K.; Nagasubramanian, K. Challenges and Opportunities in Machine-Augmented Plant Stress Phenotyping. *Trends Plant Sci.* **2020**. [CrossRef]

148. Pan-pan, H.; Wei-xu, L.; Hui-hui, L.; Zeng-xu, X. In vitro induction and identification of autotetraploid of Bletilla striata (Thunb.) Reichb.f. by colchicine treatment. *Plant Cell Tissue Organ Cult.* **2018**, *132*, 425–432. [CrossRef]

149. Ochatt, S.J.; Patat-Ochatt, E.M.; Moessner, A. Ploidy level determination within the context of in vitro breeding. *Plant Cell Tissue Organ Cult.* **2011**, *104*, 329–341. [CrossRef]

150. Ahmadi, B.; Ebrahimzadeh, H. In vitro androgenesis: Spontaneous vs. artificial genome doubling and characterization of regenerants. *Plant Cell Rep.* **2020**, *39*, 299–316. [CrossRef]

151. Santeramo, D.; Howell, J.; Ji, Y.; Yu, W.; Liu, W.; Kelliher, T. DNA content equivalence in haploid and diploid maize leaves. *Planta* **2020**, *251*, 30. [CrossRef] [PubMed]

152. Blonder, B.; Graae, B.J.; Greer, B.; Haagsma, M.; Helsen, K.; Kapás, R.E.; Pai, H.; Rieksta, J.; Sapena, D.; Still, C.J.; et al. Remote sensing of ploidy level in quaking aspen (Populus tremuloides Michx.). *J. Ecol.* **2020**, *108*, 175–188. [CrossRef]

 *agriculture*

Article

A Hybrid CFS Filter and RF-RFE Wrapper-Based Feature Extraction for Enhanced Agricultural Crop Yield Prediction Modeling

Dhivya Elavarasan [1], Durai Raj Vincent P M [1,*], Kathiravan Srinivasan [1] and Chuan-Yu Chang [2,*]

[1] School of Information Technology and Engineering, Vellore Institute of Technology (VIT), Vellore 632 014, India; dhivya.e2017@vitstudent.ac.in (D.E.); kathiravan.srinivasan@vit.ac.in (K.S.)

[2] Department of Computer Science and Information Engineering, National Yunlin University of Science and Technology, Yunlin 64002, Taiwan

* Correspondence: pmvincent@vit.ac.in (D.R.V.P.M.); chuanyu@yuntech.edu.tw (C.-Y.C.)

Received: 14 July 2020; Accepted: 7 September 2020; Published: 11 September 2020

Abstract: The innovation in science and technical knowledge has prompted an enormous amount of information for the agrarian sector. Machine learning has risen with massive processing techniques to perceive new contingencies in agricultural development. Machine learning is a novel onset for the investigation and determination of unpredictable agrarian issues. Machine learning models actualize the need for scaling the learning model's performance. Feature selection can impact a machine learning model's performance by defining a significant feature subset for increasing the performance and identifying the variability. This paper explains a novel hybrid feature extraction procedure, which is an aggregation of the correlation-based filter (CFS) and random forest recursive feature elimination (RFRFE) wrapper framework. The proposed feature extraction approach aims to identify an optimal subclass of features from a collection of climate, soil, and groundwater characteristics for constructing a crop-yield forecasting machine learning model with better performance and accuracy. The model's precision and effectiveness are estimated (i) with all the features in the dataset, (ii) with essential features obtained using the learning algorithm's inbuilt 'feature_importances' method, and (iii) with the significant features obtained through the proposed hybrid feature extraction technique. The validation of the hybrid CFS and RFRFE feature extraction approach in terms of evaluation metrics, predictive accuracies, and diagnostic plot performance analysis in comparison with random forest, decision tree, and gradient boosting machine learning algorithms are found to be profoundly satisfying.

Keywords: correlation filter; crop yield prediction; hybrid feature extraction; machine learning; recursive feature elimination wrapper; precision agriculture

---

1. Introduction

The advancement in science and machine learning has accounted for a colossal amount of information in the agrarian field subjecting to examination and incorporating procedures such as crop yield forecasting, investigation of plant diseases, enhancement of crops, etc. Machine learning has ascended with enormous processing strategies to conceive new opportunities in multi-disciplinary agrarian innovations. Though machine learning strategies handle immense sums and variations of information, accomplishing a superior performing model is a pivotal plan that needs to be focused. Further, this actualizes the need for scaling the learning model's performance. Feature extraction is a technique for determining a significant subgroup of features utilizing various statistical measures for model construction [1]. It can impact a machine learning model's performance by enlisting a substantial feature subset for boosting the performance and categorizing the variability. The most prevailing feature selection measures are the filter methods, which are generally faster, and the wrapper methods that are more reliable but computationally expensive.

Together with colossal data advances and improved measure reinforcement, machine learning has risen to determine, assess, and envision intensive information techniques in an agriculture operative environment. Exuberant upgrades in machine learning have tremendous potential results. Many researchers and authorities in present agribusiness are looking at their speculation at an increasingly prevalent scale, helping to accomplish progressively exact and steady forecasts. Precision agriculture is also known as "site-specific agriculture", an approach to deal with farm management utilizing information technology. Precision agriculture assures that the crops and soil receive precisely what they require for optimum health and profitability [2]. Present-day agrarian frameworks can discover significantly more machine learning methods to use enhancements more efficiently and adjust to different natural changes [3–5]. The objective of precision agriculture is to guarantee productivity, manageability, and conservation of the environment [6]. Machine learning in precision agriculture endows a crop management system that assists in yield forecasting in crops, crop disease management, distinguishing the crop weeds, acknowledging crop assortments, forecasting of agricultural climate, and many others. In the machine learning procedure, insignificant features in preparing a dataset will decline the forecasting efficiency [7]. Due to the extensive increase in the data amount, a pre-processing strategy such as feature selection grows into an essential and demanding step when using a machine learning technique [8,9].

1.1. Background

Feature selection is characterized as the way towards removing the excessive and unessential features using statistical measures from a dataset, to embellish the learning algorithm. It is an active explorative area in artificial intelligence applications. The predominant aim of feature extraction is to achieve an appropriate subgroup of features for defining and delineating a dataset. In machine learning, feature selection strategy gives us a method for lowering calculation time [10], enhancing forecasting results, and improving perception of the data. In other words, feature selection is an extensively used pre-processing procedure for higher-dimensional data. It incorporates the following objectives:

- Enhancing forecasting precision
- Lowering the dimensions
- Removing superfluous or insignificant features
- Improving the data interpretability
- Enhancing the model interpretability
- Decreasing the volume of the required information.

The feature extraction processes can be classified based on various standards, as depicted in Figure 1. Depending on the training data employed, they are grouped as supervised, unsupervised, and semi-supervised. Based on their inter-relationship with the learning models, they are classified as a filter, wrapper, and hybrid models. Depending on the search strategies, it is organized as a forward increase, backward deletion, and random models [11]. Additionally, considering the output type, they are classified as feature ranking and subset selection models. For higher-dimensional features, this issue cannot be resolved by consolidating all potential outcomes.

Filter techniques can recognize and eliminate insignificant features, but they cannot expel repetitive features because they do not consider conceivable dependencies between features [12]. The filter method evaluates the significance of a feature subgroup entirely depending on the intrinsic characteristics of data such as correlation, variance, F measure, entropy, the ratio of information gain, and mutual information [13,14].

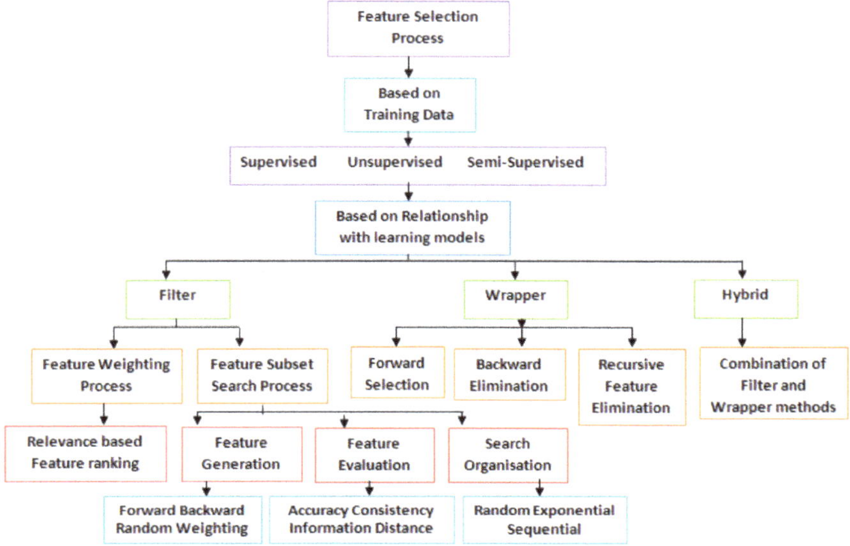

Figure 1. Overview of feature selection methods for machine learning algorithms.

The wrapper-based feature selection method [15] encompasses a feature selection algorithm over an induction algorithm. The wrapper approach is mainly helpful in solving issues such as generating a fitness function when it cannot be efficiently expressed with an accurate analytical equation. Various search algorithms such as forward and backward elimination passes, best-first search, recursive feature elimination [16] can be utilized to discover a subset of features by means of augmenting or limiting the corresponding objective function. Wrapper techniques are usually identified by the immense caliber of the selected features; however, they have a higher computational cost.

Another intermittently investigated methodology for feature selection is hybrid methods. They involve strategies endeavored to have an acceptable compromise between the computational effort and efficiency [17–19]. These methods encompass those techniques that integrate both filter and wrapper methods. It accomplishes the balance of precision and computing time.

1.2. Literature Review

It has been analyzed that the agriculturist's income rises or falls depending on the outcomes they acquire from their harvests. In an interest to enhance and support the process of determination and resolution, it is vital to perceive the definite prevailing association between the crop yield and numerous factors impacting it [20]. The factors are present in a higher level of intricacy in time and space, and the decisions are to be perceived considering the effects of soil [21], climate [22], water availability [23], landscapes, and several others that are concerned in assisting the crop yield [24]. Generally, a considerable part of agribusiness-based frameworks cannot be delineated by the essential stage-wise condition or by a definite equation, particularly at the stage when the dataset is convoluted, strident, deficient, or assorted. Structuring of these frameworks is complicated and imminent but has exceptional importance for analysts for forecasting and simulation.

Feature selection methods have been enforced in prediction and classification problems to choose a reduced list of features, which makes the algorithm to perform faster and produce precise results. Some specific issues are continuously handled with an extraordinary number of features. In the literature, some hybrid feature selection methods for agrarian frameworks combing both the filter and wrapper approaches are proposed. Muhammed et al. have proposed the identification and categorization of citrus diseases in the plant [25] depending on the improved accentuated segmentation and feature extraction. The procedure encompasses two phases: (1) detection of lesion spots in plants and (2) classification of citrus disease. The optimal features are defined by enforcing a hybrid feature extraction process, which includes the principal component analysis score, entropy, and skewed covariance vector. Yu et al. explored a new procedure of "reduced redundancy improved relevance" framework-based feature selection to choose an efficient wavelength spectrum for the hyperspectral images of cotton plants, enabling the categorization of foreign substances in cotton plants [26]. Prediction of moisture content between wood chips using the least square Support Vector Machine (SVM) kernel feature selection method has been endorsed by Hela et al. [27].

Wenbin et al. defined an efficient mutual information-based feature selection algorithm integrating information theory and rough sets [28]. The evaluation function can choose candidate features that comprise of high pertinence concerning the class and low redundancy among the selected characteristic features, in such a way that the redundancy is removed. Hosein et al. introduced a new feature extraction method, which is the combination of an advanced ant colony optimization algorithm (ACO) with an adaptive fuzzy inference system (ANFIS) [29]. It enabled them to choose the best subgroup of features from the various observed soil characteristics that leverage the soil cation exchange capacity (CEC), which is a valuable property representing the soil fertility status. Feature selection is highly essential for dimensionality reduction in the case of hyperspectral images. Ashis et al. endorsed a supervised hybrid feature extraction procedure combining the Self-adaptive differential evolution (SADE) algorithm with a fuzzy K-neighbor classifier wrapper [30] for the hyperspectral remote sensed images of agricultural data over Indiana, Kennedy Space Center of Florida, and Botswana. Somayeh et al. proposed a hybrid Genetic Algorithm—Artificial Neural Network feature extraction method to identify the significant features for the pistachio endocarp lesion problem [31]. Pistachio endocarp lesion (PEL) is one of the most significant causes of the damage of the pistachio plant. The study was framed to identify the biotic and abiotic agents that impact the existence of PEL.

The works discussed until now attempts to consider the advantages of the filter and wrapper methods and associate them appropriately. In addition, each proposed strategy utilizes its own selection procedures and assessment measures. As observed, the hybrid-based feature extraction procedures have been examined limited for agrarian datasets, and the current processes involve constraints either in their assessment measures or the number of characteristic features processed. Concerning the reasons as mentioned above, a new hybrid-based feature extraction process, unlike the other hybrid measures, is proposed, which uses a correlation-based filter stage—CFS and the random forest recursive feature elimination wrapper stage—RF-RFE. CFS can effectively screen redundant, noisy, and irrelevant features. CFS also enhances performance and reduces the size of improved knowledge structures efficiently than other filter measures. It is also computationally inexpensive, which provides a better feature subset to ease the performance of wrapper, which is usually computationally expensive. RF-RFE based wrapper, though computationally expensive, gives high-quality feature outputs. Since in the proposed approach, the RF-RFE wrapper is combined with the filter approach, the computational time is reduced. Another advantage of RF-RFE is that it does not demand any reconciliation to develop competing results.

1.3. Aim of the Paper

In this paper, a new hybrid-based feature extraction procedure combining correlation type filter CFS and a recursive feature elimination-based wrapper RF-RFE is developed. The proposed technique is applied to the paddy crop dataset to determine a prime collection of features for forecasting crop yield. Until now, the hybrid feature selection combination of CFS filter and RF-RFE wrapper has not been enforced to recognize the significant subset of features for yield prediction in crop development. The empirical results determine that the proposed method selects significant features amongst other algorithms by removing those that do not contribute to enhanced prediction results. The remainder of the paper is systemized as follows. Section 2 explains the methodology for the proposed hybrid feature abstraction method depending on the CFS RF-RFE wrapper, the data considered for the study along with the significant agrarian parameters and the details about the various machine learning models experimented and the evaluation metrics used. Section 3 demonstrates the experimental framework and outcomes of the developed hybrid feature extraction process on the agricultural dataset with various machine learning methods. Section 4 presents a discussion of results and future works. Finally, Section 5 winds up with the conclusion of the proposed work.

2. Materials and Methods

2.1. Proposed Hybrid Feature Selection Methodology

The two predominant feature extraction processes in machine learning are filter and wrapper methods. Wrappers frequently give better outcomes than filter processes, as feature extraction is advanced for the specific learning algorithm that is utilized [32]. Due to this, wrappers are very expensive to run and can be obstinate for substantial databases comprising numerous features. Moreover, as the feature selection process is tightly connected with the learning algorithm, wrappers are less frequently used than filters. In general, filters execute rapidly than the wrapper; as a result, filters portray a vastly improved possibility of scaling to databases with a substantial number of features. Filters can afford the same benefit as wrappers. When an enhanced precision for a specific learning algorithm is recommended, a filter can provide a smart beginning feature subset for a wrapper. In other words, the wrapper will be provided with a reduced feature set by the filter, thus helping the wrapper to scale efficiently for bigger datasets. The hybrid approach, which is an association of wrapper and filter methods, utilizes the ranking information from the filter method.

Further, this enhances the search in the optimization algorithm, which is used by the wrapper methods. This method exploits the advantage of both the wrappers and the filters. By connecting these two methods, we can enhance the predictive efficiency of pure filter methods and curtail the execution duration of pure wrapper methods. In this section, the proposed feature extraction procedure is explained, which conforms to the hybrid CFS filter—RF-RFE wrapper approach. A framework representing the proposed approach is explained in Figure 2.

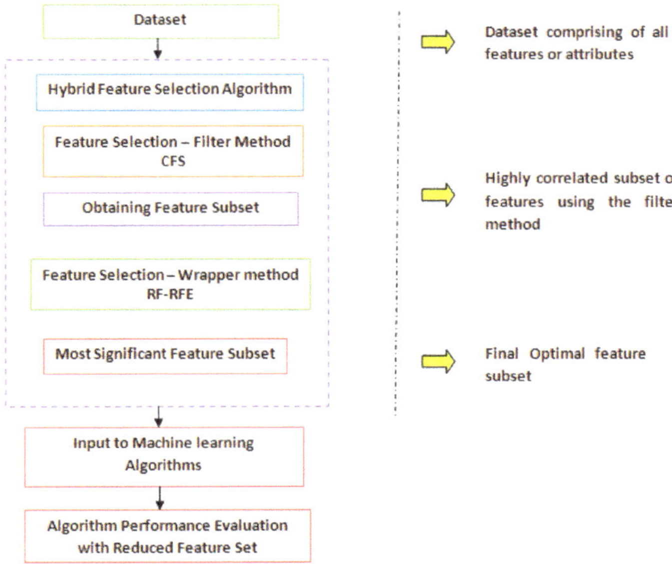

Figure 2. A framework of the proposed hybrid feature selection approach.

Generally, the hybrid filter-wrapper feature selection method comprises typically of two phases:

- The initial phase utilizes the filter method to minimize the size of the feature set by discarding the noisy insignificant features.
- The final phase utilizes the wrapper method to identify the ideal characteristic feature subgroup from the reserved feature set.

In the first step during the filter stage, the features are arranged depending on a correlation-based heuristic evaluation function. This process objective is to distinguish the characteristic features that are persistent with the information framework. The features are categorized based on their significance. With the purpose of confining the exploration into the space of all conceivable feature subgroups, this process permits a decent estimate of features as a beginning for the next step. In the second phase, i.e., the wrapper phase, the objective is to assess the features examining them as a subgroup rather than in the explicit case. Then, they are enforced to random forest-based recursive feature elimination selection process. Figure 3 explains the proposed hybrid feature selection system architecture. The following subsections delineate each phase of the proposed strategy.

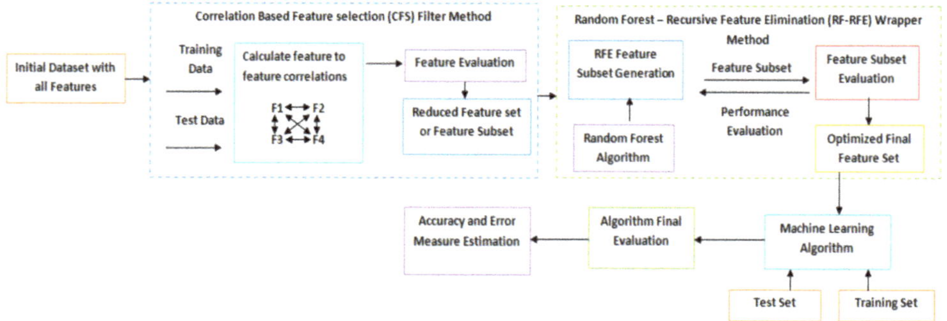

Figure 3. Architecture diagram of proposed hybrid CFS and RF-RFE feature selection approach.

2.1.1. Filter Stage—Correlation Based Feature Selector (CFS)

A filtering process assesses the quality of feature subsets depending on statistical measurements as evaluation criteria. In machine learning, one of the processes of selecting features for forecasting results can be attained based on the correlation among the features, and that such a feature selection strategy can be useful to regular machine learning algorithms. A feature is beneficial if it is conforming to a class or predicts the class [32]. A characteristic feature (X_i) is observed to be pertinent if and only if there prevail some probability (P) x_i and y such that $P(X_i = x_i) > 0$ as in Equation (1),

$$P(Y = y \mid X_i = x_i) \neq P(Y = y) \tag{1}$$

Experimental proof from the feature selection literature demonstrates that in addition to the insignificant features, superfluous features need to be removed as well. A feature is recognized as superfluous if it is exceedingly associated with one or more other features.

Moreover, this resulted in a hypothesis for feature extraction, which is a useful, acceptable characteristic feature subgroup that incorporates features that are significantly associated with class but dissociated with one other. In this scenario, the features are specific tests that measure characteristics identified with the variable of importance. For instance, a precise forecast of an individual's achievement in a subject can be obtained from a composite of various tests estimating a wide assortment of qualities rather than an individual test, which estimates a restricted set of qualities. In a given feature set if the association among the individual feature and an extrinsic variable is recognized, and the inter-relation among every other pair of the features is given, then the association among the complicated test comprising of the total features and the extrinsic variable can be determined from Equation (2),

$$r = \frac{\sum_{i=1}^{n}(x_i - \bar{x})(y_i - \bar{y})}{\sqrt{\sum_{i=1}^{n}(x_i - \bar{x})^2}\sqrt{\sum_{i=1}^{n}(y_i - \bar{y})^2}} \tag{2}$$

The above equation defines the Pearson's correlation coefficient. Where x_i and $\bar{x}$ defines the observed and average values of the features considered. y_i and $\bar{y}$ defines the observed and average values of the dataset class. If a group of n features has just been chosen, the correlation coefficient can be utilized to assess the connection between the group and the class, incorporating inter-correlation among the features. The significance of the feature group increases with the correlation between the features and classes. Additionally, it diminishes with an increasing inter-correlation. These thoughts have been examined in the literature on decision making and aggregate estimation [33]. Defining the aggregated correlation coefficient among the features and output variables as $r_{ny} = p(X_n, Y)$ and the aggregate among varying features as $r_{nn} = p(X_n, X_n)$. The group correlation coefficient calculating the relevance of the feature subset is given as follows in Equation (3),

$$J(X_n, Y) = \frac{n r_{ny}}{\sqrt{n + (n-1)r_{nn}}} \tag{3}$$

This shows that the association between a group and an external feature is an operation of the total number of individual characteristic features in the group. The formula from [34] is obtained from the Pearson's correlation coefficient by standardizing all the variables. It has been utilized in the correlation-based feature selection algorithm enabling the addition or deletion of one feature at a time. The following pseudo-code in Algorithm 1 explains the selection procedure using CFS filter.

Algorithm 1 CFS filter-based feature selection method

SELECT FEATURES
INPUT:
D_{train}—Training dataset
    P—The predictor
    n—Number of features to select
OUTPUT:
    F_x—Selected feature set
BEGIN:
    $F_0 = \varnothing$
    $x = 1$
    while $|F_x| < n$ do
        if $|F_x| < n - 1$ then
            $F_x = $ CFS (F_{x-1}, D_{train}, P)
        else
            Add the best-ranked feature f' to F_{x-1}
        end if
        $x = x + 1$
    end while
END

Predicting the feature importance based on correlation-based filters defines the following conclusions:

- The higher the correlation among the individual and the extrinsic variable, the higher is the correlation among the combination and external variables.

- The lower the inter-correlation among the individual and the extrinsic variable, the lower is the correlation among the combination and extrinsic variable.

For efficient prediction, it is obvious to remove the redundant features from the dataset. If another feature manages an existing feature's forecasting ability, then it can be removed safely. Further, to improvise the forecasting performance of the system, the reduced feature set obtained is passed on the next step of wrapper-based feature selection.

2.1.2. Wrapper Stage—Random Forest Recursive Feature Elimination (RF-RFE)

Wrapper methods use forecasting accuracy to validate the feature subset. Wrappers use the learning machine as a black box in scoring the feature subsets depending on their forecasting ability. Recursive feature elimination [34] is fundamentally a recursive process that ranks features based on a significance measure. RFE is a feature ranking procedure depending on a greedy algorithm. As per the standard of feature ranking, in all iterations, RFE will start eliminating from the full feature set the least significant features one after the other to obtain the most significant features. The recursion is required since, for a few processes, the pertinent significance of specific characteristic features can vary considerably on assessing beyond an alternate subgroup of features in the course of step-wise elimination. This is concerned primarily with profoundly correlated features. Depending on the order in which the features are discarded, the final feature set is constructed. The feature selection procedure itself comprises just acquiring the initial n features from this ranking.

Random forest falls under the ensemble-based prediction or categorization process. Substantially it grows several distinct prediction trees and utilizes them together as a combined predictor. The final prediction of a given dataset is determined by implementing an absolute rule among the choices of the respective predictors. Further, to create unassociated and distinctive insights, every tree is developed utilizing just a smaller dataset of the preparation set. Besides, to maximize the dissimilarity among the trees, the algorithm includes random contingency in the pursuit of optimal splits [35]. The wrapper stage of the proposed hybrid approach depends on the extent of the significance of the variable provided by the random forest. For each individual tree in the random forest, there exists a subgroup of the feature set which is not utilized at the time of training since every tree is developed on a bootstrap sample. These subgroups are generally termed as out-of-bag, which gives the unbiased measure of predicting the errors.

Random forest evaluates the significance of characteristic features infiltrating the framework as follows:

- In all iterations, each feature is shuffled, and over this shuffled data set, an out-of-bag estimation of the forecasting error is made.
- Naturally, when trying to alter this way, the insignificant features will not change the prediction error, inverse to the significant features.
- The corresponding loss in efficiency among the actual and the shuffled datasets is accordingly associated with the efficiency of the shuffled features.

The following pseudo-code in Algorithm 2 explains the feature selection using RF-RFE.

In the RF-RFE approach, the proportion of characteristic feature significance is connected with the recursive feature elimination algorithm. RF-RFE wrapper model is developed based on the perception of building a model (here the model is random forest) frequently and select either the best or worst operating feature. Removing the feature aside and recurring the process with the remaining features. This operation is carried out until all the features in the dataset are consumed. The features are then ranked depending on the order in which they are eliminated. In other words, this performs a greedy optimization search to determine the best performing feature subset. The following section describes the dataset and various agronomical factors impacting crop yield.

Algorithm 2 RF-RFE wrapper-based feature selection method

INPUT:
    $D_0 = [d_1, d_2, \ldots, d_n,]$—Training dataset
    $F = [f_1, \ldots, f_n]$—Set of n features
    Ranking Method $M(D, F)$
    $S = [1, 2 \ldots m]$—Subset of features
OUTPUT:
    Final ordered feature set F_s
BEGIN:
    $S = [1, 2 \ldots m]$
    $F_s = []$
    while $S \neq []$ do
    Repeat for x in $\{1 : n\}$
      Ranking feature set utilizing $M(D, F)$
      $S(f^*) \leftarrow$ F's last ranked feature
      $F_s(n - x + 1) \leftarrow S(f^*)$
      $S(F_s) \leftarrow S(F_s) - S(f^*)$
    end while
END

2.2. Significant Agrarian Parameters and Dataset Description

Machine learning has surged collectively with enormous intelligence progress and better methods to create unique contingencies to determine, assess, and appreciate information pervasive techniques for agrarian frameworks. Machine learning algorithms need an appropriate amount of data for efficient processing. Data with befitting attributes simplifies the effort of examining uniformity by eliminating the features that are superfluous or excess concerning the learning objective. This section explains in detail the various agrarian factors to be considered for yield prediction along with the study area and dataset description.

2.2.1. Agronomical Variables Impacting Yield of Crops

There are a variety of factors identified for crop yield and the vulnerabilities required for their development. The most crucial factors that affect crop yield are the climate, soil productiveness, and groundwater characteristics. These factors can epitomize an enormous risk to farmers when they are not checked and supervised precisely. Moreover, considering the ultimate objective to maximize the yield of the crop and curtail the hazard, it is vital to see explicitly the factors that affect crop yield. The following factors play a crucial role in crop enhancement.

Climate

A preeminent and the most overlooked variable that affects crop yield is climate. Climatic conditions extend past just wet and dry. While rainfall is the indispensable segment of the atmosphere, there exist few other distinct perspectives to recognize such as wind speed, humidity, temperature, and the widespread prevalence of pests during certain climatic conditions [36]. Conflicting patterns in climate lead to an excessive risk to crop and may prompt favorable conditions for specific weeds to grow.

Soil Productivity

There exist several nutrient supplements such as nitrogen, potassium, phosphorus that constitutes plant macronutrients and magnesium, zinc, calcium, iron, sulfur, etc. that constitutes plant micronutrients [37]. Every one of them is proportionally fundamental to the crop yield, regardless of the way that they are required in differing amounts. The accountability of nutrients for crop yield is indispensable for crop growth enhancement, protein formation, photosynthesis, and so forth. The unavailability of these nutrients can reduce the crop yield by conversely impacting the relevant growth factors.

Groundwater Characteristics and Availability

Water receptivity specifically influences the crop yield, and yield efficiency can change comprehensively given the varying precipitation pattern, utilizing both amount and time span. Almost no measure of rainfall can result in crop death; at the same time, ample precipitation can cause adverse effects [38]. The essential and synthetic parameters of groundwater perform an essential role in surveying the quality of water. The hydrochemical analysis uncovers the nature of water that is appropriate for irrigation, agriculture, industrial use, and drinking purposes.

2.2.2. Crop Dataset and Study Area Description

The crop data required for the proposed study is obtained from the various village blocks including Arcot, Sholinghur, Ponnai, Ammur, Kalavai, and Thimiri of the Vellore district of Tamil Nadu in India. The crop considered for the study is paddy. This district lies between 12°15′ to 13°15′ north latitude and 78°20′ to 79°50′ east longitude. Paddy is one of the significant economic crops grown in this district, and hence this district is examined for analysis. Unlike the regular soil and climatic parameters, the dataset comprises of distinctive climatic, soil, groundwater characteristics together with the fertilizer amount absorbed by the plants of the experimented region. Table 1 explains, in brief, the various parameters used for the experimental study. The data is observed for a time span of 20 years. The dataset contains paddy yield utilizing the region cultivated (in hectares), production of paddy (in tonnes), and crop yield obtained (kilogram/hectare). The data relevant to environmental aspects such as precipitation, air temperature, potential evapotranspiration, evapotranspiration of reference crop, and exceptional climatic features such as frost frequency of ground, diurnal temperature range, wind speed, humidity has been used which is obtained from the Indian water portal metadata tool. The data of soil and groundwater properties comprises of soil pH, topsoil density, amount of macronutrients existing in the soil, and the distinct groundwater characteristics such as type of aquifer, transmissivity, rock layer permeability, water conductivity, and the number of micronutrients existing in the groundwater before and after the monsoon period. Unlike the standard parameter set, the proposed work includes considering all the parameters from various aspects, including climate, hydrochemical properties of groundwater, soil, and fertilizer amount, to construct an efficient feature subset enhancing the prediction of crop yield achieving better precision than the traditional approach.

Table 1. List of dataset parameters and their description.

S. No	Parameter Name	Description	Units
1	Net cropped area	The total geographic area on which the crop has been planted at least once during a year	Integer (hectare)
2	Gross cropped area	Total area planted to crops during all growing seasons of the year	Integer (hectare)
3	Net irrigated area	The total geographic area that has acquired irrigation throughout the year	Integer (hectare)
4	Gross irrigated area	The total area under crops that have received irrigation during all the growing seasons of the year.	Integer (hectare)
5	Area rice	Total area in which the rice crop is planted	Integer (hectare)
6	Quantity rice	Total rice production in the study area	Integer (ton)
7	Yield rice	The total quantity of rice acquired	Integer (ton)
8	Soil type	Type of the soil in the study area considered 1—Medium black soil type 2—Red soil type	Integer
9	Land slope	A rise or fall of the land surface	Integer (meters)
10	Soil pH	Acidity and alkalinity measure in the soil.	Integer
11	Topsoil depth	The outermost soil layer rich in microorganisms and organic matter	Integer (meters)
12	N soil	The nitrogen amount present in the soil	Integer (kilogram/hectare)
13	P soil	The phosphorus amount present in the soil	Integer (kilogram/hectare)
14	K soil	The potassium amount present in the soil	Integer (kilogram/hectare)
15	QNitro	Amount of nitrogen fertilizers utilized	Integer (kilogram)
16	QP_2O_5	Amount of phosphorus fertilizers utilized	Integer (kilogram)
17	QK_2O	Amount of potassium fertilizers utilized	Integer (kilogram)
18	Precipitation	Rain or water vapor condensation from the atmosphere	Integer (millimeter)
19	Potential evapotranspiration	Quantity of evaporation occurring in an area in the presence of a sufficient water source	Integer (millimeter/day)
20	Reference crop evapotranspiration	The evapotranspiration rate from a crop reference surface that is not short of water	Integer (millimeter/day)
21	Ground frost frequency	Number of days referring to the condition when the upper layer soil temperature falls below the water freezing point	Integer (number of days)
22	Diurnal temperature range	Difference between the daily maximum and minimum temperature	Integer (°C)
23	Wet day frequency	The number of days in which a quantity of 0.2 mm or more of rain is observed.	Integer (number of days)
24	Vapor pressure	The pressure administered by water vapor with its condensed phase in thermodynamic equilibrium	Integer (hectopascal)
25	Maximum temperature	The highest temperature of air recorded	Integer (°C)
26	Minimum temperature	The lowest temperature of air recorded	Integer (°C)
27	Average temperature	The average temperature of air recorded	Integer (°C)
28	Humidity	The quantity of water vapor in the atmosphere	Integer (percentage)
29	Wind speed	The rate at which the air blows	Integer (miles/hour)
30	Aquifer area percentage	Percentage of an area enclosed by a body of permeable rock that can transmit or contain groundwater.	Integer (percentage)

Table 1. *Cont.*

S. No	Parameter Name	Description	Units
31	Aquifer well yield	Amount of water pumped from a well in an aquifer area	Integer (liters/minute)
32	Aquifer transmissivity	The water quantity that can be disseminated horizontally by a full saturated thickness of the aquifer	Integer (meter^2/day)
33	Aquifer permeability	A measure of the rock property, which defines how fluids can flow through it.	Integer (meter/day)
34	Pre–electrical conductivity	Average pre-monsoon electrical conductivity of groundwater	Integer (siemens/meter)
35	Post-electrical conductivity	Average post-monsoon electrical conductivity of groundwater	Integer (siemens/meter)
36	Groundwater pre-calcium	Average pre-monsoon calcium level in groundwater	Integer (milligram/Liters)
37	Groundwater post-calcium	Average post-monsoon calcium level in groundwater	Integer (milligram/Liters)
38	Groundwater pre-magnesium	Average pre-monsoon magnesium level in groundwater	Integer (milligram/Liters)
39	Groundwater post-magnesium	Average post-monsoon magnesium level in groundwater	Integer (milligram/Liters)
40	Groundwater pre-sodium	Average pre-monsoon sodium level in groundwater	Integer (milligram/Liters)
41	Groundwater post-sodium	Average post-monsoon sodium level in groundwater	Integer (milligram/Liters)
42	Groundwater pre-potassium	Average pre-monsoon potassium level in groundwater	Integer (milligram/Liters)
43	Groundwater post-potassium	Average post-monsoon potassium level in groundwater	Integer (milligram/Liters)
44	Groundwater pre-chloride	Average pre-monsoon chloride level in groundwater	Integer (milligram/Liters)
45	Groundwater post-chloride	Average post-monsoon chloride level in groundwater	Integer (milligram/Liters)

S. No—Serial Number.

The following section describes the various machine learning models and evaluation metrics used for the assessment of the proposed feature selection method.

2.3. Machine Learning Models and Evaluation Metrics

The proposed CFS filter RF-RFE wrapper hybrid statistical feature selection algorithm is tested by implementing it with the following machine learning algorithms, namely:

- Random forest
- Decision tree
- Gradient boosting.

2.3.1. Machine Learning

Decision trees are an information-based supervised machine learning algorithm [39]. They are a tree structure similar to a flow diagram, where every interior node indicates an analysis performed on the attributes, branch depicts the output of the test, and the label of a class is defined by every terminal node [40]. The significant objective of the decision tree is to identify the distinct features that represent the essential data concerning the target element, and then the dataset is split along these features such that the sub dataset's target value is as pure as possible. For evaluating the proposed hybrid feature selection method, a decision tree regressor with a maximum depth of four and best splitter value is constructed. Random forest is an ensemble-based [41] supervised machine learning algorithm, which combines several decision trees. The random forest algorithm is not biased, as there are several trees, and each is prepared on a subgroup of data. For evaluating the proposed hybrid feature selection with random forest, the regressor model is instantiated with 550 estimator decision trees and 40 random states. Boosting algorithms are a subclass of ensemble algorithms and one of the most widely used algorithms in data science [42], converting weak learners to strong learners. Gradient boosting [43] sequentially trains several models, and every new model consistently reduces the loss function of the entire procedure utilizing the gradient descendant process. The principal objective of this algorithm is to build a new base learner, which is maximally correlated with the loss function's negative gradient, which is related to the entire ensemble.

The algorithm's predictive performance with the crop data set is observed using the proposed hybrid feature selection method. A machine learning model's efficiency is defined by assessing a model against various performance metrics or using various measures of evaluation. The various performance metrics examined for the assessment of the developed work are:

- Mean Square Error (MSE),
- Mean Absolute Error (MAE),
- Root Mean Squared Error (RMSE),
- Determination Coefficient (R^2),
- Mean Absolute Percentage Error (MAPE).

2.3.2. Metrics of Evaluation

Evaluation metrics define the performance of the model. A significant aspect of the evaluation measure is their capability to differentiate among the results of various learning models. The various performance metrics used for evaluation for this study are explained in this subsection.

Mean absolute error: Given an array of predictions, mean absolute error calculates the average importance of the errors [44]. It is the arithmetical mean of the absolute variation between the actual observation and the forecasted observation and is defined as follows in Equation (4).

$$MAE = \frac{1}{n} \sum\nolimits_{j=1}^{n} \left| y_j - y_j' \right| \tag{4}$$

Here n is defined as the size of the sample, y_j depicts the original target measure and y'_j defines the forecasted target measure.

Mean Squared Error: Mean squared error is a significant criterion to determine the estimator's performance. Further, this defines how close a regressor line is to the dataset points [45]. The formula for calculating mean squared error is defined as follows in Equation (5).

$$MSE = \frac{1}{n} \sum_{j=1}^{n} \left(y_j - y'_j\right)^2 \tag{5}$$

Root mean square error: It is an estimation of the residuals or forecasted error's standard deviation [46]. To be more precise, it explains how well the information is concentrated on the best fit line. The formula for calculating the root mean squared error is defined as follows in Equation (6).

$$RMSE = \sqrt{\sum_{j=1}^{n} \frac{(y_j - y'_j)^2}{n}} \tag{6}$$

Determination Coefficient: The statistical measure R-squared or determination coefficient is utilized to determine the accuracy of the fit of the regression framework [47]. To be more defined determination coefficient defines how the developed framework is superior to the baseline framework. It is defined in Equation (7) as follows:

$$R^2 = \left(\frac{n(\sum xy) - (\sum x)(\sum y)}{\sqrt{[n \sum x^2 - (\sum x)^2][n \sum y^2 - (\sum y)^2]}} \right)^2 \tag{7}$$

Mean Absolute Percentage Error: This represents how far the model's prediction deviates from its corresponding output. It is the average of the percentage errors. It is the sum of the individual absolute errors divided by each period individually. It is defined in Equation (8) as follows:

$$MAPE = \frac{1}{n} \sum_{j=1}^{n} \frac{y_j - y'_j}{y_j} \tag{8}$$

2.3.3. Cross-Validation

During the development of the deep learning models, the dataset is subjectively split into training and test set, where the maximum amount of information is taken as a training set. Despite the fact the test dataset is small, there exists a possibility of leaving out some critical data that may have upgraded the model. In addition, there exists a concern for significant data variance. To deal with this issue, k-fold cross-validation is utilized. It is a process that is utilized to evaluate the deep learning models by re-sampling the training data for upgrading the performance. Arbitrarily splitting a time series data for cross-validation may results in a temporal dependency problem, as there is a specific dependence on past observation, and also data leakage from response to the lag variable will undoubtedly occur. In such cases, forward-chaining cross-validation is performed for time series data. It performs by beginning with a small subset of information originally for training, predict for the following data, and deciding the precision of the predicted data. The same predicted data points are encased as a part of the subsequent training data subset, and the respective data points are predicted. For the proposed approach, five-fold forward chaining cross-validation is implemented. The cross-validation is implemented using python's Sklearn machine learning library. The results of cross-validation are tabulated in Table 2.

The dataset is normalized using min-max scaling. The training and the test sets are split using the train_test_split function of Sklearn. The size of the training and the test dataset is determined by the test_size parameter, which is set as 0.3 for the experiment, indicating that 70% of the data is reserved for training, and 30% of the data is fixed for testing. The best value of "K" for cross-validation is determined using the cross_val_score function. K defines the number of groups the given data is to be split into. The dataset is split into five subsets. The error metric determined is the R^2 score, which is affixed in every iteration and attains the optimal value determining the overall model accuracy.

Table 2. Dataset testing with 5-fold forward chaining cross-validation.

Model Data Subset	Training Data (In Years)	Test Data (In Years)	Correlation Measure Value	R^2 Score
1	1996–2000	2001–2003	0.77	0.82
2	1996–2003	2004–2006	0.84	0.86
3	1996–2006	2007–2009	0.61	0.70
4	1996–2009	2010–2012	0.76	0.84
5	1996–2012	2013–2016	0.86	0.89

The following section illustrates in detail the experimental framework of the CFS and RF-RFE based hybrid feature selection method for various machine learning frameworks such as gradient boosting, random forest, and decision tree.

3. Results

This section briefs about the experimental results obtained utilizing the proposed hybrid statistical feature extraction procedure over the existing machine learning algorithms. Feature selection is an automated process of choosing the most relevant attributes or significant features from a dataset that enhances a predictive model's performance. The proposed CFS filter RF-RFE wrapper hybrid statistical feature selection algorithm is tested by implementing it with the following machine learning algorithms, namely:

- Random forest
- Decision tree
- Gradient boosting.

Some of the machine learning algorithms comprise of a beneficial inbuilt method termed as feature importance. These methods are generally utilized for forecasting, for observing the most useful variables on the model. This information can be used to engineer new features, eliminate the noisy feature data, or to continue with the existing models. This measure is used as one of the reference values for evaluating the developed hybrid feature extraction framework. The evaluation of the model is done in three phases.

- In the first phase, the models are constructed using all the features or variables in the dataset, and the prediction results are validated using various statistical evaluation measures.
- In the second phase, models are constructed using the algorithm inbuilt 'feature_importances' methods, where only significant features alone are selected, and the prediction results are evaluated.
- In the third phase, the models are built utilizing the developed hybrid feature extraction method. The most significant features as per the proposed approach are selected, and the prediction results are evaluated.

3.1. Machine Learning Algorithms Performance Estimation in Terms of Evaluation Metrics

For validating the proposed hybrid feature selection method, a gradient boosting tree with 500 regression estimators, and a learning rate of 0.01 is constructed. The efficiency and accuracy measures for all the experimented models are determined with:

- All the features,
- Selective features obtained through inbuilt feature importance method and
- Features obtained through proposed hybrid feature selection methods.

The evaluation metrics are used to define the executing model's performance. The residuals which are obtained during the experiments are the variations between the predicted and actual values. By observing the residual spread magnitude, the efficiency and the precision of the model are defined. The evaluation measures obtained through the developed hybrid feature extraction process is found to be better than the other experimented methods, which are depicted in Tables 3–5.

Table 3. Performance metric evaluation of machine learning models with all the dataset features.

Algorithm Name	The Performance Measure with All Features in the Dataset				
	MAE	MSE	RMSE	R^2	MAPE (%)
Random Forest	0.203	0.082	0.286	0.53	21.3
Decision Tree	0.481	0.378	0.614	0.48	48
Gradient Boosting	0.334	0.208	0.456	0.41	33

Table 4. Performance metric evaluation of machine learning models with algorithm inbuilt feature importance method.

Algorithm Name	The Performance Measure with Algorithm Inbuilt Feature Importance Method				
	MAE	MSE	RMSE	R^2	MAPE (%)
Random Forest	0.202	0.08	0.284	0.59	20
Decision Tree	0.356	0.225	0.474	0.50	35
Gradient Boosting	0.323	0.196	0.443	0.45	31

Table 5. Performance metric evaluation of machine learning models with the proposed hybrid feature selection algorithm.

Algorithm Name	Performance Measure with CFS Filter and RF-RFE Wrapper Feature Selection Method				
	MAE	MSE	RMSE	R^2	MAPE (%)
Random Forest	0.194	0.07	0.265	0.67	19
Decision Tree	0.341	0.182	0.426	0.55	33
Gradient Boosting	0.306	0.187	0.433	0.48	29

3.2. ML Algorithms Performance Estimation in Terms of Accuracy

Assessment of the model's efficiency is a significant model enhancement procedure. It empowers in analyzing the ideal framework for representing the information and executing the information for fore coming iterations. Accuracy measure analyses the relativity of the forecasted value to the original value. It is the rate of accurately predicted model predictions. Table 6 represents the experimented model accuracy measures with all the features in the dataset, with particular features obtained through the algorithm in-built feature_importance method and with the features obtained through the proposed hybrid feature extraction procedure. The outcomes delineate the fact that the models perform with better accuracy when tested with the proposed hybrid CFS-filter RF-RFE wrapper feature selection algorithm.

Figure 4 graphically defines the performance metric results of the machine learning models with all the features, with selective features obtained through algorithm inbuilt feature_importance method and with the features obtained through the proposed hybrid feature selection method.

Figures 5–7 graphically represent the machine learning models' accuracy using all the features in the dataset for the specific features obtained through the feature importance method and the features obtained through the proposed hybrid feature selection method. Figure 5a depicts the accuracy measure of the gradient boosting algorithm using the features obtained by implementing the proposed CFS RF-RFE feature selection method, which is 85.41%. Figure 5b defines the 84.4% accuracy measure attained using the gradient boosting algorithm's inbuilt feature selection method. Figure 5c describes the accuracy measure attained by the gradient boosting algorithm using all the features of the dataset, which is 83.71%.

Table 6. Machine Learning Model Accuracy with the proposed hybrid feature selection algorithm.

Algorithm Name	Model Accuracy with All Features in the Dataset (%)	Model Accuracy with Algorithm Inbuilt Feature Importance Method (%)	Model Accuracy with CFS Filter and RF-RFE Wrapper Feature Selection Method (%)
Random Forest	90.84	90.94	91.23
Decision Tree	77.05	80.75	82.58
Gradient Boosting	83.71	84.4	85.41

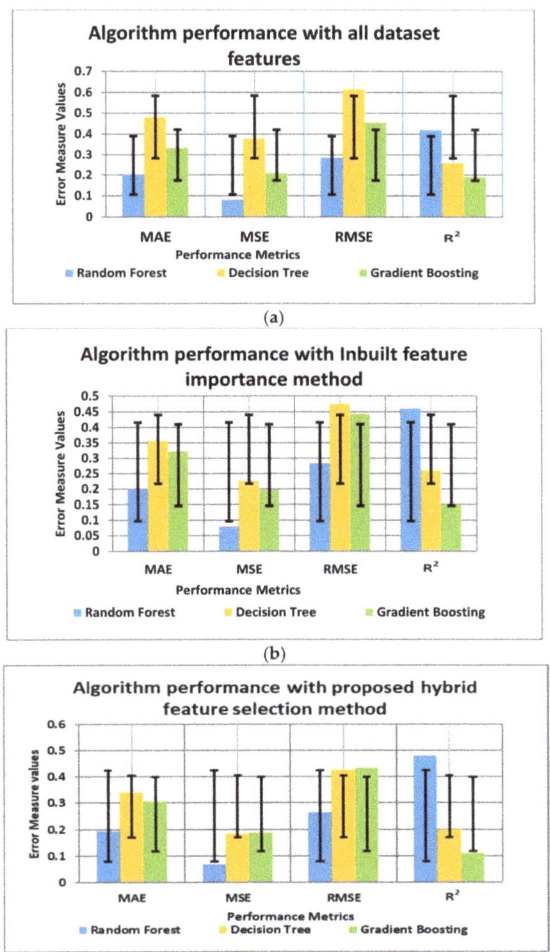

Figure 4. Performance metric results of the machine learning models with (**a**) all dataset features, (**b**) selective features through algorithm in-built feature importance method, (**c**) features obtained through the proposed hybrid feature selection method.

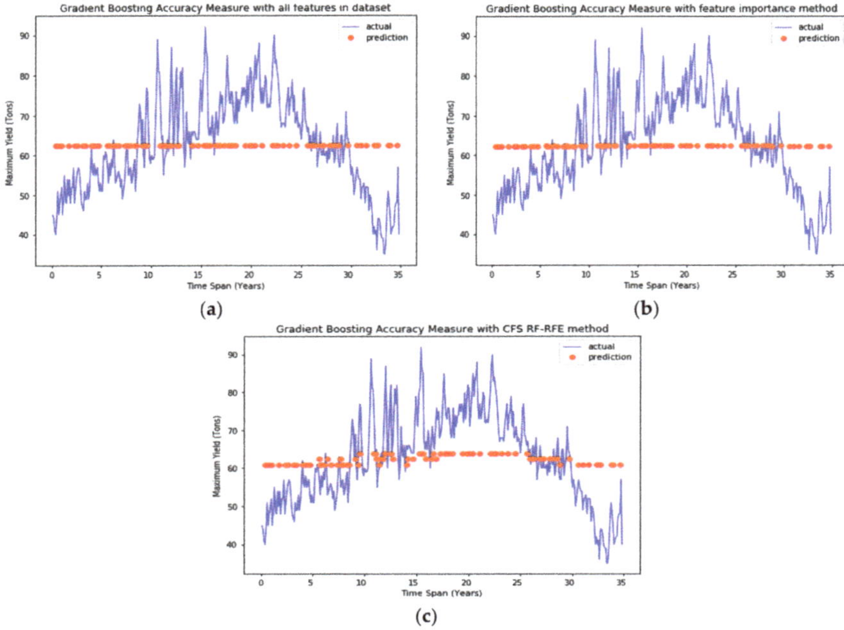

Figure 5. Gradient boosting model accuracy measure using: (**a**) proposed CFS RF-RFE feature selection method, (**b**) algorithm in-built feature importance method, (**c**) all the features in the dataset.

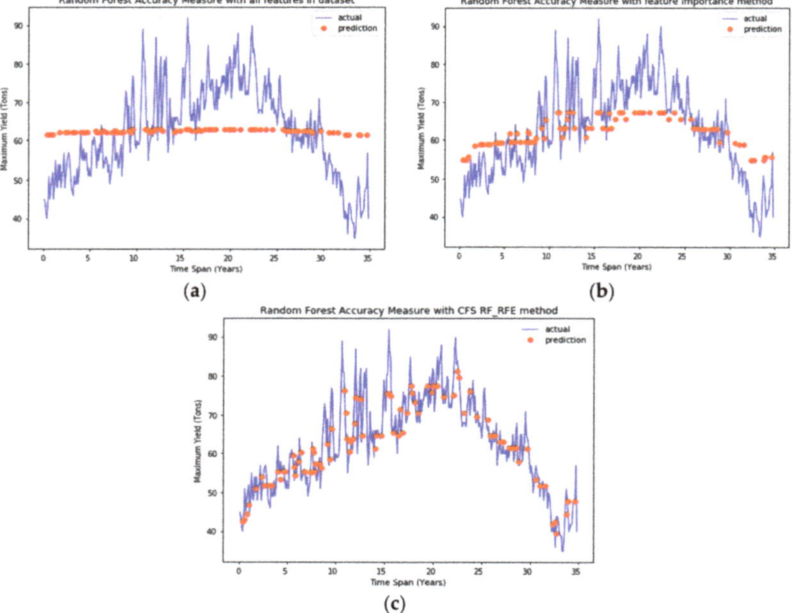

Figure 6. Random forest model accuracy measure using: (**a**) proposed CFS RF-RFE feature selection method, (**b**) algorithm in-built feature importance method, (**c**) all the features in the dataset.

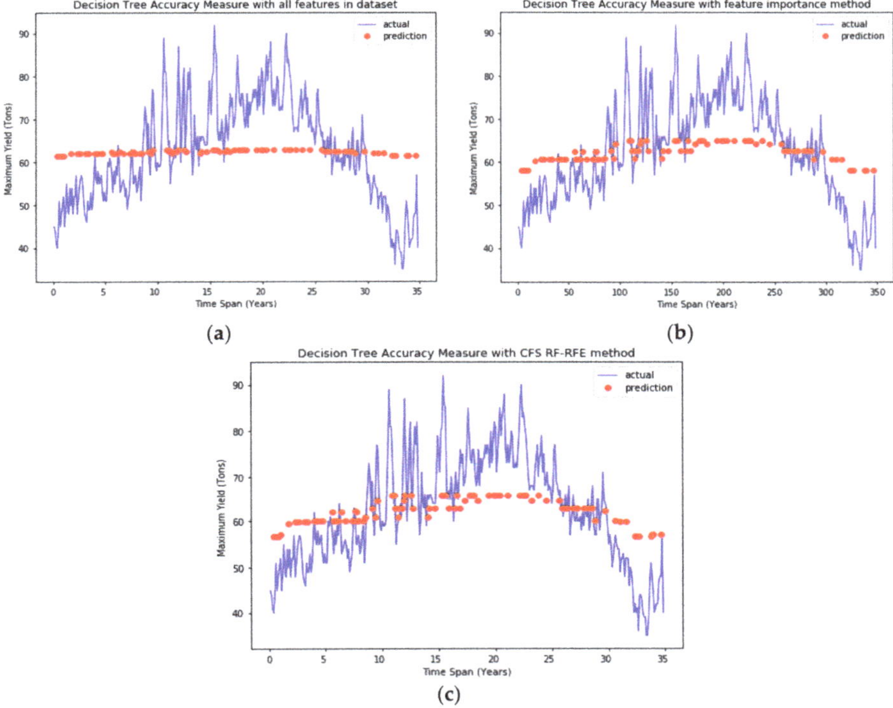

Figure 7. Decision-tree model accuracy measure using: (**a**) proposed CFS RF-RFE feature selection method, (**b**) algorithm in-built feature importance method, (**c**) all the features in the dataset.

Figure 6a depicts the accuracy measure of the random forest algorithm using the features obtained by implementing the proposed CFS RF-RFE feature selection method, which is 91.23%. Figure 6b defines the 90.94% accuracy measure attained using the random forest algorithm's inbuilt feature selection method. Figure 6c describes the accuracy measure attained by the random forest algorithm using all the features of the dataset, which is 90.84%.

Figure 7a depicts the accuracy measure of the decision tree algorithm using the features obtained by implementing the proposed CFS RF-RFE feature selection method, which is 82.58%. Figure 7b defines the 80.75% accuracy measure attained using the decision tree algorithm's inbuilt feature_selection method. Figure 7c describes the accuracy measure attained by the decision tree algorithm using all the features of the dataset, which is 77.05.

3.3. Regression Performance Analyses—Diagnostic Plots

Concerning validation of the regression results of the machine learning models, using the features from the developed hybrid feature extraction process, the regression diagnostic plots [48] are constructed. Regression diagnostic plots enhance the exploratory performance of the regression model through a set of accessible procedures to evaluate the legitimacy of the model. This assessment might be an investigation of the model's hidden statistical hypothesis or evaluation of model structure by considering plans that have less or diverse illustrative factors. They also assist in investigating subgroups of perceptions, searching for samples that are either ineffectively represented by the model, such as the outliers or those having a comparatively massive impact on the regression model forecasts. Residuals are generally leftovers of the resultant variable after fitting a model to data. However, residuals could indicate how ineffectively a model represents the data. They also uncover unexplained patterns in the information by the experimented model. Utilizing these statistics, we can review if the regression hypotheses are met and also enhance the model in an explorative manner. The diagnostic plots represent residuals in four different ways, which are presented in Figure 8. This section compiles the results obtained from the various machine learning models using the proposed hybrid CFS filter and RF-RFE wrapper feature selection method and also evaluated the forecasting models against various error measures. The following section discusses the results and future scope.

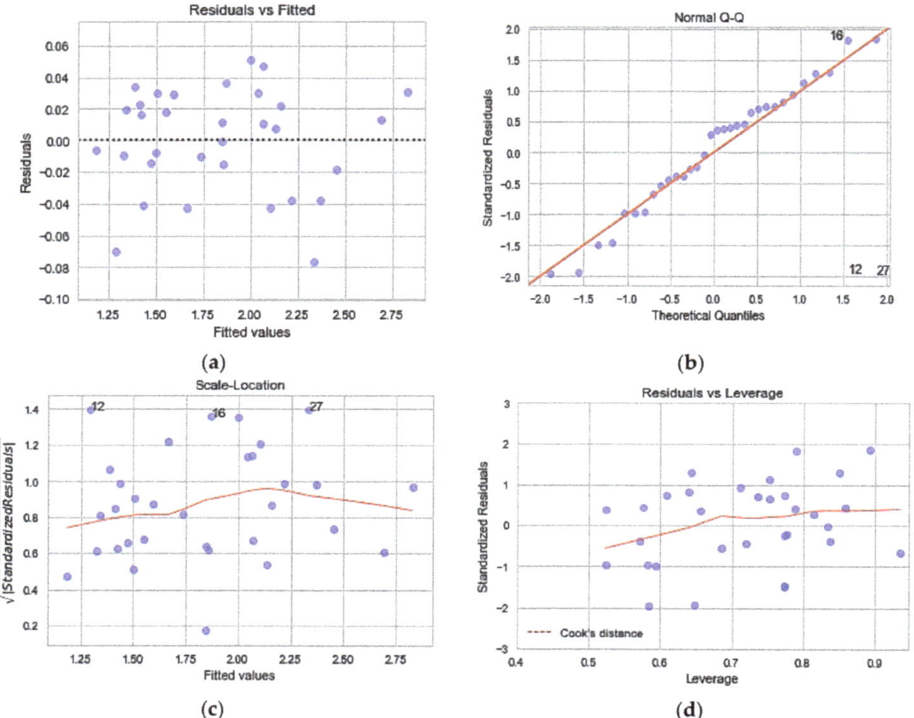

Figure 8. Residual diagnostic plots for regression analysis: (**a**) residuals vs. fitted plot, (**b**) normal Q–Q plot, (**c**) scale-location plot, (**d**) residuals vs. leverage plot.

4. Discussion

As a point, this section discusses the results obtained from the proposed model and also it briefs the future scope of the current study.

The residuals vs. fitted graph facilitates to observe the non-linear residual patterns. There can be a non-linear relation among the actual and the predictor variable, and such patterns could appear in these plots in case if the model does not catch them initially. The evenly distributed residuals about the horizontal line without any definite patterns demonstrate non-linear relationships. Figure 8a shows that the model data has met the linear regression assumptions well. There exists no distinctive data pattern referring to the linear spread of data.

The Q–Q plot analyses if the residuals follow a normal distribution with minimum deviation. It is better if the residuals interlined well on the straight line with minimum deviation. If the residual tends to possess a higher magnitude than expected from a normal distribution, then the p-values and confidence intervals fail to sufficiently account for complete data variability. Figure 8b depicts that the residuals are almost carefully plotted to the diagonal line indicating the normal distribution of the residual. The scale location plot observes if residuals are dispersed evenly within the range of the predictor. It enables us to verify the hypothesis of equal variance, i.e., homoscedasticity [49]. It is better to have a horizontal line with arbitrarily distributed points. Figure 8c indicates that the residuals are spread randomly. The residuals vs. leverage points enable to identify the most influential data. All the outliers cannot be influential i.e., they may or may not create much importance to the regression line. Cook's distance enables to create a margin. The outliers with the highest Cook's distance score or those occurring outside the cooks' distance are the influential outliers. Figure 8d delineates that there exist no influential outliers. Thus, the regression diagnostic plots define the enhanced model performance with the developed hybrid feature extraction process. A final reduced set of parameters after the proposed hybrid feature extraction process is listed in Table 7.

In addition to the feature extraction methods, an exploratory data analysis process, namely factor analysis is carried out to identify the influential variables or latent variables. It assists in data interpretation by decreasing the number of variables. Factor analysis is a linear statistical model that explains the variance among the observed variables, and the unobserved variables are called factors. Factors are associated with multiple observed variables comprising of similar response patterns. It is a process of investigating whether the variables of interest $x_1, x_2 \ldots x_n$ are linearly related to the minimal number of factors $f_1, f_2 \ldots f_n$. The primary objective of factor analysis is to minimize the observed variables and identify the unobserved variables. Moreover, this can be achieved by utilizing the factor extraction or factor rotation. Further, the proposed work factor analysis is implemented in python using the factor_analyzer package. Before implementing the factor analysis, it is necessary to assess the factorability of the dataset. Besides, this is determined using the Kaiser–Meyer–Olkin (KMO) test, which measures the data suitability for factor analysis. It defines the adequacy for the entire model and every observed variable. The KMO value varies from 0 to 1, where less than 0.1 is considered inadequate. The overall KMO for the crop dataset is observed to be 0.82, indicating its effectiveness in proceeding for factor analysis. The number of factors is defined based on the scree plot using the eigenvalues. The scree plot process defines a straight line for every factor and its eigenvalue. The variables whose eigenvalues are greater than one are considered as factors.

From the scree plot in Figure 9, it is observed that there are 32 factors whose eigenvalues are greater than 1. These factors define a cumulative variance of 57%. Factor analysis explores massive datasets and determines underlying associations, defining the group of inter-related variables. However, more than one interpretation can be made from the same data factors. This method generates 32 decisive factors that are close to the number of features determined by our proposed feature extraction method. The overall performance and the comparative results represent the fact that the proposed feature extraction process produces enhanced performance results than the other feature extraction process. Hence improves the predictive capability of the frameworks and their efficiency with lower error measures of MAE, MSE, and RMSE and higher value of determination coefficient. The diagnostic plots also result in delineating the enhanced exploratory performance of the models.

Table 7. List of dataset parameters after the proposed hybrid feature selection algorithm.

S. No	Final Set of Parameters	Description	Normal Acceptable Level	Units
1	QK$_2$O	Amount of potassium fertilizers utilized	15–20	Integer (kilogram/hectare)
2	Quantity rice	Total production of rice in the study area	2.37–2.5	Integer (ton/hectare)
3	QNitro	Amount of nitrogen fertilizers utilized	15–20	Integer (kilogram/hectare)
4	QP$_2$O$_5$	Amount of phosphorus fertilizers utilized	2–3	Integer (kilogram/hectare)
5	Vapor pressure	The pressure administered by water vapor with its condensed phase in thermodynamic equilibrium	23.8–41.2	Integer (hectopascal)
6	Gross cropped area	Total area planted to crops during all growing seasons of the year	195–220	Integer (hectare)
7	Net irrigated area	Total geographic area that has acquired irrigation throughout the year	80–110	Integer (hectare)
8	Ground frost frequency	Number of days referring to the condition when the upper layer soil temperature falls below the water freezing point	5–7	Integer (number of days)
9	Diurnal temperature range	Difference between the daily maximum and minimum temperature	90–130	Integer (°C)
10	Net cropped area	Total geographic area on which the crop has been planted at least once during a year	175–200	Integer (hectare)
11	Precipitation	Rain or water vapor condensation from the atmosphere	1400–1800	Integer (millimeter/year)
12	Gross irrigated area	Total area under crops that have received irrigation during all the growing seasons of the year.	90–116	Integer (hectare)
13	Average temperature	The average air temperature recorded in a particular location	21–25	Integer (°C)
14	Wet day frequency	The number of days in which a quantity of 0.2 mm or more of rain is observed.	45–55	Integer (number of days)
15	Area rice	Total area planted for rice crop	35–40	Integer (hectare)
16	Potential evapotranspiration	Quantity of evaporation occurring in an area in the presence of a sufficient water source	27–35	Integer (millimeter/day)
17	Reference crop Evapotranspiration	The evapotranspiration rate from a crop reference surface that is not short of water	25–30	Integer (millimeter/day)
18	Maximum temperature	The highest temperature of air recorded	21–37	Integer (°C)
19	Humidity	The quantity of water vapor in the atmosphere	60–80	Integer (percentage)
20	Wind speed	The rate at which the air blows	40–50	Integer (miles/hour)
21	Minimum temperature	The lowest temperature of air recorded	16–20	Integer (°C)
22	K soil	The potassium amount present in the soil	≥42	Integer (ton/hectare)
23	N soil	The nitrogen amount present in the soil	≥48	Integer (kilogram/hectare)
24	P soil	The phosphorus amount present in the soil	≥30	Integer (kilogram/hectare)
25	Aquifer area percentage	Percentage of the area covered by the aquifer. An aquifer is a body of permeable rock that can contain or transmit groundwater.	55–60	Integer (percentage)

Table 7. *Cont.*

S. No	Final Set of Parameters	Description	Normal Acceptable Level	Units
26	Aquifer permeability	A measure of the rock property which determines how easily water and other fluids can flow through it. Permeability depends on the extent to which pores are interconnected.	25–30	Integer (meter/day)
27	Pre–electrical conductivity	Average pre-monsoon electrical conductivity of groundwater	55–60	Integer (siemens/meter)
28	Post-electrical conductivity	Average post-monsoon electrical conductivity of groundwater	65–70	Integer (siemens/meter)
29	Groundwater pre-magnesium	Average pre-monsoon magnesium level in groundwater	68–73	Integer (milligram/litres)
30	Groundwater post-magnesium	Average post-monsoon magnesium level in groundwater	60–65	Integer (milligram/litres)
31	Groundwater pre-sodium	The average pre-monsoon sodium level in groundwater	150–170	Integer (milligram/litres)
32	Groundwater Post-Sodium	Average post-monsoon sodium level in groundwater	190–200	Integer (milligram/litres)
33	Groundwater pre-potassium	Average pre-monsoon potassium level in groundwater	15–20	Integer (milligram/litres)
34	Groundwater post-potassium	Average post-monsoon potassium level in groundwater	20–25	Integer (milligram/litres)
35	Groundwater pre-chloride	Average pre-monsoon chloride level in groundwater	320–325	Integer (milligram/litres)
36	Groundwater post-chloride	Average post-monsoon chloride level in groundwater	330–340	Integer (milligram/litres)
37	Yield rice	The total quantity of rice acquired	2.0–2.5	Integer (ton)
38	Soil PH	Acidity and alkalinity measure in the soil.	6–7	Integer
39	Topsoil depth	The outermost soil layer rich in microorganisms and organic matter	0.5–0.75	Integer (meters)

S. No—Serial Number.

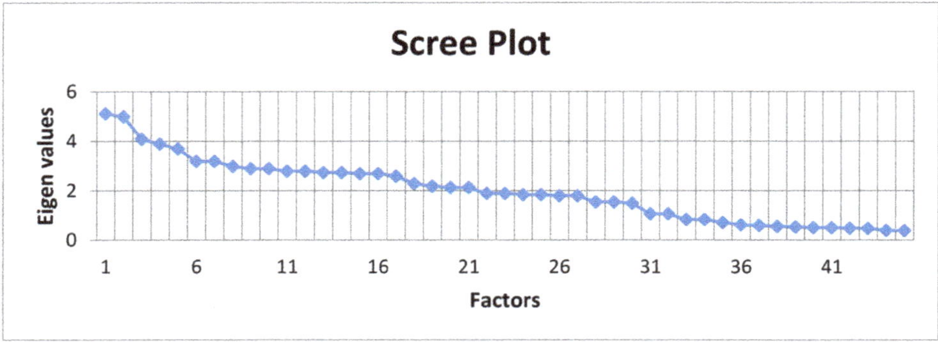

Figure 9. Scree plot defining the number of factors for factor analysis.

Future Scope

However, in this study, we have considered a varied set of parameters, including climatic, soil, and groundwater factors for forecasting crop yield. In the future, we can consider more extrinsic variables related to pesticides and weed infestations. Further improvement of the statistical filter-based selection using the adaptive prototype-based selection for improved performance can be considered. A more defined hybrid feature selection measure with the combination of deep learning wrappers with less complexity can be considered as an exciting area of research. As a part of the future work, we can consider building an ensemble feature selection model by combining the CFS Filter approach with the artificial neural network (ANN) based wrapper approach for agricultural applications. In another future work, we can also consider building a stacked generalization model-based crop yield forecasting model, using the features extracted through this proposed approach. Further, this stacked generalization model can be built by using ANN as the meta-learner. The following section concludes the paper.

5. Conclusions

Agriculture is a predominant sector among the most arduous departments incorporating the outcome of the analytical evaluation. Undoubtedly within an explicit sector, circumstances are consistently varying, starting with one sector to another. There exists unstable weather, diverse soil characteristics, persisting crop diseases, and pest infestations that influence crop yield and precision agriculture. There is an overwhelming capacity for machine learning to reform agribusiness by integrating various factors to forecast yield. Machine learning models secure a high degree to analyze the factual information, translate the data achieved, giving more in-depth knowledge into the process. To streamline the predictive model's learning process and for the efficient representation of the dataset, feature selection using various statistical measures is a crucial and significant stage. In this paper, a hybrid feature extraction process to address the feature extraction problem in machine learning models is proposed. A real-time dataset of soil, water, and climatic parameters from the Indian water portal and directorate of rice development Patna is used for the current study. The models are constructed in predicting the paddy crop yield for the interesting study area based on the climate, soil, and hydrochemical properties of groundwater. A list of 45 features was considered for model construction. Out of them, the most significant features for foreseeing the yield of crops in an interesting study area is determined using a hybrid feature extraction strategy. The proposed hybrid statistical feature extraction method is a mixture of CFS and RF-RFE wrapper, respectively. The filter method is initially implemented using correlation measures to eliminate the superfluous and non-essential features, which results in a reduced subgroup of features. These essential features obtained can be subjected to the construction of an intelligent agrarian model for the crop prediction procedure.

The advantage of the CFS filter among the other filter methods is the significantly shorter computation time. A wrapper method is then enforced on the reduced subgroup of features to find the feature set with high predictive accuracy. One of the essential highlights of the RF-RFE wrapper is that it does not need any fine-tuning to obtain competing results. Experimental results also confirm that the developed hybrid feature extraction method is superior to the other existing inbuilt feature selector methods. In addition, the efficiency of the results with fewer error measures shows improved prediction accuracy of the machine learning models.

Author Contributions: Conceptualization, D.R.V.P.M.; Funding acquisition, C.-Y.C.; Investigation, K.S.; Methodology, D.E. and C.-Y.C. All authors have read and agreed to the published version of the manuscript.

Funding: This research was partially funded by the "Intelligent Recognition Industry Service Research Center" from The Featured Areas Research Center Program within the framework of the Higher Education Sprout Project by the Ministry of Education (MOE) in Taiwan. Grant number: N/A and the APC was funded by the aforementioned Project.

Acknowledgments: We thank the India water portal for providing the meteorological data relevant to climatic factors from their MET data tool. The MET data tool provides district wise monthly and the annual mean of each metrological indicator values. We also thank the Joint Director of Agriculture, Vellore, Tamil Nadu, India, for providing the details regarding the soil and groundwater properties for the respective village blocks.

Conflicts of Interest: The authors declare that they have no conflict of interest.

References

1. Hamzeh, S.; Mokarram, M.; Haratian, A.; Bartholomeus, H.; Ligtenberg, A.; Bregt, A.K. Feature selection as a time and cost-saving approach for land suitability classification (Case Study of Shavur Plain, Iran). *Agriculture* **2016**, *6*, 52. [CrossRef]

2. Monzon, J.P.; Calviño, P.A.; Sadras, V.O.; Zubiaurre, J.B.; Andrade, F.H. Precision agriculture based on crop physiological principles improves whole-farm yield and profit: A case study. *Eur. J. Agron.* **2018**, *99*, 62–71. [CrossRef]

3. Rehman, T.U.; Mahmud, M.S.; Chang, Y.K.; Jin, J.; Shin, J. Current and future applications of statistical machine learning algorithms for agricultural machine vision systems. *Comput. Electron. Agric.* **2019**, *156*, 585–605. [CrossRef]

4. Chlingaryan, A.; Sukkarieh, S.; Whelan, B. Machine learning approaches for crop yield prediction and nitrogen status estimation in precision agriculture: A review. *Comput. Electron. Agric.* **2018**, *151*, 61–69. [CrossRef]

5. Elavarasan, D.; Vincent, D.R.; Sharma, V.; Zomaya, A.Y.; Srinivasan, K. Forecasting yield by integrating agrarian factors and machine learning models: A survey. *Comput. Electron. Agric.* **2018**, *155*, 257–282. [CrossRef]

6. Cisternas, I.; Velásquez, I.; Caro, A.; Rodríguez, A. Systematic literature review of implementations of precision agriculture. *Comput. Electron. Agric.* **2020**, *176*, 105626. [CrossRef]

7. Saikai, Y.; Patel, V.; Mitchell, P.D. Machine learning for optimizing complex site-specific management. *Comput. Electron. Agric.* **2020**, *174*, 105381. [CrossRef]

8. Guyon, I.; Elisseeff, A. An introduction to variable and feature selection. *J. Mach. Learn. Res.* **2003**, *3*, 1157–1182.

9. Liu, J.; Lin, Y.; Lin, M.; Wu, S.; Zhang, J. Feature selection based on quality of information. *Neurocomputing* **2017**, *225*, 11–22. [CrossRef]

10. Chandrashekar, G.; Sahin, F. A survey on feature selection methods. *Comput. Electr. Eng.* **2014**, *40*, 16–28. [CrossRef]

11. Cai, J.; Luo, J.; Wang, S.; Yang, S. Feature selection in machine learning: A new perspective. *Neurocomputing* **2018**, *300*, 70–79. [CrossRef]

12. Bommert, A.; Sun, X.; Bischl, B.; Rahnenführer, J.; Lang, M. Benchmark for filter methods for feature selection in high-dimensional classification data. *Comput. Stat. Data. Anal.* **2020**, *143*, 106839. [CrossRef]

13. Macedo, F.; Oliveira, M.R.; Pacheco, A.; Valadas, R. Theoretical foundations of forward feature selection methods based on mutual information. *Neurocomputing* **2019**, *325*, 67–89. [CrossRef]

14. Mielniczuk, J.; Teisseyre, P. Stopping rules for mutual information-based feature selection. *Neurocomputing* **2019**, *358*, 255–274. [CrossRef]

15. Kohavi, R.; John, G.H. Wrappers for feature subset selection. *Artif. Intell.* **1997**, *97*, 273–324. [CrossRef]

16. Chen, G.; Chen, J. A novel wrapper method for feature selection and its applications. *Neurocomputing* **2015**, *159*, 219–226. [CrossRef]

17. Jin, C.; Jin, S.W.; Qin, L.N. Attribute selection method based on a hybrid BPNN and PSO algorithms. *Appl. Soft Comput.* **2012**, *12*, 2147–2155. [CrossRef]

18. Wang, F.; Liang, J. An efficient feature selection algorithm for hybrid data. *Neurocomputing* **2016**, *193*, 33–41. [CrossRef]

19. Pourpanah, F.; Lim, C.P.; Wang, X.; Tan, C.J.; Seera, M.; Shi, Y. A hybrid model of fuzzy min–max and brain storm optimization for feature selection and data classification. *Neurocomputing* **2019**, *333*, 440–451. [CrossRef]

20. Holzman, M.E.; Carmona, F.; Rivas, R.; Niclòs, R. Early assessment of crop yield from remotely sensed water stress and solar radiation data. *ISPRS J. Photogramm. Remote. Sens.* **2018**, *145*, 297–308. [CrossRef]

21. Helman, D.; Lensky, I.M.; Bonfil, D.J. Early prediction of wheat grain yield production from root-zone soil water content at heading using Crop RS-Met. *Field Crop. Res.* **2019**, *232*, 11–23. [CrossRef]

22. Ogutu, G.E.O.; Franssen, W.H.P.; Supit, I.; Omondi, P.; Hutjes, R.W.A. Probabilistic maize yield prediction over East Africa using dynamic ensemble seasonal climate forecasts. *Agric. Meteorol.* **2018**, *250–251*, 243–261. [CrossRef]

23. Chatterjee, S.; Dey, N.; Sen, S. Soil moisture quantity prediction using optimized neural supported model for sustainable agricultural applications. *Sustain. Comput. Inform. Syst.* **2018**. [CrossRef]

24. Dash, Y.; Mishra, S.K.; Panigrahi, B.K. Rainfall prediction for the Kerala state of India using artificial intelligence approaches. *Comput. Electr. Eng.* **2018**, *70*, 66–73. [CrossRef]

25. Sharif, M.; Khan, M.A.; Iqbal, Z.; Azam, M.F.; Lali, M.I.U.; Javed, M.Y. Detection and classification of citrus diseases in agriculture based on optimized weighted segmentation and feature selection. *Comput. Electron. Agric.* **2018**, *150*, 220–234. [CrossRef]

26. Jiang, Y.; Li, C. mRMR-based feature selection for classification of cotton foreign matter using hyperspectral imaging. *Comput. Electron. Agric.* **2015**, *119*, 191–200. [CrossRef]

27. Daassi-Gnaba, H.; Oussar, Y.; Merlan, M.; Ditchi, T.; Géron, E.; Holé, S. Wood moisture content prediction using feature selection techniques and a kernel method. *Neurocomputing* **2017**, *237*, 79–91. [CrossRef]

28. Qian, W.; Shu, W. Mutual information criterion for feature selection from incomplete data. *Neurocomputing* **2015**, *168*, 210–220. [CrossRef]

29. Shekofteh, H.; Ramazani, F.; Shirani, H. Optimal feature selection for predicting soil CEC: Comparing the hybrid of ant colony organization algorithm and adaptive network-based fuzzy system with multiple linear regression. *Geoderma* **2017**, *298*, 27–34. [CrossRef]

30. Ghosh, A.; Datta, A.; Ghosh, S. Self-adaptive differential evolution for feature selection in hyperspectral image data. *Appl. Soft. Comput.* **2013**, *13*, 1969–1977. [CrossRef]

31. Sadr, S.; Mozafari, V.; Shirani, H.; Alaei, H.; Pour, A.T. Selection of the most important features affecting pistachio endocarp lesion problem using artificial intelligence techniques. *Sci. Hortic.* **2019**, *246*, 797–804. [CrossRef]

32. Kohavi, R.; John, G.H. Wrapper Approach. In *Feature Extraction, Construction and Selection*; Liu, H., Motoda, H., Eds.; Springer US: New York, NY, USA, 1998; Volume 453.

33. Robert, H.M. Methods for aggregating opinions. In *Decision Making and Change in Human Affairs*; Jungermann, H., De Zeeuw, G., Eds.; Springer: Dordrecht, The Netherlands, 1977; Volume 16.

34. Isabelle, G.; Jason, W.; Stephen, B. Vladimir vapnik gene selection for cancer classification using support vector machines. *Mach. Learn.* **2002**, *46*, 389–422. [CrossRef]

35. Elavarasan, D.; Vincent, P.M.D. Crop Yield Prediction Using Deep Reinforcement Learning Model for Sustainable Agrarian Applications. *IEEE Access* **2020**, *8*, 86886–86901. [CrossRef]

36. Park, S.; Im, J.; Jang, E.; Rhee, J. Drought assessment and monitoring through blending of multi-sensor indices using machine learning approaches for different climate regions. *Agric. Meteorol.* **2016**, *216*, 157–169. [CrossRef]

37. Elavarasan, D.; Vincent, D.R. Reinforced XGBoost machine learning model for sustainable intelligent agrarian applications. *J. Intell. Fuzzy. Syst.* **2020**, pre-press. [CrossRef]

38. Vanli, N.D.; Sayin, M.O.; Mohaghegh, M.; Ozkan, H.; Kozat, S.S. Nonlinear regression via incremental decision trees. *Pattern Recognit.* **2019**, *86*, 1–13. [CrossRef]
39. Prasad, R.; Deo, R.C.; Li, Y.; Maraseni, T. Soil moisture forecasting by a hybrid machine learning technique: ELM integrated with ensemble empirical mode decomposition. *Geoderma* **2018**, *330*, 136–161. [CrossRef]
40. Fratello, M.; Tagliaferri, R. Decision trees and random forests. In *Encyclopedia of Bioinformatics and Computational Biology*; Academic Press: Cambridge, MA, USA, 2019; pp. 374–383.
41. Herold, N.; Ekström, M.; Kala, J.; Goldie, J.; Evans, J.P. Australian climate extremes in the 21st century according to a regional climate model ensemble: Implications for health and agriculture. *Weather Clim. Extrem.* **2018**, *20*, 54–68. [CrossRef]
42. Kari, D.; Mirza, A.H.; Khan, F.; Ozkan, H.; Kozat, S.S. Boosted adaptive filters. *Digit. Signal Process.* **2018**, *81*, 61–78. [CrossRef]
43. Friedman, J.H. Stochastic gradient boosting. *Comput. Stat. Data Anal.* **2002**, *38*, 367–378. [CrossRef]
44. Ali, M.; Deo, R.C.; Downs, N.J.; Maraseni, T. Multi-stage committee based extreme learning machine model incorporating the influence of climate parameters and seasonality on drought forecasting. *Comput. Electron. Agric.* **2018**, *152*, 149–165. [CrossRef]
45. Deepa, N.; Ganesan, K. Hybrid Rough Fuzzy Soft classifier based Multi-Class classification model for Agriculture crop selection. *Soft Comput.* **2019**, *23*, 10793–10809. [CrossRef]
46. Torres, A.F.; Walker, W.R.; McKee, M. Forecasting daily potential evapotranspiration using machine learning and limited climatic data. *Agric. Water Manag.* **2011**, *98*, 553–562. [CrossRef]
47. Rousson, V.; Goşoniu, N.F. An R-square coefficient based on final prediction error. *Stat. Methodol.* **2007**, *4*, 331–340. [CrossRef]
48. Ferré, J. Regression diagnostics. In *Comprehensive Chemometrics*; Tauler, R., Walczak, B., Eds.; Elsevier: Amsterdam, The Netherlands, 2009; pp. 33–89.
49. Srinivasan, R.; Lohith, C.P. Main study—Detailed statistical analysis by multiple regression. In *Strategic Marketing and Innovation for Indian MSMEs*; India Studies in Business and Economics; Springer: Berlin/Heidelberg, Germany, 2017; pp. 69–92.

 *agriculture*

Article

Crop Growth Stage GPP-Driven Spectral Model for Evaluation of Cultivated Land Quality Using GA-BPNN

Mingbang Zhu [1], Shanshan Liu [1], Ziqing Xia [1], Guangxing Wang [2], Yueming Hu [1,3,4,5,]* and Zhenhua Liu [1]

[1]   College of Natural Resources and Environment, South China Agricultural University, Guangzhou 510642, China; zhumb90@stu.scau.edu.cn (M.Z.); ShanshanL@stu.scau.edu.cn (S.L.); xzq@stu.scau.edu.cn (Z.X.); zhenhua@scau.edu.cn (Z.L.)

[2]   Department of Geography and Environmental Resources, Southern Illinois University Carbondale (SIUC), Carbondale, IL 62901, USA; gxwang@siu.edu

[3]   Guangdong Province Engineering Research Center for Land Information Technology, South China Agricultural University, Guangzhou 510642, China

[4]   Key Laboratory of Construction Land Transformation, Ministry of Land and Resources, South China Agricultural University, Guangzhou 510642, China

[5]   Guangdong Provincial Key Laboratory of Land Use and Consolidation, South China Agricultural University, Guangzhou 510642, China

*   Correspondence: ymhu@scau.edu.cn; Tel.: +86-020-8528-8307

Received: 28 June 2020; Accepted: 30 July 2020; Published: 1 August 2020

Abstract: Rapid and accurate evaluation of cultivated land quality (CLQ) using remotely sensed images plays an important role for national food security and social stability. Current approaches for evaluating CLQ do not consider spectral response relationships between CLQ and spectral indicators based on crop growth stages. This study aimed to propose an accurate spectral model to evaluate CLQ based on late rice phenology. In order to increase the accuracy of evaluation, the Empirical Bayes Kriging (EBK) interpolation was first performed to scale down gross primary production (GPP) products from a 500 m spatial resolution to 30 m. As an indicator, the ability of MODIS-GPPs from critical growth stages (tillering, jointing, heading, and maturity stages) was then investigated by combining Pearson correlation analysis and variance inflation factor (VIF) to select the phases of CLQ evaluation. Finally, a linear Partial Least Squares Regression (PLSR) and two nonlinear models, including Support Vector Regression (SVR) and Genetic Algorithm-Based Back Propagation Neural Network (GA-BPNN), were driven to develop an accurate spectral model of evaluating CLQ based on MODIS-GPPs. The models were tested and compared in the Conghua and Zengcheng districts of Guangzhou City, Guangdong, China. The results showed that based on field measured GPP data, the validation accuracy of 30 m spatial resolution MODIS GPP products with a root mean square error (RMSE) of 7.43 and normalized RMSE (NRMSE) of 1.59% was higher than that of the 500 m MODIS GPP products, indicating that the downscaled 30 m MODIS GPP products by EBK were more appropriate than the 500 m products. Compared with PLSR ($R^2 = 0.38$ and RMSE = 87.97) and SVR ($R^2 = 0.64$ and RMSE = 64.38), the GA-BPNN model ($R^2 = 0.69$ and RMSE = 60.12) was more accurate to evaluate CLQ, implying a non-linear relationship of CLQ with the GPP spectral indicator. This is the first study to improve the accuracy of estimating CLQ using the rice growth stage GPP-driven spectral model by GA-BPNN and can thus advance the literature in this field.

Keywords: CLQ; GA-BPNN; GPP-driven spectral model; rice phenology; EBK

1. Introduction

Cultivated land quality (CLQ) has significant influence on agricultural production and resident living [1–3]. The CLQ often changes dramatically under conditions of human disturbances or environments [4]. Thus, rapid and accurate quantification of CLQ is critical. Traditionally, the assessment of CLQ is usually conducted using field measurements, which is time-consuming and costly. More importantly, this method lacks the ability to generate spatial distributions of CLQ [5–7]. Using remotely sensed data offers the potential of obtaining accurate and spatially explicit estimates of CLQ with low cost and has attracted the attention of scholars [8–12].

Current studies on CLQ evaluation using satellite imagery can be divided into two categories: traditional CLQ evaluation methods and pressure-state-response (PSR) based approaches. In the former group of CLQ evaluation methods, remote sensing data were only utilized to obtain some traditional indicators of CLQ, such as soil properties. One typical example is the study of Yang et al. (2012) in which Landsat TM images were used to derive soil organic matter, soil acidity, soil texture, and then generate the estimates of CLQ based on gradation regulations on the quality of farmland in China [13]. Instead of soil fertilizer variables, Zhao et al. (2012) utilized normalized difference vegetation index from Landsat TM imagery to evaluate CLQ [14]. However, the evaluation efficiency of CLQ using satellite image-driven evaluation methods is limited because the use of the methods is dependent on field measurements.

While, of the PSR based methods, CLQ is directly evaluated using remote sensing spectral indicators. For example, Liu et al. (2010) developed a linear model for evaluating CLQ based on predictors, including slope, sandy area ratio in a pixel, and modified soil-adjusted vegetation index [4,15]. Liu et al. (2019) generated the spatial distribution of CLQ estimates based on the Genetic Algorithm-Based Back Propagation Neural Network (GA-BPNN) model. The authors utilized five remote sensing data derived predictors, including Slope, Vegetation Index, Temperature Vegetation Dryness Index, Road Accessibility, and Patch Fractal Dimension, and found that CLQ was significantly and nonlinearly correlated with the spectral predictors [16]. Xie et al. (2018) developed a frequent pattern-growth algorithm for improving the evaluation efficiency of CLQ [17]. At present, there have been no reports about CLQ evaluation using gross primary production (GPP). Some scholars have used GPP to evaluate cultivated land productivity. Ma et al. (2018) explored the estimation of cultivated land productivity using the mean GPP from 2000 to 2018 and analyzed the change trend and amplitude of cultivated land productivity, implying that GPP provided the potential to evaluate CLQ [18]. Although the studies demonstrated the possibility of rapidly evaluating CLQ, the spectral responses between CLQ and remote sensing indicators were ignored. Moreover, the prediction accuracy of CLQ is affected by the selected spectral indicators in the PSR framework due to the limitations of image spatial and spectral resolutions.

This study aimed to propose an accurate spectral response model of CLQ based on GPP spectral indicators from the MODIS-GPPs from 2011 to 2015 at different growth stages of late rice and the corresponding temporal CLQ for mapping CLQ. Here, CLQ is defined as the farmland utilization quality grade and represents the degree of anthropogenic use and natural conditions of cultivated land [19]. The Empirical Bayes Kriging (EBK) interpolation was first employed to perform spatial downscaling transformation of the MODIS GPP images from 500 m spatial resolution to 30 m. The accurate spectral model based on GA-BPNN was then developed and validated by comparing it with a linear partial least squares regression (PLSR) and a non-linear support vector regression (SVR). The comparison of the models was made in the study area mentioned next. It is expected that the study can offer a powerful tool to rapidly and accurately estimate CLQ.

2. Materials and Methods

2.1. Study Areas

The study area (Figure 1) is situated in the Conghua and Zengcheng District of Guangzhou, Guangdong of China (22°26′–23°56′ N, 112°57′–114°03′ E). The annual average temperature is 21 °C and the annual average precipitation is about 1900 mm, concentrating between April and September [20–22]. The cultivated land area of Guangzhou in 2015 was 13,485.99 hm^2 with an annual crop yield of 440,900 tons (referencing the Statistical Communiqué of Guangzhou on the 2015 National Economic and Social Development), of which the paddy field was 11,885.16 hm^2 with the average paddy stand size of about 0.25 hm^2, accounting for 87.91% of the total cultivated land area. The cultivated land is mainly concentrated in Conghua and Zengcheng Districts. Rice is the principal crop in the study area, with an annual double rotation system (early rice: March–June and late rice: August–November).

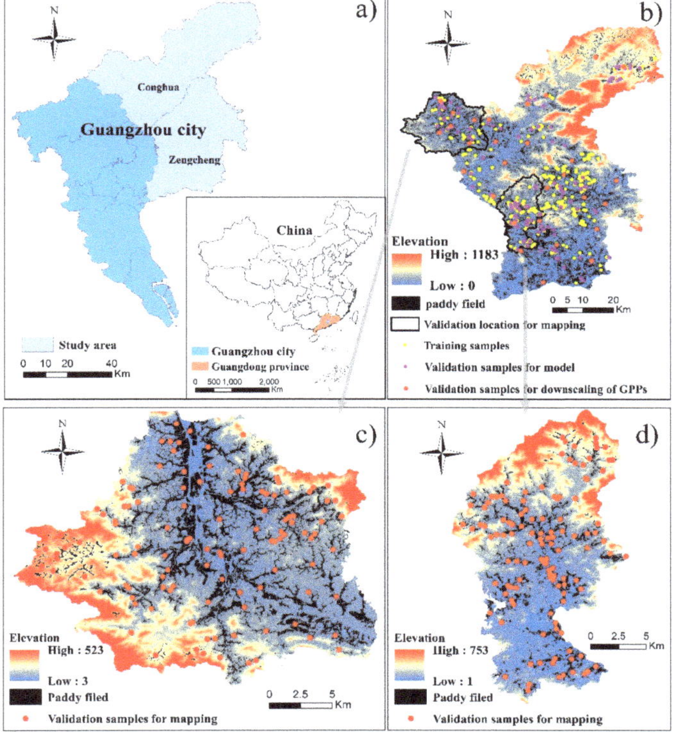

Figure 1. The study area location in Guangzhou City (**a**), and Conghua and Zengcheng District (**b**), respectively, with a total of 420 sample plots for cultivated land quality (CLQ) (training sample plots are in yellow and validation sample plots for the model in purple) and another set of 240 sample plots in red for validating multi-scale Moderate-resolution Imaging Spectroradiometer (MODIS) gross primary production (GPP) products; (**c**,**d**) the validation areas for mapping at Aotou Town of Conghua District and Zhongxin Town of Zengcheng District.

2.2. Data

In this study, to match the 500 m spatial resolution of MODIS GPP products, field sampling plots of 500 m by 500 m were designed and within each plot, biomass was taken at five locations with one placed at the plot center and the other four at the middle points from the plot center to the corners along the diagonal lines. The GPP was estimated with an empirical regression model [23]: $GPP = \frac{NPP}{0.524}$, where NPP was acquired from the following equation.

$$NPP = \frac{B \times \beta}{\alpha} \tag{1}$$

where B is the dry biomass obtained from each of the sampled plots at the heading stage and dried in a constant temperature drying oven at 110 °C. The α is the ratio of aboveground biomass to total biomass, and β is the percentage of carbon in the biomass. For cultivated land, the value of α is usually 0.8 [24], and β is 45% [25]. The rice samples were first destructively collected in the field sample plots and oven-dried at 110 °C for 50 min in the laboratory, and the temperature was then adjusted to 85 °C for 10 h until the sample weights did not change. The samples were finally taken out and weighed.

Moreover, a total of 660 sample data were extracted from the CLQ map obtained from the Guangzhou National Land Department and the CLQ values were derived using gradation regulations on farmland quality in China (Regulation for gradation on agriculture land quality GB/T 28407-2012) [26]. The plot sampled area is 30 m × 30 m, matching the 30 m spatial resolution of the downscaled MODIS products. The 660 samples were randomly divided into three groups: 294 samples in yellow for modeling (Figure 1b), 126 sample in purple for validation of the estimated CLQ (Figure 1b), and another dataset of 240 sample plots in red for assessing the accuracy of mapping CLQ at the regional scale (Figure 1c,d). In addition, 2011–2015 MODIS/Terra 8-day GPP products (MOD17A2H Version 6) at a spatial resolution of 500 m × 500 m were acquired from the Land Processes Distributed Active Archive Center (LP DAAC/NASA). The MODIS Re-projection Tool (MRT) was employed to convert the sinusoidal projection into the Albers Equal Area projection for the MODIS GPP products. The scaling factor of 0.1 was utilized to obtain the standard MODIS-GPP products [27].

According to the rice growth phases recommended by Ricepedia [28], the rice growth process can be characterized by five stages: seedling, tillering, jointing, heading, and maturity. In fact, the seedling stage was not taken into account because of the too small values of MODIS-GPPs detected and the impact of water on the spectral reflectance of rice. Thus, four growth stages were considered. Moreover, the dates of the acquired MODIS-GPP images with cloud cover less than 5% were consistent with the times of the four rice growth stages (Table 1) and the synchronous satellite-field experiment was carried out.

Table 1. Acquisition dates of MODIS-GPPs and corresponding with rice growth stages.

Growth Stage	Tillering Stage	Jointing Stage	Heading Stage	Maturity Stage
Acquisition date (m/d/y)	8/20/2011–8/27/2011	9/13/2011–9/20/2011	10/15/2011–10/22/2011	11/8/2011–11/15/2011
	8/19/2012–8/26/2012	9/12/2012–9/19/2012	10/14/2012–10/21/2012	11/7/2012–11/14/2012
	8/20/2013–8/27/2013	9/13/2013–9/20/2013	10/15/2013–10/22/2013	11/8/2013–11/15/2013
	8/20/2014–8/27/2014	9/13/2014–9/20/2014	10/15/2014–10/22/2014	11/8/2014–11/15/2014
	8/20/2015–8/27/2015	9/13/2015–9/20/2015	10/15/2015–10/22/2015	11/8/2015–11/15/2015

2.3. Methods

2.3.1. Downscaling of MODIS GPP Products Based on the EBK Interpolation

The EBK was used to downscale the MODIS GPP products from a 500 m spatial resolution to 30 m [29,30]. The EBK is superior compared with the conventional spatial downscaling methods that rely solely on the spectral data of images and do not take into account image texture and structure, as well as with the classical kriging methods that ignore the explanation of the error introduced by modeling semivariograms [31]. The EBK is a powerful non-stationary algorithm and divides an image into subsets and uses simulation to makes the process automatic for spatial interpolation [32]. In this method, the following Kriging Equation (2) was utilized to predict GPP values [30,33,34]:

$$Z_{(x_0)} = \sum_{i=1}^{n} \lambda_i \cdot Z_{(x_i)} + \sum_{i=1}^{n} s_i \cdot U_{(x_i)} \tag{2}$$

where Z_i, $(i = 1, \ldots, n)$ is the GPP value at location x_i, λ_i, $(i = 1,2,3 \ldots ,n)$, represents the kriging weight generated using the parameters of cross-variograms, and s_i, $(i = 1,2,3 \ldots ,n)$ is the kriging weight estimated on the basis of a cross-variogram between $Z_{(x_i)}$ and $U_{(x_i)}$. The n denotes the total number of observations. The variable $U_{(x)}$ was a standardized rank that was calculated as [30]:

$$U_{(x_i)} = \frac{R}{n} \tag{3}$$

where R represents a rank of the Rth order statistic of GPP on the land surface at location x_i. Downscaling the 500 m MODIS GPP products to the 30 m spatial resolution using EBK was performed using ArcGIS 10.3 Geostatistical Analyst (Environmental Systems Research Institute, Inc., Redlands, CA, USA).

2.3.2. Selecting the Phases of GPP

The important step in selecting the phases for CLQ evaluation was the determination of the relevant growth stage. In this study, Pearson product moment correlation that quantifies the linear relationship between two variables [35,36] was applied to obtain the spectral variables with the highest coefficients at the significance level of 0.05. The correlation coefficient is calculated as:

$$r_i = \frac{\sum_{n=1}^{N}\left(R_{ni} - \overline{R_i}\right)(y - \overline{y})}{\sqrt{\sum_{n=1}^{N}\left(R_{ni} - \overline{R_i}\right)^2 \sum_{n=1}^{N}(y - \overline{y})^2}}, \tag{4}$$

where r_i is the correlation coefficient between growth stage and CLQ, N is the total number of CLQ samples. R_{ni} is the ith growth stage of the nth CLQ sample, $\overline{R_i}$ is the average of the CLQ sample values in the ith growth stage, and y is the nth CLQ, $\overline{y}$ is the average value of CLQ.

Moreover, the Variance Inflation Factor (VIF) was applied to mitigate the collinearity among the GPP predictors, which is calculated as [37,38]:

$$\mathrm{VIF} = \frac{1}{1 - R_i^2} \tag{5}$$

where, R_i^2 is the determination coefficient between the ith predictor and the remaining independent variables. The larger the VIF, the greater the collinearity between the predictors. In general, the values from 0 to 10, 10 to 100, and equal to and greater than 100, respectively, imply no strong and serious collinearity.

2.3.3. Partial Least Squares Regression

In this study, to improve the evaluation of CLQ, GA-BPNN was proposed and compared with PLSR and SVR to predict CLQ using GPPs.

As a linear multivariate model, PLSR relates two data matrices, X and Y. Compared with traditional regression, however, PLSR can be used to analyze the predictors that have strong correlations [39,40], which is expressed as follows:

$$Y = X\beta + \varepsilon, \tag{6}$$

where Y is the response variable CLQ, X is the predictor GPP at different stages, β is the regression coefficient, and ε is the residual.

2.3.4. Support Vector Regression

The SVR is a widely used supervised learning method for solving the problem of regression fitting. Different from the traditional process from induction to deduction, SVR greatly simplifies the usual regression process by making efficient "transductive inference" from training samples to prediction [41–43]. Traditional regression algorithms use training samples to generate a model (a global trend) that is used to predict values of the dependent variable at unknown locations. In SVR, the values of 2011–2015 MODIS-GPPs as predictors x in this study are first plotted onto an m-dimensional feature space (m–number of predictors) and a linear model with its epsilon band-acceptable prediction surface is constructed in the feature space.

$$f(x) = \omega\varnothing(x) + b \tag{7}$$

where $f(x)$ presents CLQ estimates, $\varnothing(x)$ denotes the GPP set of nonlinear transformations, b is the error term, and ω is the weight coefficient. After preprocessing of data, the error term often has a zero mean and can be dropped. According to structural risk maximization principles, ω and b are calculated by following the objective function $R(x)$.

$$R(x) = \frac{1}{2}\|\omega\|^2 + \frac{1}{l}\sum_{i=1}^{l}\left|f(x_i) - y_i\right|_\varepsilon \tag{8}$$

where $\left|f(x_i) - y_i\right|_\varepsilon$ is the insensitive loss function, ε is the error tolerance of the insensitive loss function, l denotes the number of samples, $\|\omega\|^2$ reflects the flatness of in the m-dimensional space.

The model complexity can be simplified by minimizing $\|\omega\|^2$. By introducing (non-negative) slack variables $\xi_i, \xi_i^*, i = 1,\ldots,n$ and penalty factor (C) to derive the deviation of training samples outside the insensitive loss function, SVR is formulated by minimizing the following objective function:

$$R\left(\omega, \xi_i, \xi_i^*\right) = \tfrac{1}{2}\|\omega\|^2 + C\sum_{i=1}^{l}\left(\xi_i + \xi_i^*\right)$$
$$\text{s.t.} \begin{cases} y_i - \varnothing(x) - b \le \varepsilon + \xi_i^* \\ \omega\varnothing(x) + b - y_i \le \varepsilon + \xi_i^* \\ \xi_i, \xi_i^* \ge 0, i = 1,\ldots,n \end{cases} ,(i = 1,\ldots,n) \tag{9}$$

The above problem can be translated into the following dual problem:

$$\min_{\alpha_i^{(*)} \in R^{2l}} \tfrac{1}{2}\sum_{i,j=1}^{l}\left(\alpha_i - \alpha_i^*\right)\left(\alpha_j - \alpha_j^*\right)\left(\varnothing(x_i)\varnothing(x_j)\right) + \varepsilon\sum_{i=1}^{l}\left(\alpha_i^* + \alpha_i\right) - \sum_{i=1}^{l}y_i\left(\alpha_i^* - \alpha_i\right)$$
$$\text{s.t.} \begin{cases} \sum_{i=1}^{l}\left(\alpha_i - \alpha_i^*\right) = 0 \\ 0 \le \alpha_i^{(*)} \le C, i = 1,\ldots,l. \end{cases} \tag{10}$$

where $\alpha_i - \alpha_i^*$ is the transformation of the ω variable after introducing the Lagrange factor. The SVR function is obtained by solving the above problems:

$$f(x) = \omega\varnothing(x) + b = \sum_{i=1}^{l}\left(\alpha_i^* - \alpha_i\right)K(x_i, x) + b \tag{11}$$

where $K(x_i, x) = \varnothing(x_i)\varnothing(x)$ is kernel function.

2.3.5. Genetic Algorithm-Back Propagation Neural Network

The GA-BPNN model is developed by combining a back propagation neural network (BPNN) with the genetic algorithm optimization (GA). The BPNN has been widely used to find solutions for nonlinear relationships. The training samples are used to train the multi-layer BPNN using the error back propagation (BP) algorithm. During the process of modifying the weights, however, the standard BP algorithm ignores previous gradient direction, which often leads to the oscillation and slow convergence of the learning process [44]. The GA mimics biological evolution processes and has the capacity of finding global optimum solutions of the problems and thus can be utilized to optimize the thresholds and weights of the BPNN [45]. Therefore, the combination of BPNN and GA results in an integrated model that provides the potential of improving the efficiency and accuracy of the predictions. The flow chart of GA-BPNN with its structure [46] is shown in Figure 2.

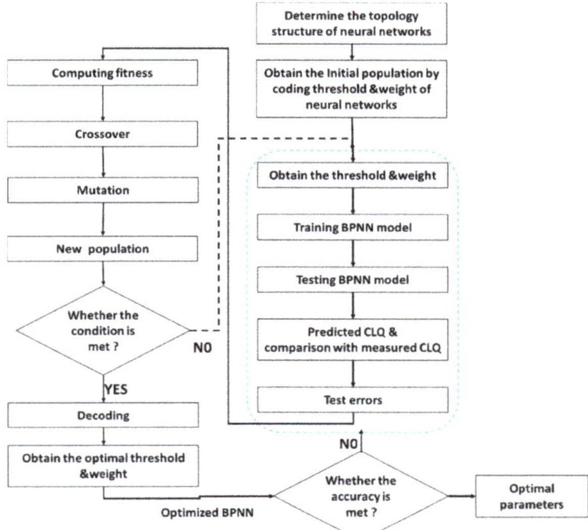

Figure 2. The Flow chart of GA-BPNN.

3. Results

3.1. Downscaling of MODIS GPPs by EBK Interpolation

To improve the accuracy of evaluating CLQ, 2011–2015 MODIS-GPPs with the 500 m spatial resolution were scaled downs to a 30 m spatial resolution using the EBK interpolation. Compared with the original standard MODIS-GPP for the same date, the downscaled MODIS-GPP on the 233rd to 289th day in 2013 shows more detailed information (Figure 3), indicating that the quality of the downscaled data is higher than the original standard data.

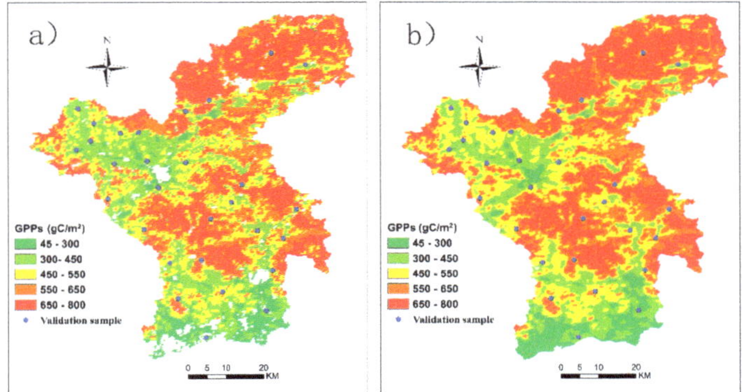

Figure 3. The spatial distribution maps of the cumulative GPP from the 233rd to 289th days in the study area in 2013: (**a**) 500 m original standard MODIS-GPPs and (**b**) 30 m MODIS-GPPs interpolated by the Empirical Bayes Kriging (EBK) method.

Table 2 shows the comparison results of MODIS-GPPs between the spatial resolutions of 500 m and 30 m based on the field observations (dry biomass) of 30 samples plots. The average MODIS-GPP was 475.09 g C/m^2 for the 500 m, 472.73 g C/m^2 for 30 m, and the average field observation was 465.49 g/m^2. Compared with the field observations, the 30 m spatial resolution MODIS GPP has a root mean square error (RMSE) of 7.43 and a normalized RMSE (NRMSE) of 1.59%, while the corresponding values for the 500 m MODIS GPP were 33.43 and 7.18%. The results imply that the downscaled MODIS-GPPs by EBK interpolation can reflect the productivity of cultivated land more accurately than the unscaled MODIS-GPPs.

Table 2. Comparison of the original 500 m spatial resolution MODIS-GPPs from the 289th day in 2013 with their downscaled 30 m products by the Empirical Bayes Kriging (EBK) method based on the dry biomass field observations of 30 sample plots.

Plot#	Field Observations	30 m MODIS-GPPs		500 m MODIS-GPPs	
		Estimates	Absolute Error (%)	Estimates	Absolute Error (%)
1	514.23	521.95	1.50	530.31	3.13
2	485.68	496.85	2.30	485.82	0.03
3	519.91	529.79	1.90	525.33	1.04
4	685.27	688.70	0.50	731.61	6.76
5	538.52	546.06	1.40	555.16	3.09
6	592.24	599.94	1.30	609.87	2.98
7	639.33	645.08	0.90	571.20	10.66
8	402.12	411.77	2.40	407.59	1.36
9	437.78	445.66	1.80	447.50	2.22
10	451.55	460.13	1.90	490.44	8.61
11	555.35	560.35	0.90	598.75	7.81
12	299.14	307.52	2.80	352.92	17.98
13	317.47	325.09	2.40	330.34	4.05
14	506.90	515.01	1.60	520.97	2.78
15	408.60	416.77	2.00	447.72	9.57
16	438.98	446.44	1.70	415.71	5.30
17	457.91	464.78	1.50	446.98	2.39
18	448.84	456.02	1.60	450.13	0.29
19	393.27	399.95	1.70	409.76	4.19
20	494.46	501.87	1.50	459.74	7.02
21	394.18	401.67	1.90	395.16	0.25
22	379.54	386.38	1.80	389.23	2.55
23	425.62	431.58	1.40	431.13	1.29
24	380.76	387.62	1.80	383.95	0.84
25	567.61	572.15	0.80	682.55	20.25
26	401.86	410.30	2.10	370.57	7.79
27	539.04	543.35	0.80	541.45	0.45
28	541.05	546.46	1.00	509.82	5.77
29	364.34	372.00	2.10	368.16	1.05
30	383.00	390.66	2.00	392.90	2.59
Mean	465.49	472.73	1.64	475.09	4.80
Stdev	91.77	91.00		97.57	
RMSE		7.43		33.43	
NRMSE (%)		1.59		7.18	

3.2. Model Comparison for CLQ Evaluation

In this study, to reduce the calculation burden, variance inflation factor (VIF) was used to perform the analysis of collinearity of the GPPs among four growth stages given a year for developing the models of evaluating CLQ based on 420 sample points. The results in Table 3 indicate that there is collinearity among the GPPs in the four growth stages. Therefore, all data at the four growth stages were selected to construct spectral models.

Table 3. The variance inflation factors (VIFs) of MODIS-GPPs at the key growth stages.

Years \ Growth Stages	Tillering	Jointing	Heading	Maturity
2011	1.971	1.981	4.611	2.874
2012	1.407	2.687	4.130	3.451
2013	1.274	4.092	7.679	4.468
2014	1.421	1.667	3.257	3.448
2015	2.073	1.699	2.655	1.026

In this study, three kinds of models including PLSR, SVR, and GA-BPNN were developed for comparing the evaluation accuracy of CLQ. For each of the years from 2011 to 2015, one PLSR model was obtained based on 294 training samples using CLQ as the response variable and GPPs at the four growth stages as the predictors. With the same training datasets for the years, the corresponding SVR and GA-BPNN models were constructed. The prediction accuracies of CLQ from all the models were assessed based on the values of RMSE, NRMSE and the coefficient of determination (R^2) between the estimated and observed CLQ values [45] according to the training and validation samples. The obtained 2011–2015 PLSR evaluation models are:

$$\hat{CLQ} = 2128.457 + 0.008 \times GPP_{Tillering/2011} + 9.388 \times GPP_{Jointing/2011} + 14.073 \times GPP_{Heading/2011} - 4.620 \times GPP_{Maturity/2011} \quad \left(R^2 = 0.38, P < 0.001 \right) \tag{12}$$

$$\hat{CLQ} = 2562.905 + 0.238 \times GPP_{Tillering/2012} + 0.428 \times GPP_{Jointing/2012} + 3.778 \times GPP_{Heading/2012} + 5.437 \times GPP_{Maturity/2012} \quad \left(R^2 = 0.39, P < 0.001 \right) \tag{13}$$

$$\hat{CLQ} = 2336.989 + 6.886 \times GPP_{Tillering/2013} + 6.834 \times GPP_{Jointing/2013} - 0.401 \times GPP_{Heading/2013} + 2.444 \times GPP_{Maturity/2013} \quad \left(R^2 = 0.40, P < 0.001 \right) \tag{14}$$

$$\hat{CLQ} = 2451.9366 + 0.586 \times GPP_{Tillering/2014} + 6.341 \times GPP_{Jointing/2014} + 0.516 \times GPP_{Heading/2014} + 6.911 \times GPP_{Maturity/2014} \quad \left(R^2 = 0.38, P < 0.001 \right) \tag{15}$$

$$\hat{CLQ} = 2328.035 + 3.458 \times GPP_{Tillering/2015} + 3.791 \times GPP_{Jointing/2015} + 5.946 \times GPP_{Heading/2015} - 4.830 \times GPP_{Maturity/2015} \quad \left(R^2 = 0.35, P < 0.001 \right) \tag{16}$$

For the development of SVR models, the support vector machine (SVM) was selected as epsilon-SVR, its loss function was set as 0.1, and the range of kernel parameter and penalty parameter was set as $(2^{-8}, 2^{8})$ [47]. Moreover, the obtained GA-BPNN models had a three-layer network and a hidden layer with 13 neuron nodes. A total of 1000 iterations was used with 10 maximum runs. Both learning rate and learning objective were 0.01. The mutation probability, crossover probability and population size were respectively 0.1, 0.3, and 10 [36]. The obtained models based on the 294 training samples are compared in Figure 4.

Based on the scattered graphs from the training samples in Figure 4, the points of estimated vs. observed CLQ are overall randomly distributed at both sides of the 1:1 lines. However, the accuracies of the predicted CLQ values vary greatly depending on the models. Overall, the estimates of CLQ from the PLSR models for 2011 to 2015 have smaller R^2 values and greater RMSE and NRMSE values, then the SVR models and the GA-BPNN models. This indicates that the GA-BPNN models performed best, implying that CLQ was nonlinearly correlated with GPPs.

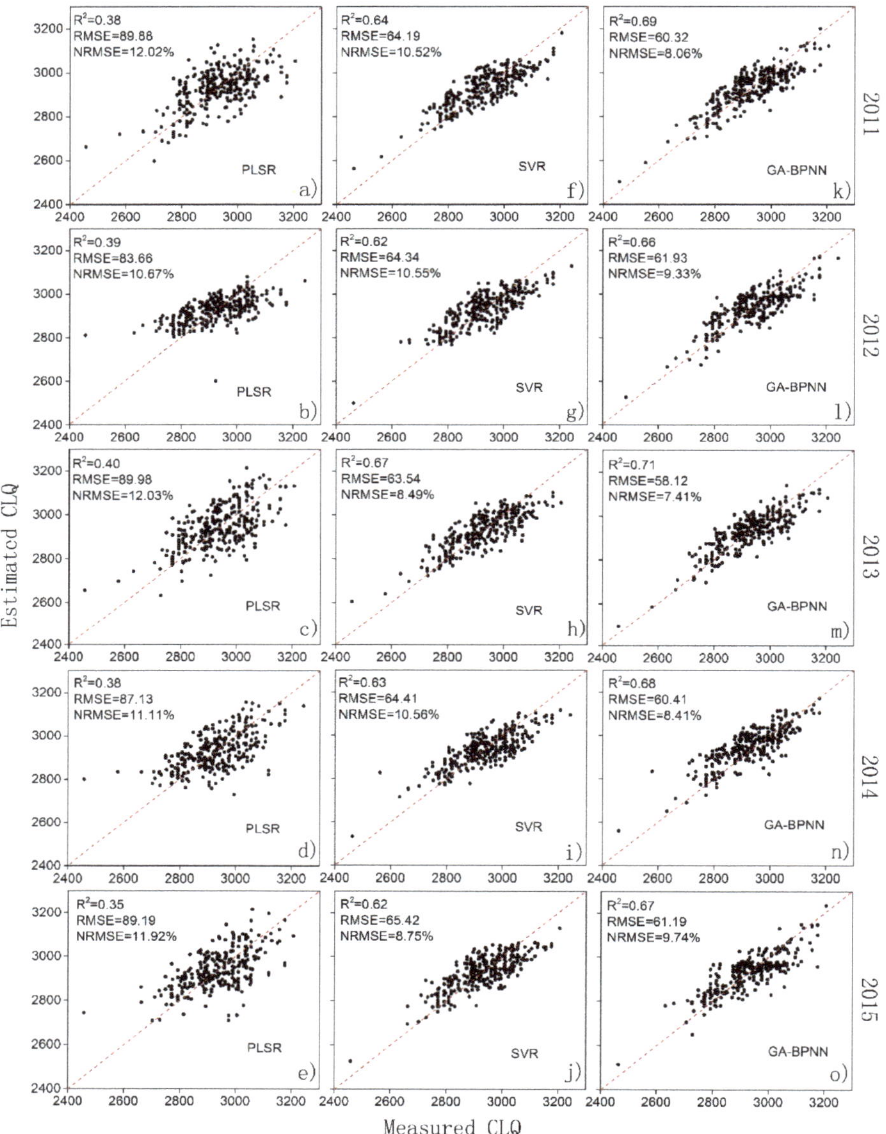

Figure 4. Scatterplots of measured versus estimated CLQ using the training dataset from 2011 to 2015: (**a–e**) partial least squares regression (PLSR) model; (**f–j**) support vector regression (SVR) model; (**k–o**) GA-BPNN model.

In addition, the predicted CLQ values from the models were validated for their accuracy using 126 validation samples in Figure 5. Given a year and a model, the points of predicted vs. measured values of CLQ were randomly placed at both sides of the 1:1 line. But, the PLSR models led to obvious overestimations and overestimations for the smaller and greater CLQ values, respectively. The overestimations were mitigated by SVR models and more mitigation was achieved by the GA-BPNN models. Among the three kinds of models, the GA-BPNN models have the smallest average RMSE of

67.37, than the SVR models with average RMSE of 73.04 and the PLSR models with average RMSE of 92.45 for years from 2011 to 2015, indicating that the GA-BPNN models have the strongest ability of predicting CLQ.

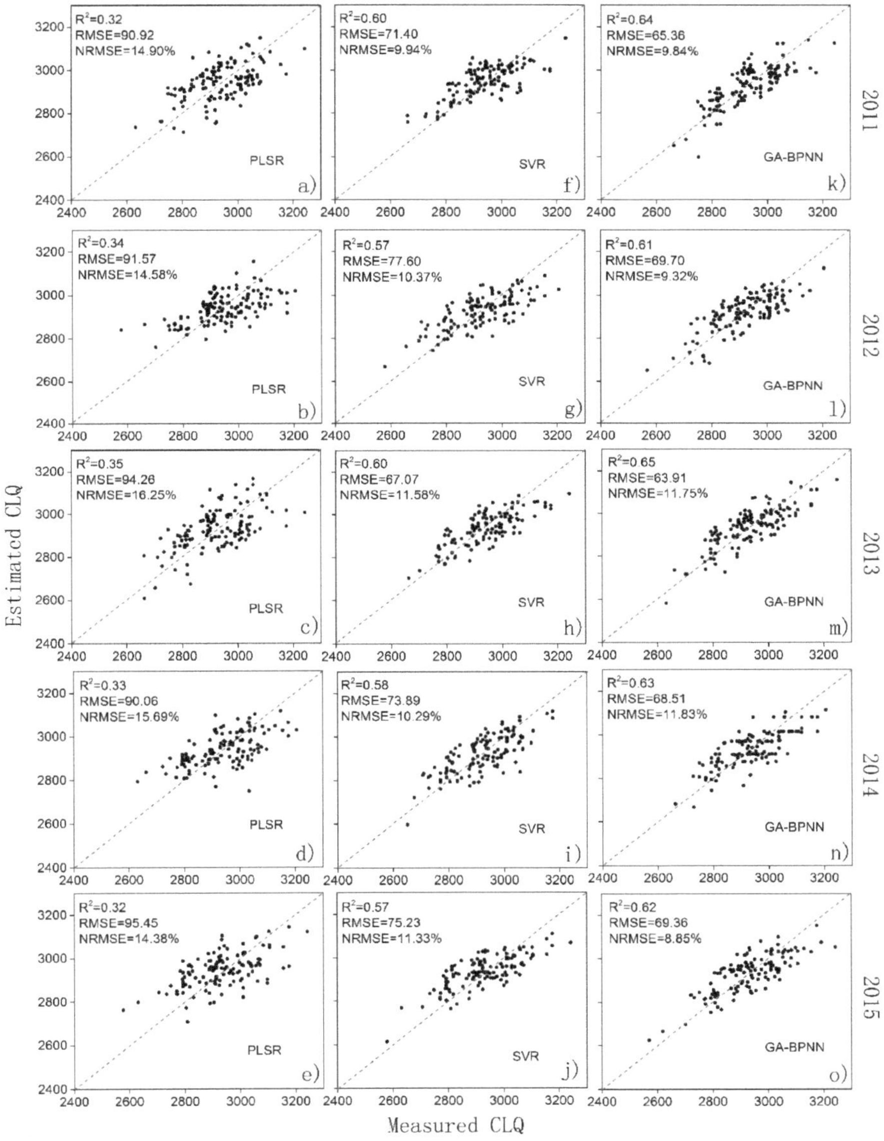

Figure 5. Scatterplots of measured versus estimated values of CLQ using the validation data set of 126 samples from 2011 to 2015: (**a–e**) PLSR model; (**f–j**) SVR model; (**k–o**) GA-BPNN model.

3.3. Mapping CLQ at the Regional Scale

The rice growth stage GPP-driven spectral model for year 2013 was used to map the CLQ for Aotou Town of the Conghua District and Zhongxin Town of the Zengcheng District to validate its capacity of predicting CLQ at the regional scale in Figure 6. The referenced values of CLQ were grouped into five classes based on the gradation regulations on agriculture land quality in China (*Regulation for gradation on agriculture land quality* GB/T 28407-2012). The R^2, RMSE, and NRMSE values of the predictions from the GA-BPNN model were calculated based on 60 sample data in Aotou and Zhongxin town, respectively (Figure 7). The prediction accuracies with RMSE of 73.32 and 104.35 and NRMSE of 10.47% and 17.75% show that the GA-BPNN model is appropriate to map CLQ at both towns.

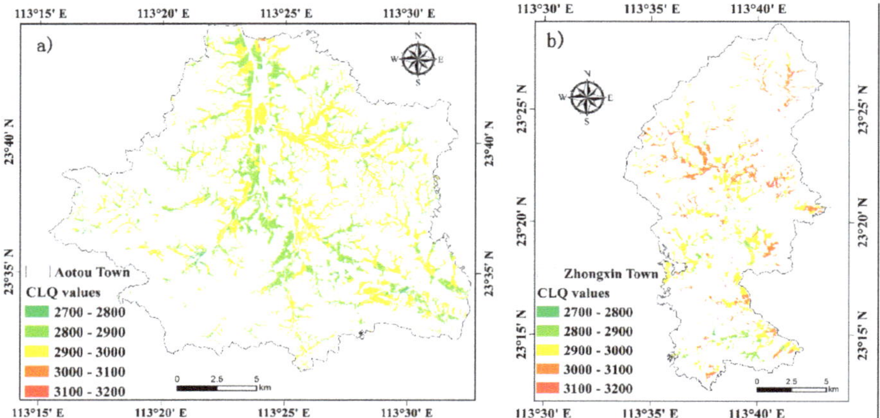

Figure 6. Spatial distributions of the predicted CLQ in 2013 using the GA-BPNN model for the study area: (**a**) Aotou Town and (**b**) Zhongxin Town.

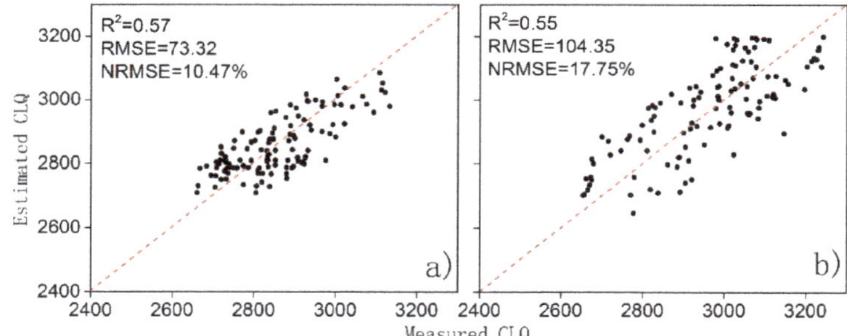

Figure 7. Measured and estimated CLQ in 2013 using the GA-BPNN model with the 120 validation sample plots for mapping: (**a**) Aotou Town and (**b**) Zhongxin Town.

4. Discussion

The CLQ implies the carrying capacity of land productivity and is critical for food supply and security. However, CLQ often changes dramatically due to human activity induced disturbances and environmental changes [4]. Thus, it is necessary to realize real-time monitoring and evaluation of CLQ in agricultural regions, especially vulnerable or urban fringe areas [48,49]. Previous studies [4,15,16] on remote sensing-based evaluation of CLQ mainly focused on retrieving spectral indicators in both the traditional evaluation system and the PSR framework system. However, it is impossible to acquire accurate CLQ data using the previous evaluation methods due to ignoring spectral relationships of spectral indicators from crop growth stages with CLQ. This study is the first attempt to propose the spectral models relating CLQ to GPP spectral indicators obtained from four growth stages of late rice phenology for evaluating CLQ. It is expected that this study can enhance the literature in this field.

Based on the comparison of the evaluation results from three kinds of models, SVR (average R^2 = 0.64 and NRMSE = 9.78%) and GA-BPNN (average R^2 = 0.69 and NRMSE = 8.59%) models performed better than the PLSR model (average R^2 = 0.38 and NRMSE = 11.55%), implying that there is obvious non-linear correlation of CLQ with GPP spectral indicator. This conclusion is consistent with the findings of previous studies [16], indicating that the non-linear models are appropriate. It was also found that the GA-BPNN models provided more accurate predictions of CLQ than the SVR and PLSR models, which was mainly attributed to the integration of BPNN with GA which has the ability of optimizing the BPNN weights and thresholds. For the SVR models, however, the kernel function and penalty factor used only referenced expert experiences and they were limited in the accuracy of CLQ evaluation [50–52].

Based on the field measured GPP values, moreover, the accuracy of the 30 m MODIS GPP generated by the EBK interpolation, with 7.43 of RMSE and 1.59% of RRMSE, was higher than those of the 500 m MODIS GPP, with 33.43 of RMSE and 7.18% of RRMSE, showing the improvements of 26% in RMSE and 5.59% in NRMSE, respectively. These results show that the downscaled 30 m MODIS GPPs were more accurate than the original 500 m spatial resolution products. Although previous studies have shown that machine learning algorithms provide potential on CLQ evaluation, with R^2 of 0.59 and NRMSE of 11.19% [16], the GA-BPNN model proposed in this study shows stronger ability for CLQ evaluation with R^2 = 0.69 and NRMSE = 8.59%, implying that the GPP spectral indicator provides a direct and effective means for estimating CLQ. The further application of the GA-BPNN model to mapping CLQ for Aotou Town and Zhongxin Town resulted in NRMSE values of 10.47% and 17.75% based on 120 validation samples. This indicated that the GA-BPNN model proposed in this study had great potential to map CLQ at a large scale.

It should be noticed that in this study the experiment was conducted only for paddy fields in which cultivated lands often have good and excellent quality. We are currently unable to verify whether the GA-BPNN model based on the relationship of CLQ with GPP can perform well in other types of cultivated lands. Therefore, in the future, we will expand the study to other kinds of cultivated lands with different grades of CLQ. Moreover, in order to further validate the GA-BPNN model for CLQ evaluations, larger sample sizes should be employed. In addition, a limited accuracy assessment using MODIS GPPs with 500 m spatial resolution was undertaken in this study. The spatial resolution of the used images will affect the evaluation accuracy of CLQ because when the rice planted areas that are smaller than the spatial resolution of the images dominate the study area, mixed pixels will exist. The GPP products from finer spatial resolution images acquired from different sensors should be tested. As the correlation coefficient method meets the assumption of normal distribution of data, it was used for determining the use of GPP from the four growth stages. In future study, more powerful methods may be introduced to obtain the growth stages of the crops. Finally, more evaluation algorithms (such as Random Forest and Deep Learning) should be attempted to improve the evaluation efficiency and accuracy for CLQ.

5. Conclusions

This study attempted to obtain an accurate spectral model for evaluation of CLQ based on the GPP spectral indicator at four important growth stages of late rice phenology by comparison of PLSR, SVR, and GA-BPNN models using the measurements of CLQ from 294 training samples and the corresponding GPP data from MOD17 products. This study was conducted in the Zengcheng and Conghua district of Guangzhou City and led to the following conclusions: (1) The downscaled 30 m spatial resolution MODIS GPP data by the EBK interpolation with NRMSE of 1.59% were more reliable than the original 500 m resolution MODIS GPP products with NRMSE of 7.18%; (2) The GA-BPNN spectral model showed the strongest prediction ability for CLQ (RMSE = 60.39) compared to the PLSR and SVM models, indicating the existence of a nonlinear relationship of CLQ with GPP spectral indicators; (3) The NRMSE values of CLQ predictions from the GA-BPNN model for two validation areas were relatively small (12.14% and 18.39%), further implying that the GA-BPNN model based on rice phenological data could be applied to accurately mapping CLQ at the regional scale. This study is the first report to provide an effective means for CLQ evaluation using a crop growth stage GPP-driven spectral model with GA-BPNN.

Author Contributions: M.Z. conceived and designed the experiments; M.Z. and S.L. analyzed the data, created the tables and figures, and finished the first version of the paper; Z.L., Y.H., Z.X., and M.Z. contributed valuable opinions during the manuscript writing; M.Z., Z.X., and G.W. revised the whole manuscript. All authors have read and agreed to the published version of the manuscript.

Funding: This research was supported by the National Natural Science Foundation of China (U1901601); National Key Research and Development Program of China (2018YFD1100103); Guangdong Province Agricultural Science and Technology Innovation and Promotion Project (No. 2019KJ102); and Guangdong Provincial Science and Technology Project of China (2017B030314155).

Acknowledgments: We gratefully acknowledge the paper writing assistance of A-Xing Zhu and Yiping Peng, as well as the experimental assistance of Hao Yang and Zhe Ni.

Conflicts of Interest: The authors declare no conflict of interest.

References

1. Tampakis, S.; Karanikola, P.; Koutroumanidis, T.; Tsitouridou, C. Protecting the productivity of cultivated lands. The viewpoints of farmers in Northern Evros. *J. Environ. Prot. Ecol.* **2010**, *11*, 601–613.

2. Xie, H.; Zou, J.; Jiang, H.; Zhang, N.; Choi, Y. Spatiotemporal pattern and driving forces of arable land-use Intensity in China: Toward sustainable land management using energy analysis. *Sustainability* **2014**, *6*, 3504–3520. [CrossRef]

3. Yan, Y.F.; Liu, J.L.; Zhang, J.B. Evaluation method and model analysis for productivity of cultivated land. *Trans. Chin. Soc. Agric. Eng.* **2014**, *30*, 204–210.

4. Liu, Y.S.; Zhang, Y.Y.; Guo, L.Y. Towards realistic assessment of cultivated land quality in an ecologically fragile environment: A satellite imagery-based approach. *Appl. Geogr.* **2010**, *30*, 271–281. [CrossRef]

5. Machin, J.; Navas, A. Land evaluation and conservation of semiarid agrosystems in Zaragoza (NE Spain) using an expert evaluation system and GIS. *Land Degrad. Dev.* **1995**, *6*, 203–214. [CrossRef]

6. Zhu, Q.; Liao, K.H.; Xu, Y.; Yang, G.S.; Wu, S.H.; Zhou, S.L. Monitoring and prediction of soil moisture spatial? Temporal variations from a hydropedological perspective: A review. *Soil Res.* **2012**, *50*, 625. [CrossRef]

7. Kalogirou, S. Expert systems and GIS: An application of land suitability evaluation. *Comput. Environ. Urban. Syst.* **2002**, *26*, 89–112. [CrossRef]

8. Qi, L.; Liu, L.; Zhao, C.; Wang, J.; Wang, J. Selection of optimum periods for extracting winter wheat based on multi-temporal remote sensing images. *Remote Sens. Technol. Appl.* **2008**, *23*, 154–160.

9. Meng, J.; Xu, J.; You, X. Optimizing soybean harvest date using HJ-1 satellite imagery. *Precis. Agric.* **2015**, *16*, 164–179. [CrossRef]

10. Becker-Reshef, I.; Justice, C.; Sullivan, M. Monitoring Global Croplands with Coarse Resolution Earth Observations: The Global Agriculture Monitoring (GLAM) Project. *Remote Sens.* **2010**, *2*, 1589–1609. [CrossRef]

11. Askari, M.S.; O'Rourke, S.M.; Holden, N.M. Evaluation of soil quality for agricultural production using visible–near-infrared spectroscopy. *Geoderma* **2015**, *243–244*, 80–91. [CrossRef]

12. Prasad, S.T.; Ronald, B.S.; Eddy, D.P. Hyperspectral Vegetation Indices and Their Relationships with Agricultural Crop Characteristics. *Remote Sens. Environ.* **2000**, *71*, 158–182.

13. Yang, J.F.; Ma, J.C.; Wang, L.C. Evaluation factors for cultivated land grade identification based on multispectral remote sensing. *Trans. Chin. Soc. Agric. Eng.* **2012**, *28*, 230–236.

14. Zhao, J.J.; Zhang, H.Y.; Wang, Y.Q.; Qiao, Z.H.; Hou, G.L. Research on the Quality Evaluation of Cultivated Land in Provincial Area Based on AHP and GIS: A Case Study in Jilin Province. *Chin. J. Soil Sci.* **2012**, *43*, 70–75.

15. Fang, L.N.; Song, J.P. Cultivated Land Quality Assessment Based on SPOT Multispectral Remote Sensing Image: A Case Study in Jimo City of Shandong Province. *Prog. Geogr.* **2008**, *27*, 71–78.

16. Liu, S.; Peng, Y.; Xia, Z.; Hu, Y.; Wang, G.; Zhu, A.-X.; Liu, Z. The GA-BPNN-Based Evaluation of Cultivated Land Quality in the PSR Framework Using Gaofen-1 Satellite Data. *Sensors* **2019**, *19*, 5127. [CrossRef]

17. Xie, X.Y.; Zheng, S.M.; Hu, Y.M.; Guo, Y.B. Study on the Method of Cultivated Land Quality Evaluation Based on Machine Learning. In Proceedings of the 2018 Fifth International Workshop on Earth Observation and Remote Sensing Applications (EORSA), Xi'an, China, 18–20 June 2018; pp. 1–5.

18. Ma, J.; Zhang, C.; Yun, W.; Lv, Y.; Chen, W.; Zhu, D. The Temporal Analysis of Regional Cultivated Land Productivity with GPP Based on 2000–2018 MODIS Data. *Sustainability* **2020**, *12*, 411. [CrossRef]

19. Xia, Z.; Peng, Y.; Liu, S.; Liu, Z.; Wang, G.; Zhu, A.-X.; Hu, Y. The Optimal Image Date Selection for Evaluating Cultivated Land Quality Based on Gaofen-1 Images. *Sensors* **2019**, *19*, 4937. [CrossRef]

20. Fu, Y.C.; Lu, X.Y.; Zhao, Y.L.; Zeng, X.T.; Xia, L.L. Assessment Impacts of Weather and Land Use/Land Cover (LULC) Change on Urban Vegetation Net Primary Productivity (NPP): A Case Study in Guangzhou, China. *Remote Sens.* **2013**, *5*, 4125–4144. [CrossRef]

21. Guangzhou Yearbook Compilation Committee. Administrative Division and Weather. In *Guangzhou Yearbook*; Guangzhou Yearbook Press: Guangzhou, China, 2010; Volume 1, pp. 4–5. (In Chinese)

22. Local Chronicles Compilation Committee of Guangzhou. Natural Geography. In *Annals of Guangzhou*; Guangzhou Press: Guangzhou, China, 1998; Volume 2, pp. 42–49. (In Chinese)

23. Chuai, X.; Guo, X.; Zhang, M.; Yuan, Y.; Li, J.; Zhao, R.; Yang, W.; Li, J. Vegetation and climate zones based carbon use efficiency variation and the main determinants analysis in China. *Ecol. Indic.* **2020**, *111*, 105967. [CrossRef]

24. Steingrobe, B.; Schmid, H.; Gutser, R.; Claassen, N. Root production and root mortality of winter wheat grown on sandy and loamy soils in different farming systems. *Biol. Fertil. Soils* **2001**, *33*, 331–339. [CrossRef]

25. Lobell, D.B.; Hicke, J.A.; Asner, G.P.; Field, C.B.; Tucker, C.J.; Los, S.O. Satellite estimates of productivity and light use efficiency in United States agriculture. *Glob. Chang. Biol.* **2002**, *8*, 722–735. [CrossRef]

26. Hu, C.Z. Theory and Method of China's Agricultural Land Classification and Gradation: On General Framework and Technical Scheme of the Agricultural Land Classification Rules. *China Land Sci.* **2012**, *26*, 4–13.

27. The Land Processes Distributed Active Archive Center (LP DAAC/NASA). Available online: https://lpdaac.usgs.gov/ (accessed on 10 August 2019).

28. CGIAR. Ricepedia. Available online: http://ricepedia.org/rice-as-a-plant/growth-phases (accessed on 10 August 2019).

29. Krivoruchko, K.; Gribov, A. Pragmatic Bayesian kriging for non-stationary and moderately non-Gaussian data. In *Mathematics of Planet Earth, Proceedings of the 15th Annual Conference of the International Association for Mathematical Geosciences, Madrid, Spain, 2–6 September 2013*; Pardo-Igúzquiza, E., Guardiola-Albert, C., Heredia, J., Moreno-Merino, L., Durán, J.J., Vargas-Guzmán, J.A., Eds.; Springer: Berlin/Heidelberg, Germany, 2014; pp. 61–64.

30. Fabijańczyk, P.; Zawadzki, J.; Magiera, T. Magnetometric assessment of soil contamination in problematic area using empirical Bayesian and indicator kriging: A case study in Upper Silesia, Poland. *Geoderma* **2017**, *308*, 69–77. [CrossRef]

31. Liu, Z.; Huang, R.; Hu, Y.; Fan, S.; Feng, P. Generating high spatiotemporal resolution LAI based on MODIS/GF-1 data and combined kriging-cressman interpolation. *Int. J. Agric. Biol. Eng.* **2016**, *9*, 120–131.

32. Krivoruchko, K.; Butler, K. *Unequal Probability-Based Spatial Sampling*; Esri: Redlands, CA, USA, 2013.

33. Goovaerts, P. Kriging and Semivariogram Deconvolution in the Presence of Irregular Geographical Units. *Math. Geosci.* **2008**, *40*, 101–128. [CrossRef]

34. Omre, H. Bayesian kriging-merging observations and qualified guesses in kriging. *Math. Geol.* **1987**, *19*, 25–39. [CrossRef]

35. Song, Y.Q.; Zhao, X.; Su, H.Y.; Li, B.; Hu, Y.M.; Cui, X.S. Predicting Spatial Variations in Soil Nutrients with Hyperspectral Remote Sensing at Regional Scale. *Sensors* **2018**, *18*, 3086. [CrossRef]

36. Zhao, L.; Hu, Y.M.; Zhou, W.; Liu, Z.H.; Pan, Y.C.; Shi, Z.; Wang, L.; Wang, G.X. Estimation Methods for Soil Mercury Content Using Hyperspectral Remote Sensing. *Sustainability* **2018**, *10*, 2474. [CrossRef]

37. Zheng, J.H. *Statistical Dictionary*; China Statistics Press: Beijing, China, 1995.

38. Liu, Z.; Lu, Y.; Peng, Y.; Zhao, L.; Wang, G.; Hu, Y. Estimation of Soil Heavy Metal Content Using Hyperspectral Data. *Remote Sens.* **2019**, *11*, 1464. [CrossRef]

39. Wold, S.; Sjostrom, M.; Eriksson, L. PLS-regression: A basic tool of chemometrics. *Chemom. Intell. Lab. Syst.* **2001**, *58*, 109–130. [CrossRef]

40. Wold, H. Nonlinear estimation by iterative least squares procedure. In *Research Papers in Statistics*; David, F.N., Ed.; Wiley: Hoboken, NJ, USA, 1966; pp. 414–444.

41. Vapnik, V. *The Nature of Statistical Learning Theory*, 2nd ed.; Springer: Berlin/Heidelberg, Germany, 1999.

42. Cherkassky, V.; Ma, Y. Selection of Meta-parameters for Support Vector Regression. In Proceedings of the International Conference on Artificial Neural Networks—ICANN 2002, Madrid, Spain, 28–30 August 2002; Dorronsoro, J.R., Ed.; Lecture Notes in Computer Science; Springer: Berlin/Heidelberg, Germany, 2002; Volume 2415, pp. 687–693.

43. Wang, X.; Wang, Z.Q.; Jin, G.; Yang, J. Land reserve prediction using different kernel based support vector regression. *Trans. Chin. Soc. Agric. Eng.* **2014**, *30*, 204–211.

44. Ye, H.; Ma, Y.; Dong, L.M. Land Ecological Security Assessment for Bai Autonomous Prefecture of Dali Based Using PSR Model–with Data in 2009 as Case. *Energy Procedia* **2011**, *5*, 2172–2177. [CrossRef]

45. Saleh, S.M.; Ibrahim, K.H.; Magdi, E.M. Study of genetic algorithm performance through design of multi-step LC compensator for time-varying nonlinear loads. *Appl. Soft Comput.* **2016**, *48*, 535–545. [CrossRef]

46. Yang, Z.; Zhou, Q.; Wu, X.; Zhao, Z.Y.; Tang, C.; Chen, W.G. Detection of Water Content in Transformer Oil Using Multi Frequency Ultrasonic with PCA-GA-BPNN. *Energies* **2019**, *12*, 1379. [CrossRef]

47. Chang, C.C.; Lin, C.J. LIBSVM: A library for support vector machines. *ACM Trans. Intell. Syst. Technol.* **2011**, *2*, 1–27. [CrossRef]

48. Xiao, L.; Yang, X.; Cai, H.; Zhang, D. Cultivated Land Changes and Agricultural Potential Productivity in Mainland China. *Sustainability* **2015**, *7*, 11893–11908. [CrossRef]

49. Xu, M.G.; Lu, C.G.; Zhang, W.J.; Li, L.; Duan, Y.H. Situation of the quality of arable land in China and improvement strategy. *Chin. J. Agric. Resour. Reg. Plan.* **2016**, *37*, 8–14.

50. Murilo, V.F.; Luiz, C.B.; Antonio, P.B. Width optimization of RBF kernels for binary classification of support vector machines: A density estimation-based approach. *Pattern Recognit. Lett.* **2019**, *128*, 1–7.

51. Zhang, X.Y.; Liu, Y.C. A Performance Analysis of Support Vector Machines with Gauss Kernel. *Comput. Eng.* **2003**, *29*, 22–25.

52. Qiu, Y.H.; Wu, C.M.; Pu, G.L. Summary of genetic algorithms research. *Appl. Res. Comput.* **2008**, *10*, 2911–2916.

Article

Corn Grain Yield Estimation from Vegetation Indices, Canopy Cover, Plant Density, and a Neural Network Using Multispectral and RGB Images Acquired with Unmanned Aerial Vehicles

Héctor García-Martínez [1], Héctor Flores-Magdaleno [1,*], Roberto Ascencio-Hernández [1], Abdul Khalil-Gardezi [1], Leonardo Tijerina-Chávez [1], Oscar R. Mancilla-Villa [2] and Mario A. Vázquez-Peña [3]

[1] Colegio de Postgraduados, Carretera México-Texcoco Km. 36.5, Montecillo, Texcoco 56230, Mexico; hector.garcia@colpos.mx (H.G.-M.); ascenciohr@colpos.mx (R.A.-H.); kabdul@colpos.mx (A.K.-G.); tijerina@colpos.mx (L.T.-C.)

[2] Centro Universitario de la Costa Sur, Universidad de Guadalajara, Avenida Independencia Nacional 151, Autlán C.P. 48900, Jalisco, Mexico; oscar.mancilla@academicos.udg.mx

[3] Departamento de Irrigación, Universidad Autónoma Chapingo, Carretera México-Texcoco, km 38.5, Chapingo C.P. 56230, Mexico; mavazquez.coahuila@gmail.com

* Correspondence: hectorfm27@gmail.com; Tel.: +52-5545113621

Received: 26 May 2020; Accepted: 2 July 2020; Published: 8 July 2020

Abstract: Corn yields vary spatially and temporally in the plots as a result of weather, altitude, variety, plant density, available water, nutrients, and planting date; these are the main factors that influence crop yield. In this study, different multispectral and red-green-blue (RGB) vegetation indices were analyzed, as well as the digitally estimated canopy cover and plant density, in order to estimate corn grain yield using a neural network model. The relative importance of the predictor variables was also analyzed. An experiment was established with five levels of nitrogen fertilization (140, 200, 260, 320, and 380 kg/ha) and four replicates, in a completely randomized block design, resulting in 20 experimental polygons. Crop information was captured using two sensors (Parrot Sequoia_4.9, and DJI FC6310_8.8) mounted on an unmanned aerial vehicle (UAV) for two flight dates at 47 and 79 days after sowing (DAS). The correlation coefficient between the plant density, obtained through the digital count of corn plants, and the corn grain yield was 0.94; this variable was the one with the highest relative importance in the yield estimation according to Garson's algorithm. The canopy cover, digitally estimated, showed a correlation coefficient of 0.77 with respect to the corn grain yield, while the relative importance of this variable in the yield estimation was 0.080 and 0.093 for 47 and 79 DAS, respectively. The wide dynamic range vegetation index (WDRVI), plant density, and canopy cover showed the highest correlation coefficient and the smallest errors (R = 0.99, mean absolute error (MAE) = 0.028 t ha^{-1}, root mean square error (RMSE) = 0.125 t ha^{-1}) in the corn grain yield estimation at 47 DAS, with the WDRVI index and the density being the variables with the highest relative importance for this crop development date. For the 79 DAS flight, the combination of the normalized difference vegetation index (NDVI), normalized difference red edge (NDRE), WDRVI, excess green (EXG), triangular greenness index (TGI), and visible atmospherically resistant index (VARI), as well as plant density and canopy cover, generated the highest correlation coefficient and the smallest errors (R = 0.97, MAE = 0.249 t ha^{-1}, RMSE = 0.425 t ha^{-1}) in the corn grain yield estimation, where the density and the NDVI were the variables with the highest relative importance, with values of 0.295 and 0.184, respectively. However, the WDRVI, plant density, and canopy cover estimated the corn grain yield with acceptable precision (R = 0.96, MAE = 0.209 t ha^{-1}, RMSE = 0.449 t ha^{-1}). The generated neural network models provided a high correlation coefficient between the estimated and the observed corn grain yield, and also showed acceptable errors in the yield estimation. The spectral information registered through remote sensors mounted on unmanned aerial vehicles and its processing in vegetation indices, canopy cover, and plant density allowed the characterization and estimation of corn grain yield. Such information is very useful for decision-making and agricultural activities planning.

Keywords: vegetation indices; UAV; neural network; corn plant density; corn canopy cover; yield prediction

---

1. Introduction

Corn is one of the main food crops for the population, and together with wheat and rice, it is one of the most important cereals in the world. According to the United States Department of Agriculture (USDA), in 2019, a global area of 192.21 million hectares was estimated, with China and the United States being the countries with the largest sown area [1]. In Mexico, 7.4 million hectares of corn was sown in 2018, with a national production of 27.7 million tons and a mean yield of 3.83 t ha^{-1} [2]. For the Mexican population, it constitutes the basis for alimentation, providing energy and proteins [3]. In Mexico, 86% of the surface is cultivated under rainfed conditions in plots of less than 5 ha. Some of these producers use the agroecosystem called "milpa", in which several species (beans, pumpkins, and others) are grown on the same plot of corn [4]. Meanwhile, according to Food and Agriculture Organization (FAO), the world population will increase by 35% by 2050, reaching 9100 million inhabitants, mainly in developing countries [5]. To feed this population, food production must sustainably increase by 70%, considering the safety and conservation of natural resources. Some studies indicate that corn yield could decrease in the coming years as a result of anthropogenic climate change [6–9]. There are three main impacts of climate change on agriculture: (a) deterioration in crop yields; (b) effects on production, consumption, and commercialization; and (c) effects on per capita caloric consumption and child nutrition [10]. Corn crop yield is directly related to many factors like the environment, management practices, genotype, and their interactions [11]. The influence of regional climate patterns and large-scale meteorological phenomena can have a significant impact on agricultural production [12]. Genotypes have improved significantly over the years, and important technological developments have been made in the machinery used in management practices. Under these circumstances, yield prediction can be important data for food production, making well-informed and timely economic and management decisions. Correct early detection of problems associated to crop yield factors can help increase the yields and subsequent incomes of farmers. Accurate, objective, reliable, and timely predictions of crop yields in large areas are fundamental to help guarantee the adequate food supply of a nation and assist responsible politicians to make plans and set prices for imports/exports [13].

Yields in corn crops vary spatially and temporally in the plots, according to the conditions present in each site such as weather, altitude, variety, planting density, available irrigation or the amount of rain (water supply), the available nutrients for the plant (soil plus fertilizer), and the date of sowing, which are the main factors that influence the yield of a plot. In recent decades, corn yield has increased as a result of genetic improvement and agronomic management. Increases in plant density and the use of synthetic fertilizers have been the main factors responsible for increases in corn yields. Plant density (number of plants per unit area) is one of the components of grain yield (number of ears per unit area, number of grains per ear, grain weight) that has an impact on the final corn yield [14]; however, its accurate measurement after plant emergence is not practical in large-scale production fields owing to the amount of labor required [15].

Precision agriculture (PA) is a management concept based on observation, measurement, and response to variability of the crops in the field [16]. PA technology allows farmers to recognize variations in the fields and apply variable rate treatments with a much finer degree of precision than before. Identifying spatial and temporal variability within the field shows the potential to support crop management concepts to satisfy most of the growing environmental, economic, market, and public pressures on agriculture [17]. Remote sensing is generally considered one of the most important technologies for precision agriculture and intelligent agriculture; it can monitor many crops and vegetation parameters through images at various wavelengths [18]. With the development of unmanned aerial systems (UAS), their use in remote sensing and precision agriculture offers the possibility of obtaining field data in an easy, fast, and profitable way, resulting in images with high spatial and temporal resolution. The successful adoption of remote sensing based on unmanned aerial vehicles (UAV) depends on changes in sensitivity on vegetation indices (VI) and growth stages [19]. Processing different vegetation indices has been associated with physiological parameters of the plant, such as plant pigments, vigor, aerial biomass, yield estimation, plant physiology, and stress. Xue and Su in 2017 [20] reviewed the developments and applications of 100 vegetation indices (VIs) in remote sensing; some VIs employ a range of reflectance values in a narrow band of the electromagnetic spectrum for more precise measurements correlating them with the grain yield, providing reliable information for yield forecasting [21,22], but they require more technologically advanced sensors. As a low-cost alternative, there are VIs that are obtained from red-green-blue (RGB) images calculated from commercial cameras. These have shown in some studies their ability to predict grain yield, quantify nutrient deficiencies, and measure the impact of diseases [23–25].

The factors that affect crop yields, such as soil, climate, and management, are so complex that traditional statistics cannot give accurate results. Various machine learning techniques have been used for yield prediction, such as decision trees, self-organizing maps (SOMs), multivariate regression, support vector machines, association rule mining, and neural networks [26–31]. As a machine learning tool, the artificial neural network (ANN) is an attractive alternative to process the massive data set generated by production and research in precision agriculture [11]. The models used in machine learning relate crop yield, as an implicit function of input variables, for example, climate variables, soil, and water characteristics, which can be a very complex function, not necessarily linear. Recently, different types of neural networks have been used to predict yield in wheat, soybean, rice, corn, and tomato, using databases of genotypes, environment, management practices, and multispectral images, obtaining acceptable results [32–37].

Studies have been carried out to estimate corn yield, using data obtained from remote sensors and ANN. Fieuzal et al. [38] use multi-temporal data from satellites and radar, as well as a neural network in the estimation of maize yield with an R^2 of 0.69. On the other hand, in another study [39], polarimetric synthetic aperture radar (PolSAR) and neural network data are used in the estimation of corn biomass, obtaining good results with an R = 0.92. Han et al. [40] estimate the aerial biomass of corn from spectral information, plant height, and structural information using data from remote sensors and unmanned aerial vehicles with machine learning regression algorithms, obtaining good results (R^2 = 0.69). Michelon et al. [41] use ANN and chlorophyll readings to estimate corn productivity, resulting in a correlation coefficient of 0.73 in stage V6. Olson et al. [19] related vegetation indices and crop height with maize yield, finding high correlations; another approach uses yield data from parents to predict maize yield in plant breeding using neural networks [42]. Canopy cover, plant density, and vegetation indices calculated from data collected by remote sensors can be used to forecast corn grain yield using neural networks. As far as we know, this is the first study where canopy cover, plant density, and various vegetation indices are estimated from images obtained by UAVs to forecast corn grain yield using a neural network.

In this study, the corn grain yield was estimated through the use of a neural network for a corn crop established under different doses of nitrogen fertilization from multispectral and digital images acquired by sensors mounted on an unmanned aerial vehicle. The images were processed and vegetation indices, canopy cover, and plant density were extracted. A neural network was designed having the following input parameters: vegetation indices (normalized difference vegetation index, NDVI; normalized difference red edge, NDRE; wide dynamic range vegetation index, WDRVI; excess green, EXG; triangular greenness index, TGI; visible atmospherically resistant index, VARI), canopy cover, and plant density.

2. Materials and Methods

2.1. Study Site

An experiment was established in a plot on 5 April 2018 at the Colegio de Postgraduados, Campus Montecillo, located on the Mexico-Texcoco highway, km. 36.5 Montecillo, Texcoco, State of Mexico (19°27′40.58″ N, 98°54′8.57″ W, 1250 m above sea level). Soil texture is sandy, bulk density of 1.45 g cm^{-3}, organic matter of 1.59%, pH of 9.1, and electrical conductivity of 1.72 dS m^{-1}. Five nitrogen levels (140, 200, 260, 320, and 380 kg ha^{-1}) were evaluated in a completely randomized block experimental design with four replicates, resulting in 20 experimental units (120 m^2 per unit), using the Asgrow Albatross corn variety. A drip-band irrigation system with drippers at 20 cm separation and a flow of 1.6 L/h was designed to irrigate each experimental unit. The five nitrogen levels were applied to the irrigation water at dose intervals of 30% for the first 40 days, 50% of the dose at 40–80 days, and 20% after 80 days. The sowing was done by hand with a distance of 80 cm between rows and 30 cm between plants, with three seeds per hole. Weeds were controlled manually. A mean temperature of 16.9 °C was present during the experiment period, and a total precipitation of 470 mm occurred, with these conditions being very close to the normal conditions.

The experimental units were harvested at the end of September 2018, harvesting the entire area of the experimental unit, so the total weight harvested corresponded to an area of 120 m^2 (0.8 m separation × 6 rows × 25 m long) for all replicates of each treatment in the experiment. The harvested ears and grains were dried at approximately 14% humidity, so grain yield (t ha^{-1}) was calculated according to the following:

$$Gy = \frac{X}{120} \times 10, \tag{1}$$

Grain yield per plant (g plant^{-1}) was calculated according to the following:

$$Gyp = \frac{X \times 1000}{Np}, \tag{2}$$

where X represents grain weight per experimental area (kg m^{-2}), Gy is grain yield, Gyp is grain yield per plant, and Np is the number of plants present in the area.

2.2. Acquisition and Analysis of Data from Remote Sensors Mounted on Unmanned Aerial Vehicles

Four orthomosaics were used in this work according to Table 1, which were acquired by a 72 g Parrot Sequoia camera with a 16 Mpx RGB sensor. It also incorporates four single band sensors with a spectral bandwidth in green (530–570 nm), red (640–680 nm), red edge (730–740 nm), and near infrared (NIR) (770–810 nm) of 1.2 Mpx, and a 35 g solar sensor for real-time correction of lighting differences. The Parrot Sequoia camera was attached to a 3 DR SOLO quadcopter (3D Robotics, Berkeley, CA, USA) with flight autonomy of 20 min, with a horizontal precision range of 1–2 m. Two orthomosaics used in this work were generated from a 20 Mpx DJI RGB sensor integrated into a DJI Phantom 4 quadcopter (DJI, Shenzhen, Guangdong, China), with a load capacity of 1.388 kg and autonomy per flight of 30 min, equipped with a GPS/GLONASS satellite positioning system with a vertical precision range ±0.5 m and horizontal precision range of 1.5 m.

Table 1. Flight log, sensors, and orthomosaics on different days after sowing (DAS).

DAS	Development Stage	Date	Sensor	No. of Images	Area (m^2)	Ground Sampling Distance (GSD) (cm)
47	V8	21 May	DJI FC6310	272	10,067	0.49
47	V8	21 May	Parrot Sequoia	752	9409	2.15
79	R0	22 June	DJI FC6310	188	7720	0.49
79	R0	22 June	Parrot Sequoia	752	9409	2.15

V8: Vegetative stage—eight leaves with collar visible; R0: Reproductive stage—anthesis.

The four flights were carried out in a period from 21 May to 22 June 2018, at a flight height of 30 m. The images were acquired with 80% overlap and 80% sidelap, and the orthomosaic was generated with the Pix4D software (Pix4D SA, Lausanne, Switzerland) using structure-from-motion. The structure-from-motion processing technique searches for features in individual images that are used to relate to features that match between overlapping images, called keypoints; using these key points, camera parameters can be calibrated to external parameters (such as position, scale, and image orientation) and a point correlation (matching) is performed based on the characteristics, identifying and relating similar characteristics between images in common areas or overlapping areas. The calculated 3D position of the matched points is densified and textured with the corresponding images, from which the orthomosaic is generated by projecting each textured pixel onto a 2D plane [43,44]. Ground control points (GCP) were added for orthomosaic georeferencing according to Figure 1a. The required GCP density depends on the required project accuracy, network geometry, and the quality of images [45,46]. Six marks were placed as ground control points (GCPs) distributed in the experimental plot for each flight. A Global Navigation Satellite System (GNSS) Real Time Kinematics (RTK) V90 PLUS Hi-Target system (Hi-Target Surveying Instrument Co., Ltd., Guangzhou, China) was used to record the center of each control point with RTK precision, Horizontal: 8 mm + 1 Part Per Million (ppm) Root Mean Square (RMS), and Vertical: 15 mm + 1 ppm RMS. Table 1 shows the summary of the images acquired in each flight and the area covered for the four orthomosaics at 47 and 79 days after sowing (DAS). Moreover, the corn growth stages for each flight are shown (V8 and R0). Generated orthomosaics showed a ground sample distance (GSD) of 0.49 cm/pixel and a model RMS error of 5.6 cm for the 20 Mpx RGB sensor. The multispectral images obtained by the Sequoia camera were processed in the Pix4D software, which integrates the camera's light sensor to correct the estimated reflectance and performs a radiometric calibration using a calibration panel image [47]. The multispectral Ag template and the calibration panel by Airinov provided by the camera were used, defining the known reflectance values for each spectral band of the panel equal to 0.171, 0.216, 0.268, and 0.372 for green, red, red edge, and NIR, respectively. The obtained orthomosaics showed a ground sampling distance of 2.15 cm/pixel.

Six vegetation indices were estimated based on the generated orthomosaics. Three vegetation indices based on the reflectance of the visible spectrum (RGB): TGI (triangular greenness index), EXG (green extraction index), and VARI (visible atmospherically resistant index), according to Table 2, and three multispectral indices based on the reflectance in the near infrared, the red, and the red edge bands: NDVI (normalized difference vegetation index), NDRE (green extraction index), and WDRVI (wide dynamic range vegetation index). The calculation of the vegetation indices was done using the raster calculator module of the Open Source program licensed under the GNU—General Public License of the Geographic Information System (QGIS) software.

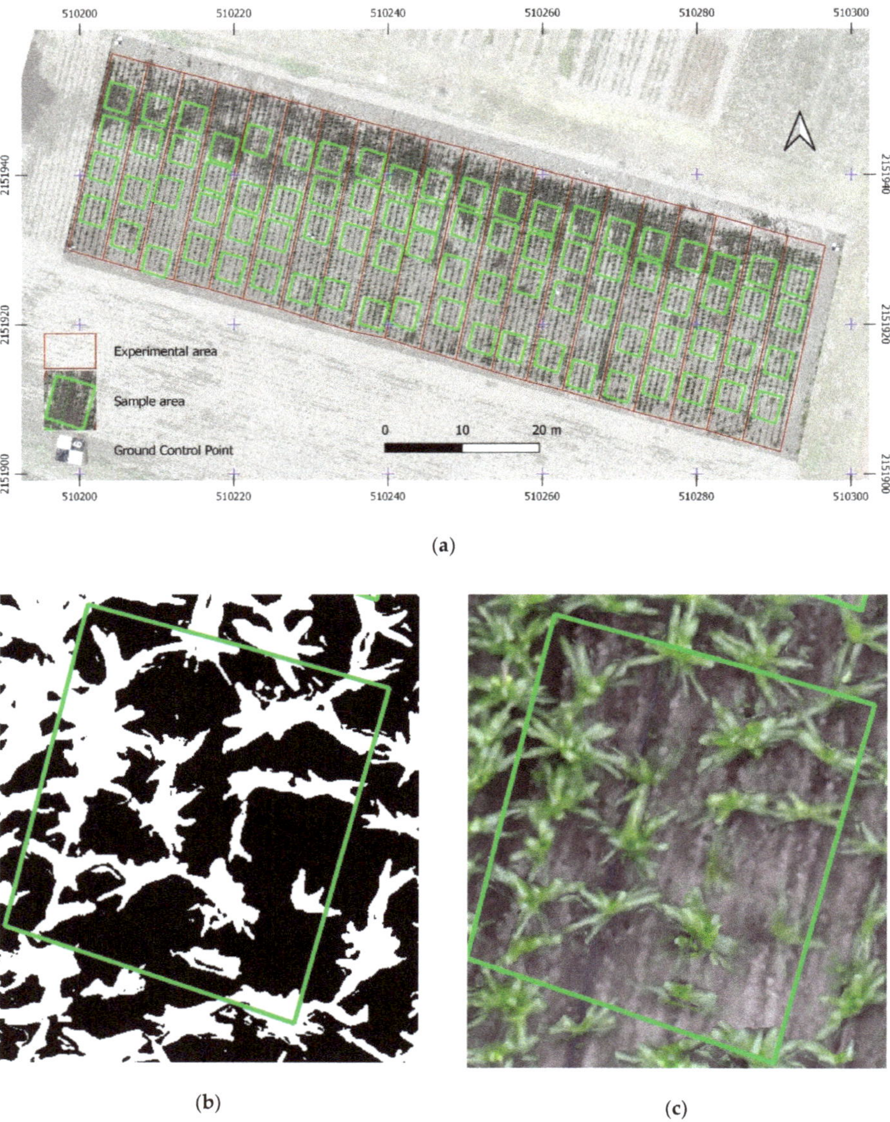

Figure 1. (**a**) Experimental plot and polygons sampled in the extraction of vegetation indices, canopy cover, and plant density; (**b**) binary image resulting from the classification of vegetation and soil; (**c**) polygon selected for the sampling of the canopy cover and plant density.

Table 2. Vegetation indices estimated from multispectral and visible images. R_{Nir}, R_{red}, R_{RE}, R_{Green}, R_{Red}, and R_{Blue} are the reflectance values for the near infrared, red, red edge, green, red, and blue bands, respectively.

Vegetation Index	Formula	Reference
Triangular greenness index	$TGI = R_{Green} - 0.39R_{Red} - 0.61R_{Blue}$	[48]
Excess green index	$EXG = 2g - r - b$	[49]
Visible atmospherically resistant index	$VARI = (R_{Red} - R_{Green})/(R_{Green} + R_{Red} - R_{Blue})$	[50]
Normalized difference vegetation index	$NDVI = (R_{Nir} - R_{Red})/(R_{Nir} + R_{Red})$	[51]
Normalized difference red edge	$NDRE = (R_{Nir} - R_{RE})/(R_{Nir} + R_{RE})$	[52]
Wide dynamic range vegetation index *	$WDRVI = (\alpha{\cdot}R_{Nir} - R_{Red})/(\alpha{\cdot}R_{Nir} + R_{Red})$	[53]

* WDRVI with an α coefficient value of 0.1 presented a good relationship with corn canopy cover [53].

The variables r, g, and b of the EXG index are normalized values of the red, green, and blue channels, respectively, according to

$$r = \frac{R_N}{R_N + G_N + B_N}, \; g = \frac{G_N}{R_N + G_N + B_N}, \; b = \frac{B_N}{R_N + G_N + B_N} \tag{3}$$

$$R_N = \frac{R}{R_{max}}, \; G_N = \frac{G}{G_{max}}, \; B_N = \frac{B}{B_{max}} \tag{4}$$

where R_N, G_N, and B_N are the normalized values of each band; R, G, and B are the non-normalized values of the red, green, and blue channels, respectively; and $R_{max} = G_{max} = B_{max}$ are the maximum digital numbers for each channel (255 on the 0–255 scale).

TGI index was normalized to have a value in the range of the others indices, between 0 and 1, according to the following:

$$Y_{nor} = \frac{Y - Y_{min}}{Y_{max} - Y_{min}}, \tag{5}$$

where Y_{nor} is the normalized index value, Y is the index value without normalizing, Y_{min} is the minimum index value, and Y_{max} is the maximum index value.

2.3. Plant Count, Determination of Plant Density, and Estimation of Canopy Cover

The corn plants present in the study area were counted digitally according to García et al. [54]. The orthomosaics were transformed from the RGB color model to the CIELab model, where we only worked on the channel a* to extract the vegetation. Corn plant samples were selected at 47 days after planting, twelve of them for each experimental area. Through normalized cross-correlation with the normxcorr2 command in Matlab (Mathworks, Natick, MA, USA), the corn plants present in the binary image were classified and, using an image component labeling technique, assigning a label (1 ... i), the pixels of each plant were grouped. The area of the plants selected in the correlation was used as a criterion, counting plants with 20% less than the minimum area of the smallest selected sample. The digitally counted plants were registered in a table. The plant density (plants m^{-2}) in each experimental area was calculated according to the following:

$$Pd = \frac{Np}{Ea}, \tag{6}$$

where Pd is the plant density in plants/m^2, Np is the number of plants present in the sampled area, and Ea is the sampled area in square meters.

The canopy cover was calculated from the generated TGI vegetation index, using the Open Source program licensed under the GNU—General Public License of the Geographic Information System (QGIS) and System for Automated Geoscientific Analyses (SAGA) [55]. In QGIS, the K-Means Clustering for Grids module was used to group the pixels into two classes: soil and vegetation, through the combined method of minimum distance and simple scaling [56]. The K-means method of image segmentation is a technique that does not use the histogram for the segmentation process, so it is not affected by the noise introduced in the image, evaluating and grouping the different pixels in similar data sets, being suitable for large data sets. On the other hand, the K-means approach in image segmentation is fast and efficient in terms of computational cost [57,58]. In the classification of pixels into two classes in QGIS, a binary raster was generated, where pixels with a value of 0 belong to soil and pixels with a value of 1 belong to vegetation, or vice versa. Thus, the estimation of the percentage of canopy cover was obtained by the following:

$$Cc = \frac{NP}{TP} \tag{7}$$

where Cc is the canopy cover, NP are the pixels corresponding to vegetation per unit area, and TP is the total pixels per unit area.

Sampling was carried out in four 11 m^2 polygons for each replicate, according to Figure 1a. In each of them, the canopy cover (Figure 1b,c) and the plant density for the two flight dates were calculated.

2.4. Vegetation Pixels' Segmentation and Extraction of the Indices Values

The images acquired through the sensors and the orthomosaics generated from them contain reflectance corresponding to soil and vegetation, but only the reflectance corresponding to vegetation was of interest to us. With the eCognition software and the object-based image analysis (OBIA) technique, multi-resolution segmentation was realized. This combines pixels or adjoining objects present in the image based on spectral and shape criteria. It also works according to the scale of the objects present; large scale results in large objects, and vice versa [59]. The homogeneity criterion measures how homogeneous or heterogeneous the object present in the image is, calculated from a combination of object color and shape properties. Color homogeneity is based on the standard deviation of the spectral colors. Shape homogeneity is based on the deviation of a compact (or smooth) shape and can have a value of up to 0.9. In the segmentation process, a 10-scale value was used and the homogeneity criteria like shape and compactness were 0.1 and 0.5, respectively. For the classification of the segmented pixels in the image, three classes present in the image (Figure 2a,b) were determined: soil, vegetation, and shadows. Supervised classification of the nearest neighbor was used to select pixel samples for each class; they were compared with the other objects present with respect to the mean and standard deviation of the supervised sample [60].

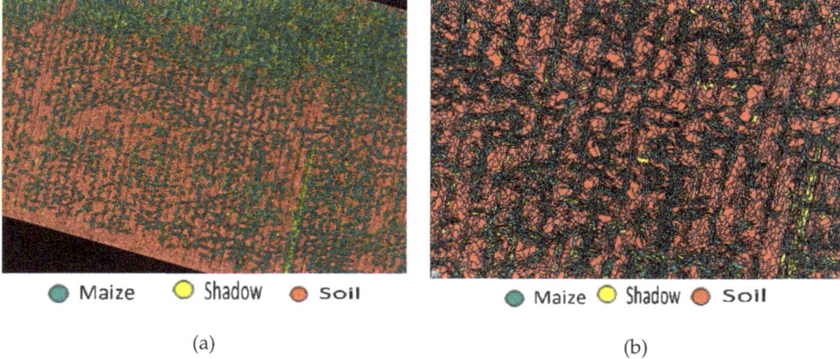

Figure 2. Segmentation and classification of objects present in the orthomosaics for the extraction of pixels belonging to and classified as corn plants: (**a**) and (**b**) processing and analysis of the crop image based on objects for three classes: corn, shadow, and soil.

Once the pixels present in the image were classified, we exported the pixels classified in the corn class (Figure 3a,b) as polygons in shape file format, with the QGIS software. The mean values of the indices calculated for each polygon of corn class were extracted through the zone statistics complement; the values were stored in a shape file. The mean values obtained from the indices were extracted for 76 sampled polygons, 4 per replicate of the 5 treatments as the average of the mean values of all the pixels classified as corn contained in each polygon. The sample polygon of each replicate on the edge of the crop was discarded to eliminate edge effects.

(**a**)   (**b**)

Figure 3. Segmentation and classification of objects present in the orthomosaics for the extraction of pixels belonging to and classified as corn plants: (**a**) image of the crop without segmentation and classification; (**b**) image of the crop with a shape of polygons generated from the segmented and classified pixels of corn plants.

2.5. Development and Training of the Feed-Forward Neural Network

A multilayer perceptron type feed-forward neural network was developed using Matlab software (Mathworks, Natick, MA, USA). A feed-forward neural network contains multiple neurons arranged in layers: input, hidden, and output. The information goes in one direction, so there are no cycles or loops [61]. The input layer receives values of the input variables (x1, x2, ..., xn); between the input layer and the hidden layer are the weights (w1, w2, ..., wn) that represent the memory of the network; and the output neurons return the output (y1, y2, .., yn) through the sum of all the inputs multiplied by the weights plus a value associated with the neuron, called bias (b) [62].

$$y = \sum_{i=0}^{i=n} W_i X_i + b, \tag{8}$$

The outputs of the neurons before passing to the other nodes are transformed through an activation function f(x) to limit the output of the neuron. The precision in the prediction by a neural network is related to the type of activation function used; the non-linear activation functions are the most used (Sigmoid, Hyperbolic Tangent, Rectified liner Unit (ReLU), Exponential Linear Unit (ELU), Softmax) [63,64], so the final output (a) will be as follows:

$$a = f\left(\sum_{i=0}^{i=n} W_i X_i + b\right), \tag{9}$$

As input parameters or variables, we used the mean value of the vegetation indices, the canopy cover, and the plant density, as well as the yield in tons per hectare as labels and output parameters. In total, 70% of the total data entered into the neural network was used for training, 15% for validation, and 15% for testing using a random partition of the entire data set. For each combination of input variables, the neural network was trained ten times using ten different random data sets. The training algorithm used was Levenberg–Marquart [65,66], which is a training function that updates the values in the weights and bias according to the Levenber–Marquart optimization. Tan-sigmoid was used as activation function, as it defines the behavior of the neurons in charge of calculating the degree or state of activation of the neuron, according to the total input function [67]. The descending gradient with momentum type was used as a learning rule, as it calculates the change in weight (W) for a given neuron. On the basis of the weights of the input neurons and the error (E), the weight W, the learning rate (LR), and the moment constant (MC) are calculated, according to the descent of the gradient with momentum. The objective of training the neural network is to minimize the resulting errors for a training data set by adjusting the weights (W).

2.6. Statistical Analysis

For the performance analysis of the estimated yields by the neural network, estimated yields were compared to the observed yields, estimating the root mean square error (RMSE), the mean absolute error (MAE), and the correlation coefficient (R) in the training, validation, testing, and total data entered into the neural network. The correlation coefficient between the observed yields and the vegetation indices, the plant density, and the canopy cover was also estimated.

$$E_s = (EY - OY), \tag{10}$$

$$MAE = \frac{1}{N} \sum_{i=1}^{N} |(E_s)_i|, \tag{11}$$

$$RMSE = \sqrt{\frac{\sum_{i=1}^{N} (EY - OY)^2}{N}}, \tag{12}$$

where N is the number of the sampled polygon, $|(E_s)_i|$ is the absolute value of E_s, EY is the estimated yield, and OY is the observed yield.

2.7. Variables' Relative Importance in the Yield Estimation Using Garson's Algorithm

Neural network models are difficult to interpret and it is difficult to identify which predictors are the most important and how they relate to the property being modeled. The weights connecting neurons in a neural network are partially analogous to the coefficients in a generalized linear model, and these weights combined with their effects on model predictions represent the relative importance of predictors in their associations with the variable being predicted [68]. Garson's algorithm discriminates the relative importance of the predictor variables for a single response variable, that is, the force with which a specific variable explains the predicted variable is determined by identifying all the weighted connections between the nodes of interest, and identifying the weights that connect to the specific input node that passes through the neural network and reaches the variable that is being predicted. This allows to obtain a unique value for each descriptor variable used in the neural network model [69,70]. To know the predictor variables' relative importance (NDVI, NDRE, WDRVI, EXG, TGI, VARI, canopy cover, and density) in the corn yield estimation, Garson's algorithm was used.

3. Results and Discussion

3.1. Corn Plant Vegetation Indices

The segmentation and classification of objects into soil and vegetation allowed to filter mean values of the vegetation indices with respect to the pixels belonging to corn plants. Table 3 shows the mean values computed through segmentation and classification of the objects present in the image for two dates in the corn crop development. Differences are observed in the mean values of the vegetation indices for 47 and 79 days after sowing. The classification of the pixels captured in the sensors into two classes allowed us to extract the values of the indices belonging to corn, so the result was a mean value of the vegetation. Contrarily, if the classification of the pixels into classes was not carried out, there would be a bias when considering pixels belonging to soil and shadows in the mean computation for the polygons taken as a sample, resulting in a mixed value of corn plants and soil pixels. Vegetation indices values increased for 79 DAS with respect to the values shown for 47 DAS, except for the VARI index, which did not present a wide variation with the crop development. Similar results were shown in the studies carried out by Gitelson et al. [53,71]. Vegetation indices did not show differences for the nitrogen treatments used; similar results are reported by Olson et al. [19].

Table 3. Mean values of the vegetation indices and yield for each replicate of the treatments under different degrees of nitrogen fertilization.

		Vegetation Index												Yield (g plant^{-1})	
		NDVI		NDRE		WDRVI		EXG		TGI		VARI			
DAS	Treatment	Mean	Std DeV	Mean	Std DeV	Mean	Std DeV	Mean	Std DeV	Mean	Std DeV	Mean	Std DeV	Mean	Std DeV
	N140	0.46	0.05	0.19	0.02	−0.55	0.04	0.40	0.03	0.40	0.03	0.12	0.02	109.3	19.6
	N200	0.48	0.04	0.19	0.01	−0.53	0.04	0.42	0.03	0.42	0.03	0.13	0.02	134.0	2.7
47	N260	0.46	0.05	0.18	0.02	−0.55	0.05	0.41	0.02	0.41	0.02	0.13	0.02	135.6	2.7
	N320	0.46	0.04	0.18	0.01	−0.55	0.04	0.42	0.04	0.42	0.04	0.13	0.02	137.1	3.6
	N380	0.47	0.05	0.19	0.02	−0.53	0.05	0.43	0.02	0.43	0.03	0.14	0.01	138.3	5.3
	N140	0.90	0.02	0.26	0.02	−0.13	0.06	0.53	0.03	0.54	0.01	0.13	0.02	109.3	19.6
	N200	0.91	0.01	0.26	0.03	−0.10	0.06	0.54	0.03	0.54	0.01	0.14	0.02	134.0	2.7
79	N260	0.90	0.01	0.25	0.02	−0.13	0.06	0.52	0.03	0.54	0.01	0.13	0.02	135.6	2.7
	N320	0.90	0.01	0.25	0.02	−0.12	0.05	0.53	0.04	0.54	0.01	0.13	0.02	137.1	3.6
	N380	0.90	0.01	0.25	0.02	−0.11	0.05	0.53	0.03	0.54	0.01	0.13	0.02	138.3	5.3

3.2. Plant Density, Canopy Cover, and Yield

Figure 4 shows the correlation coefficient at 47 and 79 days after sowing between the canopy cover, plant density, and yield; it also shows the correlation coefficient between the applied nitrogen dose and grain yield per plant. The canopy cover calculated through the TGI index is closely related to the density and grain yield of corn; the higher the density, the higher the cover and yield for 47 and 79 DAS. Therefore, corn yield has a direct response to canopy cover and planting density. For the period from 47 to 79 DAS, the canopy cover increased by an average of 15% with a standard deviation of 5%. The correlation coefficient for canopy cover was similar for both flight dates, around 0.76. Plant density, according to Figure 4c, showed a high correlation with a coefficient of 0.94; thus, at a higher density, an increase in grain yield is expected. This means that plant density explained 94% of the yield. Positive yield responses have been reported with an increase in planting density, a significant increase of 4.5–6, moderate of 6–7.5, and low of 7.5 to 9 plants per square meter [14]. The different nitrogen treatments applied to the experiment influenced grain yield per plant; as the nitrogen dose increases, there is a positive gain in grain yield per plant, as observed in Figure 4d, with mean grain weight yield per plant of 109.3, 134.0, 135.6, 137.1, and 138.3 g per plant for 140, 200, 260, 320, and 380 kg ha^{-1}, for the conditions of the experimental site and the variety used.

3.3. Vegetation Indices and Yield

For flights at 47 DAS and 79 DAS, six vegetation indices (NDVI, NDRE, WDRVI, EXG, TGI, VARI) were generated; three multispectral indices and three indices in the visible spectrum (RGB). Figure 5 shows the correlation coefficient between the vegetation indices corresponding to the corn class and the observed grain yield at 47 DAS. During the 47 DAS, the WDRVI index presented values ranging from −0.45 to −0.63 with a correlation coefficient of 0.54, thus presenting the highest correlation at this crop stage. Meanwhile, the NDRE index showed a low correlation of 0.23 with values ranging from 0.15 to 0.24. The NDVI ranged from 0.36 to 0.55. In the study carried out by Maresma et al. [72], they found that the WDRVI better explained the corn grain yield with different nitrogen treatments; on the other hand, the index is reported to be sensitive to the leaf area index (LAI) and the canopy cover [53].

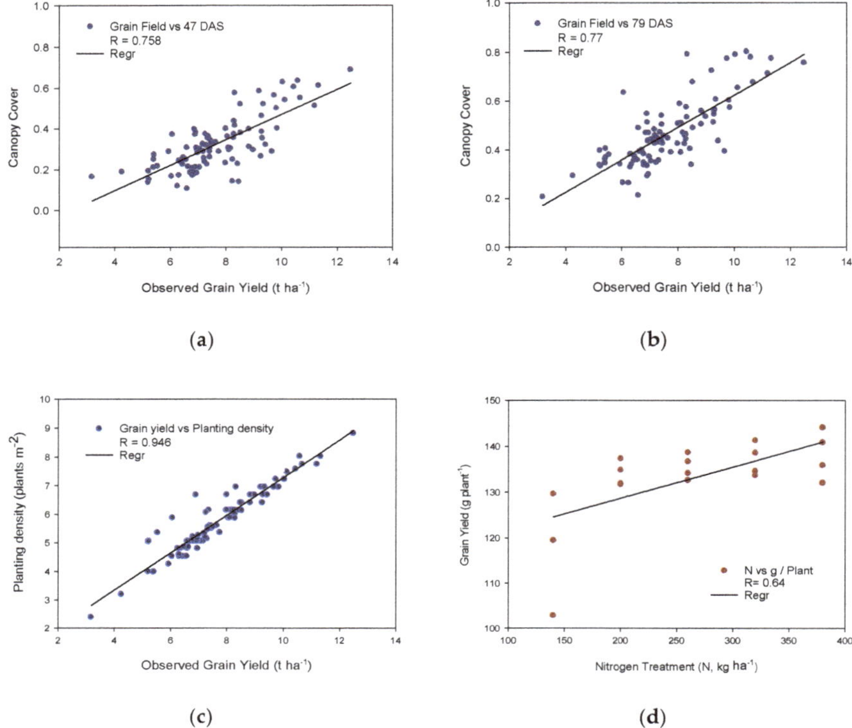

Figure 4. (**a**) Correlation between the canopy cover at 47 days after sowing (DAS) and yield; (**b**) correlation between the canopy cover at 79 DAS and yield; (**c**) correlation between plant density and yield; (**d**) correlation between the applied nitrogen dose and grain weight per plant.

At 72 DAS, the correlation coefficient of the vegetation indices with respect to the observed corn grain yield increases slightly for all the indices, which is related to the increase in canopy cover and the presence of more pixels in the corn class and fewer pixels in the soil category. In Figure 6, the NDVI, NDRE, and WDRVI showed correlation coefficients of 0.68, 0.31, and 0.65, respectively, and the indices values resulted in ranges of 0.86–0.92, 0.21–0.31, and −0.241 to 0.003, respectively. These increases in the vegetation indices are related to the increase in biomass, leaf area index (LAI), leaf chlorophyll content (LCC), canopy cover (CC), and yield [23,50,72–74]. The NDVI can explain 0.68 of the corn grain yield. Figure 6 shows a proportional relation between an increase in grain yield and an increase in the NDVI index; the same is true for the NDRE and WDRVI. Regarding the normalized EXG, TGI, and VARI indices computed in the visible spectrum (RGB) for 47 DAS, Figure 5d,e show that EXG and TGI presented low correlation coefficients, 0.22 and 0.23, respectively. The VARI in Figure 5f showed a better fit regarding to these indices, with a correlation coefficient of 0.52. In Figure 6, the values of the indices were found between 0.338–0.503, 0.335–0.501, and 0.02–0.19 for EXG, TGI, and VARI, respectively, according to the sampled polygons. In Figure 6 at 79 DAS, 0.71 of the yield can be explained by the EXG index, while VARI showed a value of 0.67. TGI showed a low correlation coefficient. The values of the indices ranged as follows: 0.47–0.60, 0.51–0.55, and 0.10–0.18 for EXG, TGI, and VARI, respectively. Some research works indicate that the VARI has a high correlation with grain yield, chlorophyll content, and the fraction of photosynthetically active radiation intercepted [75,76].

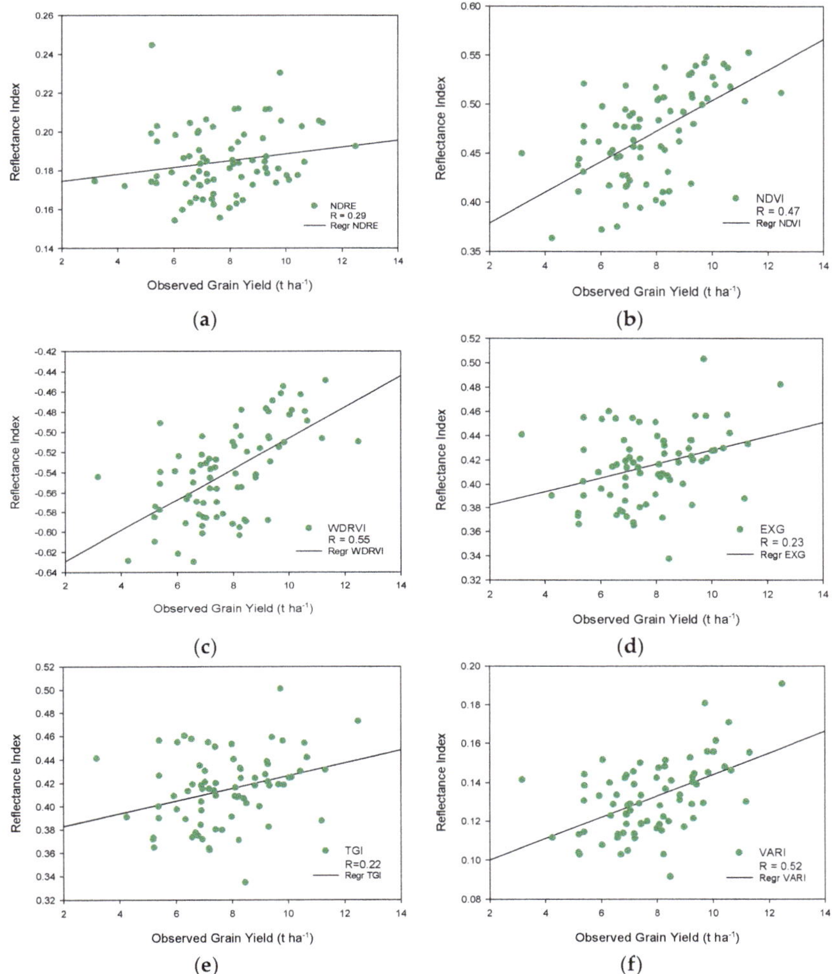

Figure 5. Correlation of vegetation indices and yield at 47 DAS. (**a**) Normalized difference red edge index (NDRE); (**b**) normalized difference vegetation index (NDVI); (**c**) wide dynamic range vegetation index (WDRVI); (**d**) excess green (ExG); (**e**) triangular greenness index (TGI); (**f**) visible atmospherically resistant index (VARI).

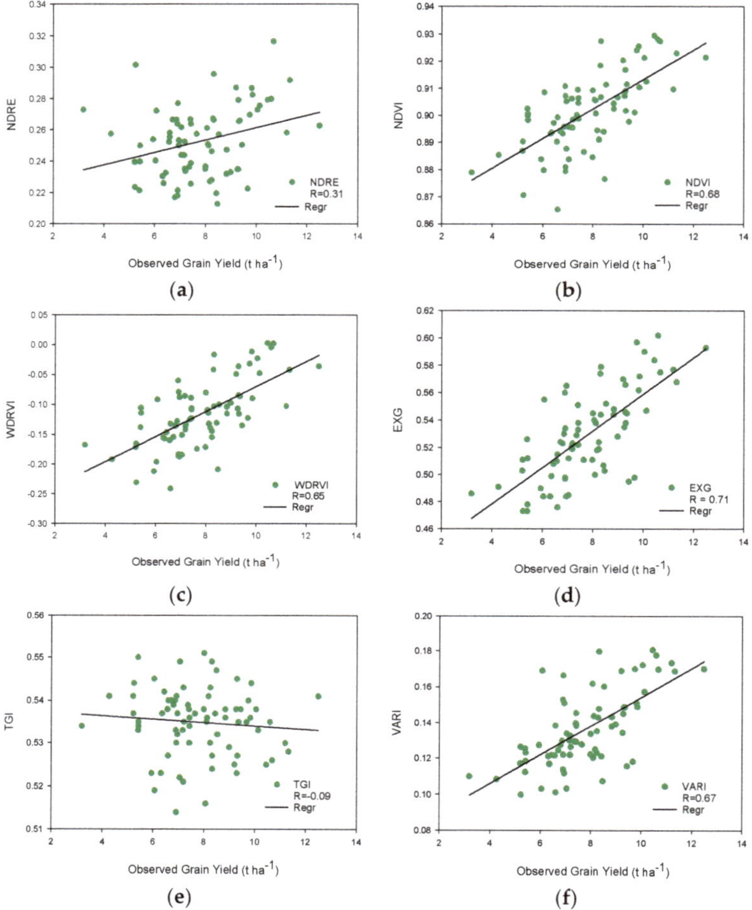

Figure 6. Correlation of vegetation indices and yield for 79 DAS for corn crops. (**a**) Normalized difference red edge index (NDRE); (**b**) normalized difference vegetation index (NDVI); (**c**) wide dynamic range vegetation index (WDRVI); (**d**) excess green (ExG); (**e**) triangular greenness index (TGI); (**f**) visible atmospherically resistant index (VARI).

3.4. Training, Validation, and Testing of the Artificial Neural Network for Estimating Yield

A feed-forward neural network was created with 2 layers and 40 neurons, where a combination of the normalized vegetation indices, plant density, and canopy cover, according to Table 4 for 47 DAS, was entered as input parameters. Another neural network for 79 DAS was created to estimate the corn grain yield of 76 sampled polygons obtained from all of the experimental treatments. The results are shown in Table 4; in column one, we have the combination of input parameters: NDVI, NDRE, WDRVI, EXG, TGI, VARI, canopy cover (C), and plant density (D). The correlation coefficient for training, validation, testing, and total entered data is also presented, as well as the mean absolute error (MAE) and the root mean square error (RMSE).

Table 4. Training, validation, and testing of the artificial neural network with different input variables from vegetation indices, plant density, and canopy cover for the estimation of corn grain yield at 47 and 79 days after sowing (DAS).

Input Variables	47 DAS						79 DAS					
	R Training	R Validation	R Test	R Total	MAE (t ha^{-1})	RMSE (t ha^{-1})	R Training	R Validation	R Test	R Total	MAE (t ha^{-1})	RMSE (t ha^{-1})
NDVI, NDRE, WDRVI, D, C	0.96	0.97	0.98	0.97	0.285	0.414	0.96	0.97	0.98	0.96	0.242	0.337
NDVI, NDRE, WDRVI, TGI, EXG, VARI, D, C	0.98	0.93	0.98	0.97	0.307	0.400	**0.97**	**0.90**	**0.99**	**0.97**	**0.249**	**0.425**
NDRE, D, C	0.97	0.99	0.95	0.97	0.256	0.365	0.97	0.88	0.97	0.96	0.278	0.437
EXG, D, C	0.97	0.98	0.99	0.98	0.252	0.354	0.95	0.97	0.99	0.96	0.280	0.431
TGI, EXG, VARI, D, C	0.97	0.94	0.96	0.96	0.331	0.470	0.92	0.97	0.96	0.94	0.292	0.562
C, D	0.97	0.93	0.95	0.96	0.298	0.441	0.96	0.92	0.95	0.96	0.298	0.441
NDVI, D, C	0.96	0.99	0.93	0.97	0.280	0.425	0.96	0.94	0.91	0.96	0.300	0.449
TGI, D, C	0.96	0.97	0.98	0.97	0.304	0.414	0.94	0.95	0.97	0.95	0.312	0.459
VARI, D, C	0.97	0.99	0.93	0.97	0.279	0.395	0.97	0.93	0.90	0.95	0.347	0.571
NDVI, NDRE, WDRVI, TGI, EXG, VARI, C	0.92	0.86	0.96	0.92	0.512	0.643	0.94	0.97	0.88	0.94	0.381	0.538
NDVI, NDRE, WDRVI, TGI, EXG, VARI	0.80	0.94	0.92	0.86	0.622	0.809	0.84	0.91	0.90	0.86	0.528	0.876
EXG, C	0.81	0.81	0.84	0.81	0.733	0.938	0.80	0.81	0.84	0.80	0.623	0.947
NDVI, NDRE, WDRVI	0.74	0.65	0.85	0.73	0.811	1.093	0.82	0.85	0.87	0.85	0.641	0.837
NDVI, C	0.83	0.82	0.90	0.84	0.597	0.883	0.80	0.86	0.88	0.82	0.649	0.884
VARI, C	0.86	0.94	0.51	0.85	0.604	0.836	0.86	0.79	0.92	0.86	0.653	0.917
NDVI, NDRE, WDRVI, C	0.81	0.93	0.83	0.84	0.618	0.874	0.87	0.87	0.62	0.84	0.672	0.856
WDRVI, C	0.87	0.62	0.95	0.87	0.584	0.784	0.82	0.64	0.56	0.79	0.689	0.971
TGI, EXG, VARI	0.69	0.64	0.82	0.67	0.908	1.189	0.83	0.72	0.72	0.80	0.720	0.951
TGI, EXG, VARI, C	0.86	0.90	0.87	0.86	0.629	0.817	0.78	0.93	0.71	0.78	0.746	1.015
WDRVI, D, C	0.99	0.99	0.99	0.99	0.028	0.125	0.96	0.99	0.89	0.96	0.209	0.449
NDRE, C	0.75	0.98	0.97	0.81	0.527	0.986	0.68	0.68	0.55	0.65	0.774	1.179
TGI, C	0.85	0.62	0.78	0.82	0.701	0.909	0.50	0.62	0.29	0.54	1.017	1.380

NDVI, normalized difference vegetation index; NDRE, normalized difference red edge index; WDRVI, wide dynamic range vegetation index; EXG, green extraction index; TGI, triangular greenness index; VARI, visible atmospherically resistant index; C, canopy cover; D, plant density (plants * m^{-2}); MAE, mean absolute error; RMSE, root mean square error.

For 47 DAS, which is shown in Table 4 and Figure 7, the input parameters WDRVI, plant density, and canopy cover showed the highest correlation coefficient and the smallest errors for the corn grain yield estimation (R = 0.99, MAE = 0.028 t ha^{-1}, RMSE = 0.125 t ha^{-1}) when the total data were used. Using the same parameters above, except plant density as input parameter in the neural network, the correlation coefficient decreased, and the errors increased (R = 0.87, MAE = 0.584 t ha^{-1}, RMSE = 0.784 t ha^{-1}), which indicates that plant density is an important parameter in estimating yield for this flight date. A combination of six vegetation indices (NDVI, NDRE, WDRVI, EXG, TGI, VARI), plant density, and canopy cover generates a model with high correlation in yield estimation (R = 0.97), with a mean absolute error of 307 kg per hectare and a root mean square error of 400 kg per hectare. Doing the same analysis, but without plant density, a lower correlation was obtained with the total data (R = 0.92, MAE = 0.512 t ha^{-1} y RMSE = 0.643 t ha^{-1}), having greater precision than the combination of WDRVI and canopy cover. On the other hand, a combination of only six vegetation indices as input parameters resulted in a correlation of 0.86 and an MAE of 0.622 t ha^{-1}, with an RMSE of 0.809 t ha^{-1} when using the total data in yield estimation. This is good if we consider that there is a decrease in computational cost in obtaining plant density and canopy cover. The EXG index, canopy cover, and plant density showed a correlation coefficient of 0.98. In general, combinations that include canopy cover presented high correlation coefficients (R ≥ 0.80), and incorporating plant density as an input parameter increased the value of the correlation coefficient (R ≥ 0.95). The multispectral indices without canopy cover and plant density as input parameters showed a good correlation (R = 0.73, MAE = 0.811 t ha^{-1}, RMSE = 1.093 t ha^{-1}) in yield estimation. The RGB indices without canopy cover and plant density showed a lower correlation than the multispectral indices, explaining 0.67 of the corn grain yield. Plant density and canopy cover showed a high correlation (R = 0.96, MAE = 0.298 t ha^{-1}, RMSE = 0.441 t ha^{-1}).

For 79 DAS, which is shown in Figure 8 and Table 4, the six vegetation indices, canopy cover, and plant density presented the highest correlation coefficient and the smallest errors (R = 0.97, MAE = 0.249 t ha^{-1}, RMSE = 0.425 t ha^{-1}) when the total data were used. The EXG index, canopy cover, and plant density also showed a good correlation coefficient and small errors (R = 0.97, MAE = 0.280 t ha^{-1}, RMSE = 0.431 t ha^{-1}) in yield estimation. Vegetation indices and canopy cover increased the correlation coefficient with respect to 47 DAS, while the indices without canopy cover and plant density maintained a similar correlation coefficient for all the data. The TGI index and canopy cover showed the lowest correlation and biggest errors (R = 0.54, MAE = 1.017 t ha^{-1}, RMSE = 1.380 t ha^{-1}) in the yield estimate. The multispectral vegetation indices and the visible vegetation indices presented a high correlation in the estimation of corn grain yield at 47 and 79 DAS for the different nitrogen doses tested, presenting slightly higher correlations at 47 DAS.

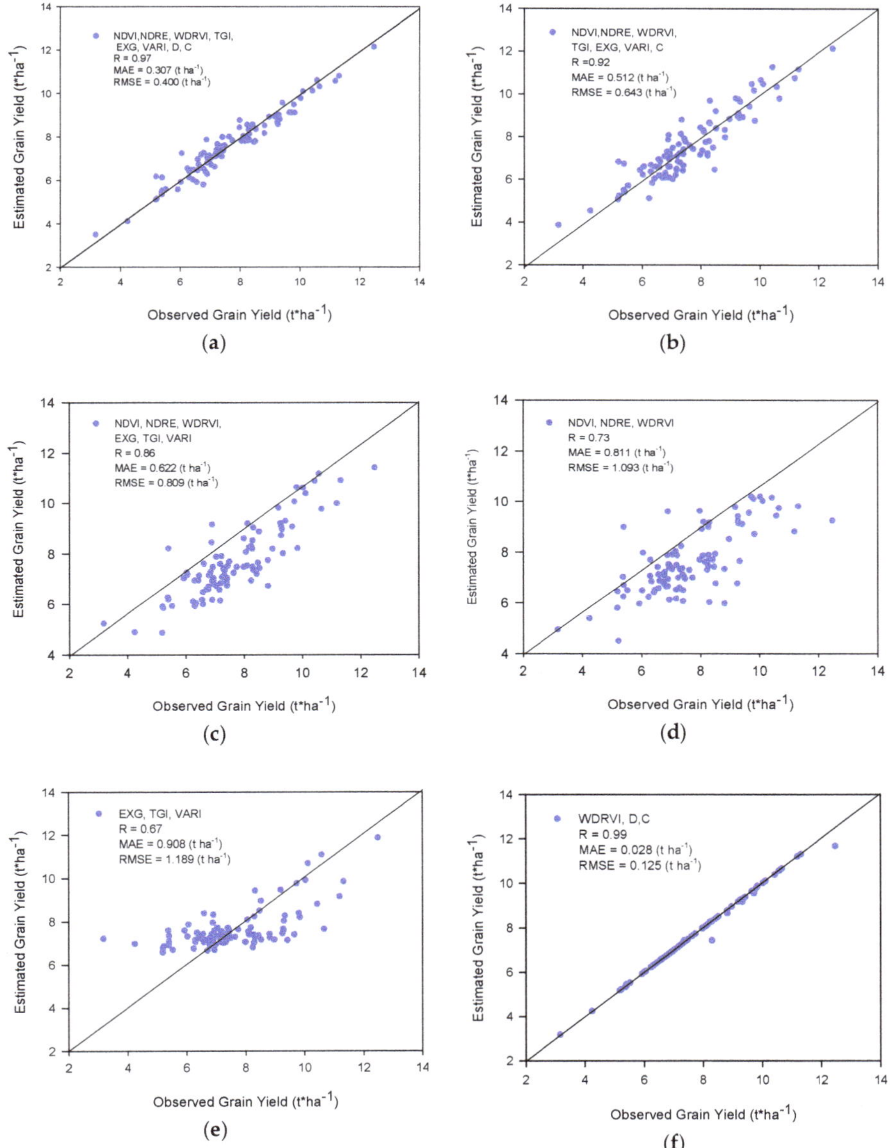

Figure 7. Yields estimated by the neural network during 47 DAS for all observed yield data. (**a**) Yield estimated with the NDVI, NDRE, WDRVI, EXG, TGI, and VARI vegetation indices, as well as planting density (D) and canopy cover (C); (**b**) yield estimated with the NDVI, NDRE, WDRVI, EXG, TGI, VARI vegetation indices, as well as C; (**c**) yield estimated with the NDVI, NDRE, WDRVI, EXG, TGI, and VARI vegetation indices; (**d**) yield estimated with the NDVI, NDRE, and WDRVI vegetation indices; (**e**) yield estimated with EXG, TGI, and VARI vegetation indices; (**f**) yield estimated with the WDRVI vegetation index, as well as D and C.

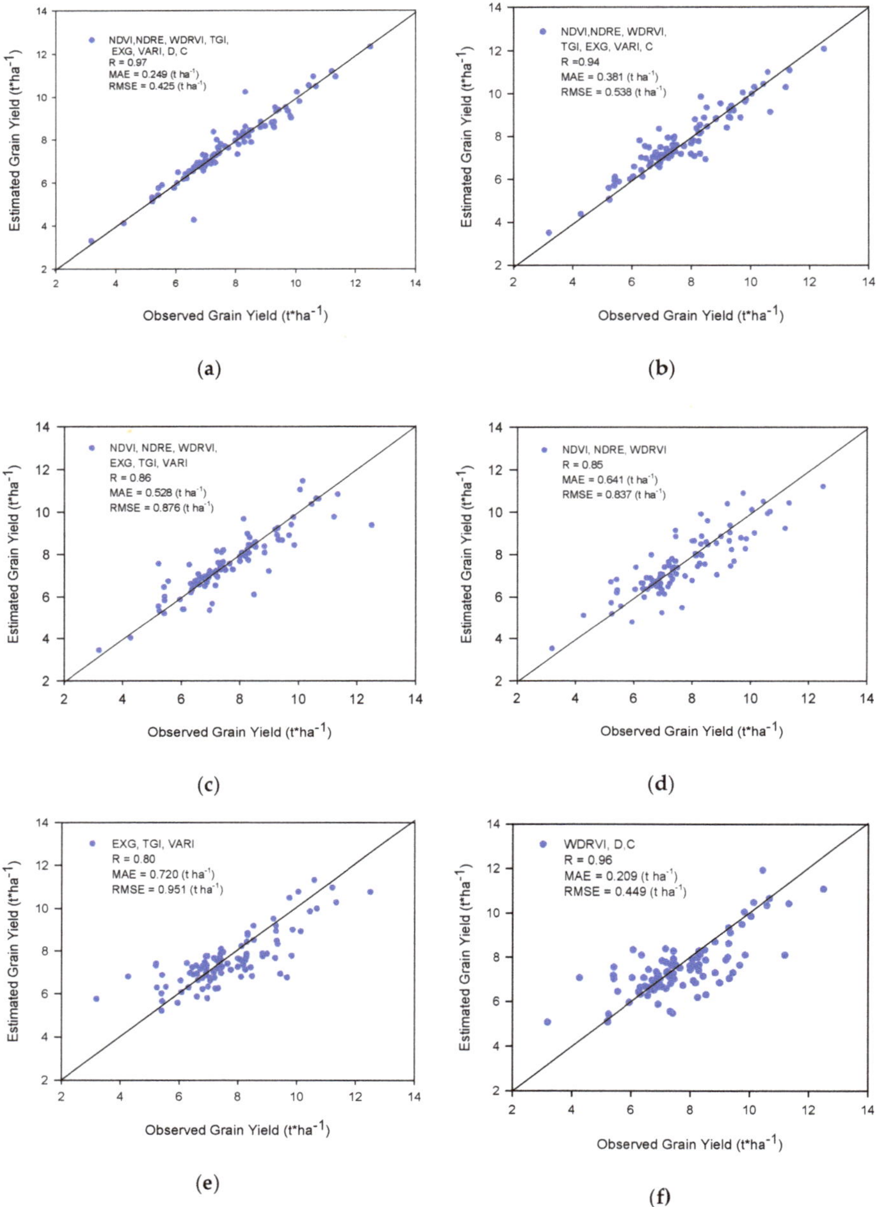

Figure 8. Yields estimated by the neural network during 79 DAS for all observed yield data. (**a**) Yield estimated with the NDVI, NDRE, WDRVI, EXG, TGI, and VARI vegetation indices, as well as plant density (D) and canopy cover (C); (**b**) yield estimated with the NDVI, NDRE, WDRVI, EXG, TGI, and VARI vegetation indices, as well as C; (**c**) yield estimated with the NDVI, NDRE, WDRVI, EXG, TGI, and VARI vegetation indices; (**d**) yield estimated with the NDVI, NDRE, and WDRVI vegetation indices; (**e**) yield estimated with EXG, TGI, and VARI vegetation indices; (**f**) yield estimated with the WDRVI vegetation index, as well as D and C.

3.5. Variables' Relative Importance in the Yield Estimation Using Garson's Algorithm

The importance of the predictor variables (NDVI, NDRE, WDRVI, EXG, TGI, VARI, density, and canopy cover) with respect to the predicted variable (yield) is shown in Figure 9; this was calculated using Garson's algorithm. The results show that the density is the most important predictor for the 47 and 79 DAS with a relative importance of 0.269 and 0.295, respectively; the WDRVI index (0.175) was the second best predictor in importance for the 47 DAS; while the NDVI index (0.184) for the 79 DAS. The VARI index was the least important predictor in the yield estimation for the 47 and 79 DAS with 0.058 and 0.031 relative importance, respectively.

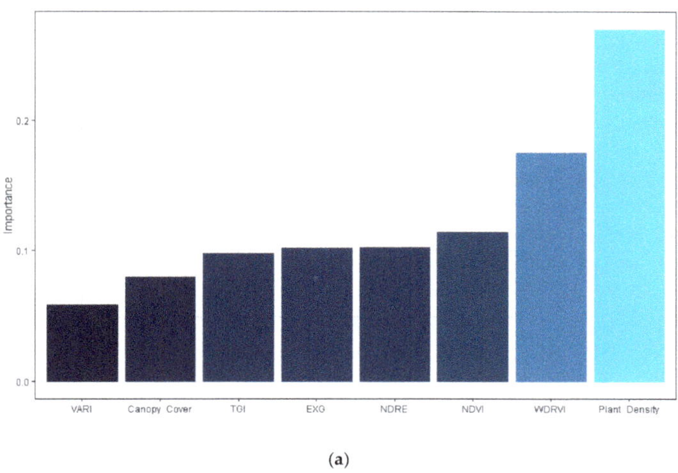

(**a**)

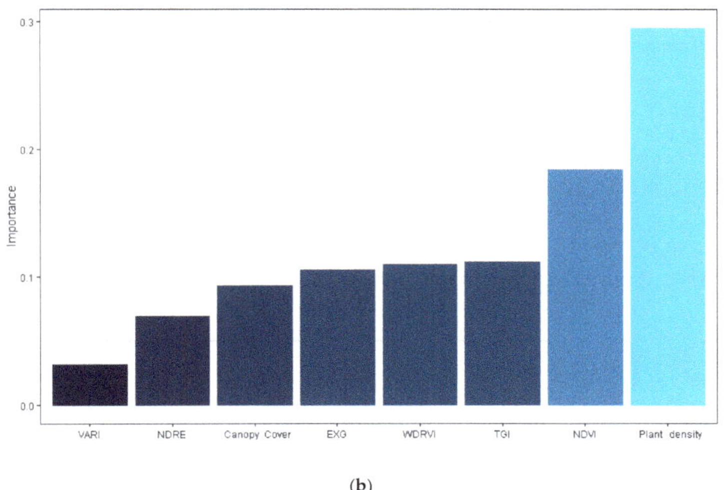

(**b**)

Figure 9. Relative importance of the predictors in the corn grain yield estimation. (**a**) Relative importance of the predictors for the 47 DAS; (**b**) relative importance of the predictors for the 79 DAS.

4. Conclusions

In the present study, the corn grain yield was estimated designing a neural network model based on vegetation indices, canopy cover, and plant density. The relative importance of the predictor variables was also analyzed. The information obtained through the digital processing of images taken by unmanned aerial vehicles allowed to monitor the crop development. The correlation coefficient between the plant density, obtained through the digital count of corn plants, and the corn grain yield was 0.94; this variable was the one with the highest relative importance in the yield estimation according to Garson's algorithm. The canopy cover, digitally estimated by object-oriented classification and using the TGI index, showed a correlation coefficient of 0.77 with respect to the corn grain yield, while the relative importance of this variable in the yield estimation was 0.080 and 0.093 for the 47 and 79 DAS, respectively. The WDRVI, plant density, and canopy cover showed the highest correlation coefficient and the smallest errors (R = 0.99, MAE = 0.028 t ha^{-1}, RMSE = 0.125 t ha^{-1}) in the corn grain yield estimation at 47 DAS, with the WDRVI and the density being the variables with the highest relative importance for this crop development date. For the 79 DAS flight, the combination of the NDVI, NDRE, WDRVI, EXG, TGI, and VARI indices, as well as plant density and canopy cover, generated the highest correlation coefficient and the smallest errors (R = 0.97, MAE = 0.249 t ha^{-1}, RMSE = 0.425 t ha^{-1}) in the corn grain yield estimation, with the density and the NDVI being the variables with the highest relative importance with values of 0.295 and 0.184, respectively. However, the WDRVI, plant density, and canopy cover estimated the corn grain yield with acceptable precision (R = 0.96, MAE = 0.209 t ha^{-1}, RMSE = 0.449 t ha^{-1}). The generated neural network models provided a high correlation coefficient between the estimated and the observed corn grain yield; it also showed acceptable errors in the yield estimation. The spectral information registered through remote sensors mounted on unmanned aerial vehicles and its processing in vegetation indices, canopy cover, and plant density allow the characterization and estimation of corn grain yield. Such information is very useful for decision-making and agricultural activities planning.

At the time of establishing agricultural crops, different techniques, tools, and management are used during the crop development, so it is desirable to carry out future trials for different climates, soils, management, and varieties to have a broader database and with greater parameters to be used in the modeling of crop yields through the use of neural networks.

Author Contributions: Conceptualization, H.F.-M.; Data curation, H.G.-M.; Formal analysis, H.G.-M.; Investigation, H.G.-M. and H.F.-M.; Supervision, H.F.-M.; Writing—original draft, H.G.-M.; Writing—review & editing, H.F.-M., R.A.-H., A.K.-G., L.T.-C., O.R.M.-V., and M.A.V.-P. All authors have read and agreed to the published version of the manuscript.

Funding: This research received no external funding.

Acknowledgments: We thank the financial support of the Colegio de Postgraduados and the National Council of Science and Technology of Mexico (CONACyT) for making this study possible.

Conflicts of Interest: The authors declare no conflict of interest.

References

1. USDA-Office of the Chief Economist. Available online: https://www.usda.gov/oce/commodity/wasde/ (accessed on 28 April 2020).
2. ASERCA. CIMA. Available online: https://www.cima.aserca.gob.mx/ (accessed on 4 March 2020).
3. Domínguez Mercado, C.A.; de Jesús Brambila Paz, J.; Carballo Carballo, A.; Quero Carrillo, A.R. Red de valor para maíz con alta calidad de proteína. *Rev. Mex. Cienc. Agríc.* **2014**, *5*, 391–403.
4. Mercer, K.L.; Perales, H.R.; Wainwright, J.D. Climate change and the transgenic adaptation strategy: Smallholder livelihoods, climate justice, and maize landraces in Mexico. *Glob. Environ. Chang.* **2012**, *22*, 495–504. [CrossRef]
5. Tarancón, M.; Díaz-Ambrona, C.H.; Trueba, I. Cómo alimentar a 9.000 millones de personas en el 2050? In Proceedings of the XV Congreso Internacional de Ingeniería de Proyectos, Huesca, Spain, 6–8 July 2011.

6. Cervantes, R.A.; Angulo, G.V.; Tavizón, E.F.; González, J.R. Impactos potenciales del cambio climático en la producción de maíz Potential impacts of climate change on maize production. *Investigación Ciencia* **2014**, *22*, 48–53.

7. Moore, F.C.; Lobell, D.B. Reply to Gonsamo and Chen: Yield findings independent of cause of climate trends. *Proc. Natl. Acad. Sci. USA* **2015**, *112*, E2267. [CrossRef] [PubMed]

8. Ruiz Corral, J.A.; Medina García, G.; Ramírez Díaz, J.L.; Flores López, H.E.; Ramírez Ojeda, G.; Manríquez Olmos, J.D.; Zarazúa Villaseñor, P.; González Eguiarte, D.R.; Díaz Padilla, G.; Mora Orozco, C.D.L. Cambio climático y sus implicaciones en cinco zonas productoras de maíz en México. *Rev. Mex. Cienc. Agríc.* **2011**, *2*, 309–323.

9. Tinoco-Rueda, J.A.; Gómez-Díaz, J.D.; Monterroso-Rivas, A.I.; Tinoco-Rueda, J.A.; Gómez-Díaz, J.D.; Monterroso-Rivas, A.I. Efectos del cambio climático en la distribución potencial del maíz en el estado de Jalisco, México. *Terra Latinoam.* **2011**, *29*, 161–168.

10. Bolaños, H.O.; Vázquez, M.H.; Juárez, G.G.; González, G.S. Cambio climático: Una percepción de los productores de maíz de temporal en el estado de Tlaxcala, México. *CIBA Rev. Iberoam. Las Cienc. Biológicas Agropecu.* **2019**, *8*, 1–26. [CrossRef]

11. Khaki, S.; Wang, L.; Archontoulis, S.V. A CNN-RNN Framework for Crop Yield Prediction. *Front. Plant Sci.* **2020**, *10*. [CrossRef]

12. Dahikar, S.S.; Rode, S.V. Agricultural Crop Yield Prediction Using Artificial Neural Network Approach. *Int. J. Innov. Res. Electr. Electron. Instrum. Control Eng.* **2014**, *2*, 683–686.

13. Li, A.; Liang, S.; Wang, A.; Qin, J. Estimating Crop Yield from Multi-temporal Satellite Data Using Multivariate Regression and Neural Network Techniques. *Photogramm. Eng. Remote Sens.* **2007**, *73*, 1149–1157. [CrossRef]

14. Assefa, Y.; Vara Prasad, P.V.; Carter, P.; Hinds, M.; Bhalla, G.; Schon, R.; Jeschke, M.; Paszkiewicz, S.; Ciampitti, I.A. Yield Responses to Planting Density for US Modern Corn Hybrids: A Synthesis-Analysis. *Crop. Sci.* **2016**, *56*, 2802–2817. [CrossRef]

15. Kitano, B.T.; Mendes, C.C.T.; Geus, A.R.; Oliveira, H.C.; Souza, J.R. Corn Plant Counting Using Deep Learning and UAV Images. *IEEE Geosci. Remote Sens. Lett.* **2019**, 1–5. [CrossRef]

16. Lindblom, J.; Lundström, C.; Ljung, M.; Jonsson, A. Promoting sustainable intensification in precision agriculture: Review of decision support systems development and strategies. *Precis. Agric.* **2017**, *18*, 309–331. [CrossRef]

17. Geipel, J.; Link, J.; Claupein, W. Combined Spectral and Spatial Modeling of Corn Yield Based on Aerial Images and Crop Surface Models Acquired with an Unmanned Aircraft System. *Remote Sens.* **2014**, *6*, 10335–10355. [CrossRef]

18. Tsouros, D.C.; Bibi, S.; Sarigiannidis, P.G. A Review on UAV-Based Applications for Precision Agriculture. *Information* **2019**, *10*, 349. [CrossRef]

19. Olson, D.; Chatterjee, A.; Franzen, D.W.; Day, S.S. Relationship of Drone-Based Vegetation Indices with Corn and Sugarbeet Yields. *Agron. J.* **2019**, *111*, 2545–2557. [CrossRef]

20. Xue, J.; Su, B. Significant Remote Sensing Vegetation Indices: A Review of Developments and Applications. *J. Sens.* **2017**. [CrossRef]

21. Peñuelas, J.; Filella, I. Visible and near-infrared reflectance techniques for diagnosing plant physiological status. *Trends Plant Sci.* **1998**, *3*, 151–156. [CrossRef]

22. Serrano, L.; Filella, I.; Peñuelas, J. Remote Sensing of Biomass and Yield of Winter Wheat under Different Nitrogen Supplies. *Crop. Sci.* **2000**, *40*, 723–731. [CrossRef]

23. Buchaillot, M.; Gracia-Romero, A.; Vergara-Diaz, O.; Zaman-Allah, M.A.; Tarekegne, A.; Cairns, J.E.; Prasanna, B.M.; Araus, J.L.; Kefauver, S.C. Evaluating Maize Genotype Performance under Low Nitrogen Conditions Using RGB UAV Phenotyping Techniques. *Sensors* **2019**, *19*, 1815. [CrossRef]

24. Kefauver, S.C.; El-Haddad, G.; Vergara-Diaz, O.; Araus, J.L. RGB picture vegetation indexes for High-Throughput Phenotyping Platforms (HTPPs). In *Remote Sensing for Agriculture, Ecosystems, and Hydrology XVII, Volume 9637*; International Society for Optics and Photonics: Touluse, France, 2015; p. 96370. [CrossRef]

25. Vergara-Diaz, O.; Kefauver, S.C.; Elazab, A.; Nieto-Taladriz, M.T.; Araus, J.L. Grain yield losses in yellow-rusted durum wheat estimated using digital and conventional parameters under field conditions. *Crop. J.* **2015**, *3*, 200–210. [CrossRef]

26. Jeong, J.H.; Resop, J.P.; Mueller, N.D.; Fleisher, D.H.; Yun, K.; Butler, E.E.; Timlin, D.J.; Shim, K.M.; Gerber, J.S.; Reddy, V.R.; et al. Random Forests for Global and Regional Crop Yield Predictions. *PLoS ONE* **2016**, *11*. [CrossRef] [PubMed]

27. Oguntunde, P.G.; Lischeid, G.; Dietrich, O. Relationship between rice yield and climate variables in southwest Nigeria using multiple linear regression and support vector machine analysis. *Int. J. Biometeorol.* **2018**, *62*, 459–469. [CrossRef] [PubMed]

28. Pantazi, X.E.; Moshou, D.; Alexandridis, T.; Whetton, R.L.; Mouazen, A.M. Wheat yield prediction using machine learning and advanced sensing techniques. *Comput. Electron. Agric.* **2016**, *121*, 57–65. [CrossRef]

29. Panda, S.S.; Panigrahi, S.; Ames, D.P. Crop Yield Forecasting from Remotely Sensed Aerial Images with Self-Organizing Maps. *Trans. ASABE* **2010**, *53*, 323–338. [CrossRef]

30. Schwalbert, R.A.; Amado, T.; Corassa, G.; Pott, L.P.; Prasad, P.V.V.; Ciampitti, I.A. Satellite-based soybean yield forecast: Integrating machine learning and weather data for improving crop yield prediction in southern Brazil. *Agric. For. Meteorol.* **2020**, *284*, 107886. [CrossRef]

31. Waheed, T.; Bonnell, R.B.; Prasher, S.O.; Paulet, E. Measuring performance in precision agriculture: CART—A decision tree approach. *Agric. Water Manag.* **2006**, *84*, 173–185. [CrossRef]

32. Ashapure, A.; Oh, S.; Marconi, T.G.; Chang, A.; Jung, J.; Landivar, J.; Enciso, J. Unmanned aerial system based tomato yield estimation using machine learning. In *Autonomous Air and Ground Sensing Systems for Agricultural Optimization and Phenotyping IV*; International Society for Optics and Photonics: Baltimore, MD, USA, 2019; Volume 11008, p. 110080O. [CrossRef]

33. Fu, Z.; Jiang, J.; Gao, Y.; Krienke, B.; Wang, M.; Zhong, K.; Cao, Q.; Tian, Y.; Zhu, Y.; Cao, W.; et al. Wheat Growth Monitoring and Yield Estimation based on Multi-Rotor Unmanned Aerial Vehicle. *Remote Sens.* **2020**, *12*, 508. [CrossRef]

34. Khaki, S.; Wang, L. Crop Yield Prediction Using Deep Neural Networks. In *Smart Service Systems, Operations Management, and Analytics*; Yang, H., Qiu, R., Chen, W., Eds.; Springer Proceedings in Business and Economics; Springer: Berlin, Germany, 2020; pp. 139–147. [CrossRef]

35. Kim, N.; Ha, K.-J.; Park, N.-W.; Cho, J.; Hong, S.; Lee, Y.-W. A Comparison between Major Artificial Intelligence Models for Crop Yield Prediction: Case Study of the Midwestern United States, 2006–2015. *ISPRS Int. J. Geo Inf.* **2019**, *8*, 240. [CrossRef]

36. Wang, A.X.; Tran, C.; Desai, N.; Lobell, D.; Ermon, S. Deep Transfer Learning for Crop Yield Prediction with Remote Sensing Data. In Proceedings of the 1st ACM SIGCAS Conference on Computing and Sustainable Societies (COMPASS '18), New York, NY, USA, 20–22 June 2018; pp. 1–5. [CrossRef]

37. You, J.; Li, X.; Low, M.; Lobell, D.; Ermon, S. Deep Gaussian Process for Crop Yield Prediction Based on Remote Sensing Data. In Proceedings of the Thirty-First AAAI Conference on Artificial Intelligence, San Francisco, CA, USA, 4–9 February 2017; Available online: https://aaai.org/ocs/index.php/AAAI/AAAI17/paper/view/14435 (accessed on 19 February 2020).

38. Fieuzal, R.; Marais Sicre, C.; Baup, F. Estimation of corn yield using multi-temporal optical and radar satellite data and artificial neural networks. *Int. J. Appl. Earth Obs. Geoinf.* **2017**, *57*, 14–23. [CrossRef]

39. Reisi-Gahrouei, O.; Homayouni, S.; McNairn, H.; Hosseini, M.; Safari, A. Crop Biomass Estimation Using Multi Regression Analysis and Neural Networks from Multitemporal L-Band Polarimetric Synthetic Aperture Radar Data. Available online: https://pubag.nal.usda.gov/catalog/6422744 (accessed on 27 February 2020).

40. Han, L.; Yang, G.; Dai, H.; Xu, B.; Yang, H.; Feng, H.; Li, Z.; Yang, X. Modeling maize above-ground biomass based on machine learning approaches using UAV remote-sensing data. *Plant Methods* **2019**, *15*, 10. [CrossRef]

41. Michelon, G.K.; Menezes, P.L.; de Bazzi, C.L.; Jasse, E.P.; Magalhães, P.S.G.; Borges, L.F. Artificial neural networks to estimate the productivity of soybeans and corn by chlorophyll readings. *J. Plant Nutr.* **2018**, *41*, 1285–1292. [CrossRef]

42. Khaki, S.; Khalilzadeh, Z.; Wang, L. Predicting yield performance of parents in plant breeding: A neural collaborative filtering approach. *PLoS ONE* **2020**, *15*, e0233382. [CrossRef] [PubMed]

43. Allan, B.M.; Ierodiaconou, D.; Hoskins, A.J.; Arnould, J.P.Y. A Rapid UAV Method for Assessing Body Condition in Fur Seals. *Drones* **2019**, *3*, 24. [CrossRef]

44. Lucieer, A.; Jong, S.M.; de Turner, D. Mapping landslide displacements using Structure from Motion (SfM) and image correlation of multi-temporal UAV photography. *Prog. Phys. Geogr. Earth Environ.* **2014**, *38*, 97–116. [CrossRef]

45. James, M.R.; Robson, S.; d'Oleire-Oltmanns, S.; Niethammer, U. Optimising UAV topographic surveys processed with structure-from-motion: Ground control quality, quantity and bundle adjustment. *Geomorphology* **2017**, *280*, 51–66. [CrossRef]

46. Franzini, M.; Ronchetti, G.; Sona, G.; Casella, V. Geometric and Radiometric Consistency of Parrot Sequoia Multispectral Imagery for Precision Agriculture Applications. *Appl. Sci.* **2019**, *9*, 5314. [CrossRef]

47. Radiometric Corrections. Support. Available online: http://support.pix4d.com/hc/en-us/articles/202559509 (accessed on 16 June 2020).

48. Hunt, E.R.; Doraiswamy, P.C.; McMurtrey, J.E.; Daughtry, C.S.T.; Perry, E.M.; Akhmedov, B. A visible band index for remote sensing leaf chlorophyll content at the canopy scale. *Int. J. Appl. Earth Obs. Geoinf.* **2013**, *21*, 103–112. [CrossRef]

49. Woebbecke, D.M.; Meyer, G.E.; Von Bargen, K.; Mortensen, D.A. Color indices for weed identification under various soil, residue, and lighting conditions. *Trans. ASAE* **1995**, *38*, 259–269. [CrossRef]

50. Gitelson, A.A.; Kaufman, Y.J.; Stark, R.; Rundquist, D. Novel Algorithms for Remote Estimation of Vegetation Fraction. *Pap. Nat. Resour.* **2002**, *80*, 76–87. [CrossRef]

51. Rouse, J.W. Monitoring Vegetation Systems in the Great Plains with ERTS. 1974. Available online: https://ntrs.nasa.gov/search.jsp?R=19740022614 (accessed on 19 February 2020).

52. Barnes, E.M.; Clarke, T.R.; Richards, S.E.; Colaizzi, P.D.; Haberland, J.; Kostrzewski, M.; Waller, P.; Choi, C.; Riley, E.; Thompson, T.; et al. Coincident detection of crop water stress, nitrogen status and canopy density using ground-based multispectral data. In Proceedings of the Fifth International Conference on Precision Agriculture and Other Resource Management, ASA–CSSA–SSSA, Bloomington, MN, USA, 16–19 July 2000; Precision Agriculture Center, University of Minnesota, ASA-CSSA-SSSA: Madison, WI, USA, 2000; pp. 16–19.

53. Gitelson, A.A. Wide Dynamic Range Vegetation Index for Remote Quantification of Biophysical Characteristics of Vegetation. *J. Plant Physiol.* **2004**, *161*, 165–173. [CrossRef] [PubMed]

54. García-Martínez, H.; Flores-Magdaleno, H.; Khalil-Gardezi, A.; Ascencio-Hernández, R.; Tijerina-Chávez, L.; Vázquez-Peña, M.A.; Mancilla-Villa, O.R. Digital Count of Corn Plants Using Images Taken by Unmanned Aerial Vehicles and Cross Correlation of Templates. *Agronomy* **2020**, *10*, 469. [CrossRef]

55. Conrad, O.; Bechtel, B.; Bock, M.; Dietrich, H.; Fischer, E.; Gerlitz, L.; Wehberg, J.; Wichmann, V.; Böhner, J. System for Automated Geoscientific Analyses (SAGA) v. 2.1.4. *Geosci. Model Dev.* **2015**, *8*, 1991–2007. [CrossRef]

56. Rubin, J. Optimal classification into groups: An approach for solving the taxonomy problem. *J. Theor. Biol.* **1967**, *15*, 103–144. [CrossRef]

57. Kumar, A.; Tiwari, A. A Comparative Study of Otsu Thresholding and K-means Algorithm of Image Segmentation. *Int. J. Eng. Technol. Res.* **2019**, *9*, 2454–4698. [CrossRef]

58. Liu, D.; Yu, J. Otsu Method and K-means. In Proceedings of the 2009 Ninth International Conference on Hybrid Intelligent Systems, Shenyang, China, 12–14 August 2009; Volume 1, pp. 344–349. [CrossRef]

59. Benz, U.C.; Hofmann, P.; Willhauck, G.; Lingenfelder, I.; Heynen, M. Multi-resolution, object-oriented fuzzy analysis of remote sensing data for GIS-ready information. *ISPRS J. Photogramm. Remote Sens.* **2004**, *58*, 239–258. [CrossRef]

60. eCognition Suite Dev RB. Available online: https://docs.ecognition.com/v9.5.0/Page%20collection/eCognition%20Suite%20Dev%20RB.htm (accessed on 21 April 2020).

61. Chandola, V.; Banerjee, A.; Kumar, V. Anomaly detection: A survey. *ACM Comput. Surv. CSUR* **2009**, *41*, 15:1–15:58. [CrossRef]

62. Torres, J. Deep Learning—Introducción Práctica con Keras. Jordi TORRES.AI. Available online: https://torres.ai/deep-learning-inteligencia-artificial-keras/ (accessed on 24 February 2020).

63. Pedamonti, D. Comparison of Non-Linear Activation Functions for Deep Neural Networks on MNIST Classification Task. ArXiv180402763 Cs Stat. Available online: http://arxiv.org/abs/1804.02763 (accessed on 16 June 2020).

64. Sharma, S. Activation functions in neural networks. *Data Sci.* **2017**, *6*, 310–316.

65. Levenberg, K. A method for the solution of certain non-linear problems in least squares. *Q. Appl. Math.* **1944**, *2*, 164–168. [CrossRef]

66. Marquardt, D.W. An Algorithm for Least-Squares Estimation of Nonlinear Parameters. *J. Soc. Ind. Appl. Math.* **1963**, *11*, 431–441. [CrossRef]

67. Vogl, T.P.; Mangis, J.K.; Rigler, A.K.; Zink, W.T.; Alkon, D.L. Accelerating the convergence of the back-propagation method. *Biol. Cybern.* **1988**, *59*, 257–263. [CrossRef]
68. Zhang, Z.; Beck, M.W.; Winkler, D.A.; Huang, B.; Sibanda, W.; Goyal, H. Opening the black box of neural networks: Methods for interpreting neural network models in clinical applications. *Ann. Transl. Med.* **2018**, *6*. [CrossRef] [PubMed]
69. Garson, G.D. Interpreting neural-network connection weights. *AI Experts* **1991**, *6*, 46–51.
70. Goh, A.T. Back-propagation neural networks for modeling complex systems. *Artif. Intell. Eng.* **1995**, *9*, 143–151. [CrossRef]
71. Gitelson, A.A.; Viña, A.; Arkebauer, T.J.; Rundquist, D.C.; Keydan, G.; Leavitt, B. Remote estimation of leaf area index and green leaf biomass in maize canopies. *Geophys. Res. Lett.* **2003**, *30*. [CrossRef]
72. Maresma, Á.; Ariza, M.; Martínez, E.; Lloveras, J.; Martínez-Casasnovas, J.A. Analysis of Vegetation Indices to Determine Nitrogen Application and Yield Prediction in Maize (*Zea mays* L.) from a Standard UAV Service. *Remote Sens.* **2016**, *8*, 973. [CrossRef]
73. Zhang, Y.; Han, W.; Niu, X.; Li, G. Maize Crop Coefficient Estimated from UAV-Measured Multispectral Vegetation Indices. *Sensors* **2019**, *19*, 5250. [CrossRef]
74. Zhang, M.; Zhou, J.; Sudduth, K.A.; Kitchen, N.R. Estimation of maize yield and effects of variable-rate nitrogen application using UAV-based RGB imagery. *Biosyst. Eng.* **2020**, *189*, 24–35. [CrossRef]
75. Zhou, X.; Zheng, H.B.; Xu, X.Q.; He, J.Y.; Ge, X.K.; Yao, X.; Cheng, T.; Zhu, Y.; Cao, W.X.; Tian, Y.C. Predicting grain yield in rice using multi-temporal vegetation indices from UAV-based multispectral and digital imagery. *ISPRS J. Photogramm. Remote Sens.* **2017**, *130*, 246–255. [CrossRef]
76. Fernández, E.; Gorchs, G.; Serrano, L. Use of consumer-grade cameras to assess wheat N status and grain yield. *PLoS ONE* **2019**, *14*. [CrossRef]

 *agriculture*

Article

Modeling the Dynamic Response of Plant Growth to Root Zone Temperature in Hydroponic Chili Pepper Plant Using Neural Networks

Galih Kusuma Aji [1,2,*], Kenji Hatou [3] and Tetsuo Morimoto [3]

[1] The United Graduate School of Agricultural Sciences, Ehime University, Matsuyama, Ehime 790-8577, Japan
[2] Department of Bioresources Technology and Veterinary, Universitas Gadjah Mada, Daerah Istimewa Yogyakarta 55281, Indonesia
[3] Faculty of Agriculture, Ehime University, Matsuyama, Ehime 790-8577, Japan; hatou@ehime-u.ac.jp (K.H.); morimoto@agr.ehime-u.ac.jp (T.M.)
* Correspondence: ajikun@ugm.ac.id

Received: 25 May 2020; Accepted: 16 June 2020; Published: 17 June 2020

Abstract: One of the essential factors in the root zone environment that affects plant growth is temperature. Determining the optimal root zone temperature condition in a hydroponic system during cultivation could lead to an improvement in plant growth. An optimal control strategy can be determined by identifying the eco-physiological process using a dynamic model. However, it is difficult to develop a dynamic model of the responses of plant growth to root zone temperature because the eco-physiological processes of plants are quite complicated. We propose an intelligent approach that can deal with this complex system. Non-linear autoregressive with exogenous input (NARX) neural networks were used to develop a dynamic model of the responses of plant growth to root zone temperature. The responses of chili pepper plant growth as affected by root zone temperature were measured during 60 days of cultivation inside a growth chamber using a non-destructive and continuous system based on a load cell. Five datasets of dynamic responses of plant growth were obtained for system identification. The results suggest that the application of a neural network is useful for modeling the dynamic response of plant growth to root zone temperature in hydroponic cultivation, with promising performance.

Keywords: plant growth; dynamic response; root zone temperature; dynamic model; NARX neural networks; hydroponics

---

1. Introduction

In recent years, there has been growing interest in adopting hydroponic systems for crop production worldwide. In combination with protected agriculture such as greenhouses and plant factories, about 3.5% of the worldwide area has adopted hydroponic systems for crop production [1]. As a soilless technology, the benefits of adopting a hydroponic system for growing plants are flexibility and accuracy in controlling the root zone environment. By using a hydroponic system, the root zone environmental factors that influence crop growth and development can be controlled in optimal conditions during each cultivation stage [1]. Thus, the crops can grow to their maximum potential. Given its potential advantages, hydroponic culture is categorized as an intensive method of crop production. In comparison with soil culture, hydroponic systems can offer a higher yield, yet lower water and land usage [1].

Among the root zone environmental factors, temperature is one of the most important factors that influence plant growth and development [2]. It has been reported that the mechanisms of nutrient and water absorption processes within the roots are mainly regulated by root zone temperature

(RZT) [3–5]. Therefore, RZT has been widely used as the determining factor for promoting plant growth in hydroponic systems. For instance, during cold and hot air temperatures in winter and summer, instead of completely controlling the air temperature, RZT control is used as a more economical and wise solution to maintain plant growth under greenhouse cultivation [6,7]. A better understanding of optimal RZT in hydroponic cultivation could lead to the improvement of plant growth.

Studies on environmental control technology for plant production systems [8–11] suggest that the optimal plant conditions during cultivation vary between growth stages. This is because the physiological status of the plant is changing and is remarkably affected by changes in environmental factors. This control approach is mainly used in modern protected agriculture that applies real-time control techniques such as optimal and adaptive control as its control strategy [9,12,13]. For realizing the optimal strategy, an exact dynamic model of an eco-physiological process is necessary. This is because by predicting and simulating the behavior of the eco-physiological process of a plant using a dynamic model, an optimal control strategy can be determined easily [14]. This means that the model's accuracy plays a major part in the performance of optimal control strategies. Moreover, a dynamic model is also used to better understand the process behavior and synthesis of the control system [15].

In order to control the growth of plants using the RZT, a dynamic model of the process is necessary. By constructing a dynamic model of the process, the optimal control strategy of RZT in hydroponic systems can be determined. In recent years, various studies on eco-physiological modeling and environmental control technology for plant production systems in protected cultivation have been intensively conducted [8,9,11,13,14,16,17]. In our previous study, we examined and identified the response of the leaf water content of plants as affected by changes in the RZT within a short period of time [18]; however, the dynamic response of plant growth as affected by change in the RZT has not been identified. A better understanding of this system is necessary for a comprehensive study on plant growth control in plant production systems.

It is well known that there are two main methods for the model-building of a system. First is theoretical modeling, which applies a mathematical method from a well-defined system based on theoretical analysis, e.g., derived from physical and chemical laws [19]. The other approach is experimental modeling, which is also known as system identification, in which a model is constructed based on the measurement of input–output signals of a system. One major advantage of system identification is that it can be applied for an unknown process [19]. The building of a dynamic model of the eco-physiological process of a plant, however, is difficult to conduct successfully using theoretical modeling. This is because most of the eco-physiological processes of plants have strong non-linearity, time delay, and time-varying parameters, which can be characterized as a complex system with uncertain parameters [8,14,20–22]. This unknown process is also termed a black box system.

One useful approach to identifying a black box system is artificial neural networks (ANNs) [23]. ANNs are an information-processing approach inspired by the biological neural system. Artificial neural networks can identify a complex system without requiring prior knowledge of the relationships among the parameters within the system by learning from its input–output signals [24–26]. With these capabilities, artificial neural networks are useful in handling uncertainties and non-linear relationships [23]. Moreover, artificial neural networks are a general-purpose approach to dealing with such a complex system. Not only effective for non-linear regression, artificial neural networks are also useful for other applications such as classification, clustering, pattern recognition, and forecasting. Further, analysis using artificial neural networks does not require restrictive assumptions about the data (non-parametric), which makes this approach flexible to use [23]. These properties make artificial neural networks superior to other statistical methods in the identification of a complex system [26]. Given this ability, artificial neural networks have high potential to successfully identify the complex eco-physiological process of plants. In the agricultural sector, the artificial neural network technique has been extensively developed in various applications [9,27,28]. Hence, system identification using the artificial neural networks mentioned above may be useful in constructing a model of the dynamic response of plant growth to RZT.

The present study is an attempt to apply an artificial neural network to model the dynamic responses of plant growth as affected by change in the RZT in hydroponic chili pepper plants. The dynamic model proposed here was developed using system identification based on artificial neural networks in a single input–single output (SISO) system. The model estimates the output variable of dynamic response in plant growth using the input variable of RZT.

2. Materials and Methods

2.1. Plant Materials

Pepper (*Capsicum annum* L.) is an important commodity, and it is one of the most popular crops grown in protected agriculture all over the world [29]. Moreover, peppers are well known as temperature-sensitive cultivars [30–33]. Hence, in this study, a pepper cultivar was used.

Chili pepper seeds (*Capsicum annum* L. cv. Takanotsume Togarashi; Takii Seed Ltd., Kyoto, Japan) were sown in seedling trays and germinated at a controlled room temperature of 26 °C. After germination, at 15 days after seeding (DAS), the seedlings were transplanted to soilless media of water-soaked polyurethane sponge blocks (2.3 × 2.3 × 2.7 cm) and grown in a greenhouse at day/night temperatures of 25/20 °C under ambient light. Seedlings were watered as needed and supplied with nutrient solution (Otsuka liquid fertilizer; OAT Agrio Co., Ltd., Tokyo, Japan) at 1 dS m^{-1}.

2.2. Experimental Design

At 35 DAS, the plants were transferred to a growth chamber (2.5 × 2.5 × 2 m; (NK System; Nippon Medical and Chemical Instruments Co., Ltd., Osaka, Japan). During observations, the growth chamber was set up consistently with a 12 h photoperiod of 270 µmol m^{-2} s^{-1} PPFD (photosynthetic photon flux density) as measured at the base of the growth chamber using a T&D TR-74i illuminance ultraviolet (UV) recorder (T&D Corporation, Matsumoto, Japan), day/night temperature 25/20 ± 1°C, relative humidity 55/70 ± 5%, and nutrient solution 2.3 ± 0.2 dS m^{-1}. Meanwhile, the dissolved oxygen level of the nutrient solution was maintained with the application of an air bubble generator.

To obtain adequate information about the dynamic response of plant growth to RZT, five random patterns of RZT were applied in the range 15–37 °C. Each pattern consisted of three sample plants planted in an independent deep flow technique (DFT) hydroponic system. A total of 15 plants were used for this experiment. To control the RZT, each pattern was equipped with an automatic independent water temperature control system using a NETC-3 thermostat (Newmarins Co. Ltd., Fukuoka, Japan). The thermostat regulated the nutrient solution temperature by controlling a ceramic water heater (Power Safe PRO, Nisso, Japan) and cooling water circulator (FCW-10 Fine Circulators; TGK Co. Ltd., Tokyo, Japan), which circulated cooling water through a copper pipe inside the hydroponic system. To supply the nutrient solution inside the container, an automatic system periodically pumped each container with nutrient solution. The control system is depicted in Figure 1.

2.3. Measurement of Plant Growth

Dynamic changes in plant development were monitored through changes in the fresh weight of the plants during observation. In a controlled environment, the fresh weight of a plant provides useful information on the status of the plant and is also an indicator of plant growth [34–36]. In this study, an automated non-destructive plant weight measurement system for pepper plants was developed. The measurement system consisted of a CZL635 micro load cell (loads up to 5 kg and 0.05% precision; Phidgets Inc., Calgary, Canada), load cell support structure, and plant holder, as illustrated in Figure 1. However, because the plant roots were in the nutrient solution, the weight of the plant roots was balanced by buoyant force. Thus, only the weight of the plant shoots was measured. To record the change in plant weight during the experiment, data were sampled at five-minute intervals and stored on a microcomputer (Raspberry Pi 3 Model B+; Raspberry Pi Foundation, Cambridge, UK).

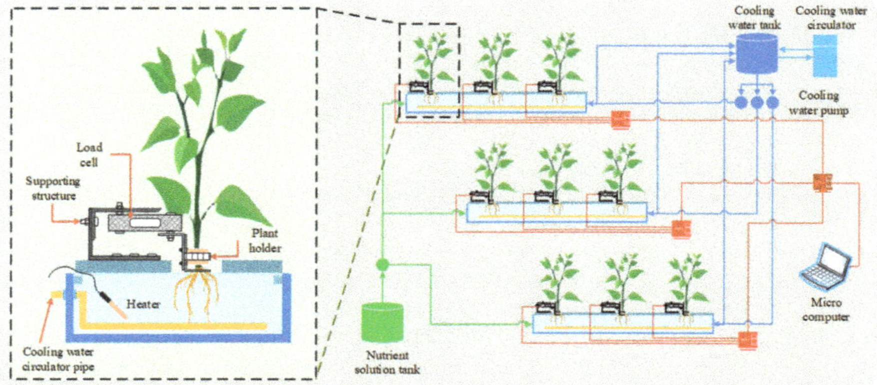

Figure 1. Root zone temperature control system and plant growth measuring system using a load cell sensor.

The dynamic response of plant growth was determined as the growth rate of plant weight, which measures the sensitivity of the change in plant weight (ΔW, [g]) with respect to the change in time (Δt, [d]), as shown in the following equation:

$$WR(t) = \frac{\Delta W}{\Delta t} \tag{1}$$

2.4. System Identification Method

2.4.1. Data Preprocessing

Sensors produce a large number of data points, which can be an advantage in modeling; therefore, these data need to be reliably preprocessed. In supervised learning models, unreliable data could lead to wrong outputs, which affect model performance [37]. Data preprocessing methods were applied in this study to avoid unreliability and to organize data for model development.

Data that deviate considerably from the remaining observations can be defined as outliers. In this study, to detect and remove outliers from time series data, the Hampel Identifier method [38] was applied. The missing data then need to be replaced to maintain completeness and the trend of time series data. Thus, the missing observation data were replaced by the moving average interpolation of their latest neighboring time series data.

The response of plant growth to changes in environmental conditions can be categorized as a long response that can be identified daily. Thus, resampling or aggregation of time series data was needed. In this step, data were resampled on a daily basis using the averaging procedure and then smoothed using the Savitzky–Golay filter [39]. All data preprocessing was performed using the Matlab® Signal Processing Toolbox™ R2019a (MathWorks® Inc., Natick, MA, USA).

2.4.2. Dynamic Neural Networks for System Identification

In the present study, system identification was considered as a single input–single output system with unknown parameters. Figure 2 shows a block diagram of the SISO system. For the control system of plant growth, the output variable of plant growth response (*WR*) was estimated from the input variable of RZT (*RT*).

Artificial neural networks are gaining attention as a general solution for handling non-linear system problems due to their universal approximation capabilities [23]. However, specific neural network architectures are required to solve certain problems optimally. Thus, in recent years, many variants of artificial neural network architecture have been developed. For identification of a dynamic system, in this study, a non-linear autoregressive with exogenous input (NARX) network [40] is considered. NARX networks are a particular class of artificial neural networks characterized by a feedback procedure on the output which can identify any non-linear dynamic system. NARX networks have been employed in various applications of dynamic systems. NARX networks also showed better performance in the prediction of long-term dependencies [41]. Moreover, in particular, for constructing a dynamic model in control system problems, NARX has been widely employed with excellent performance [18,40,42–45].

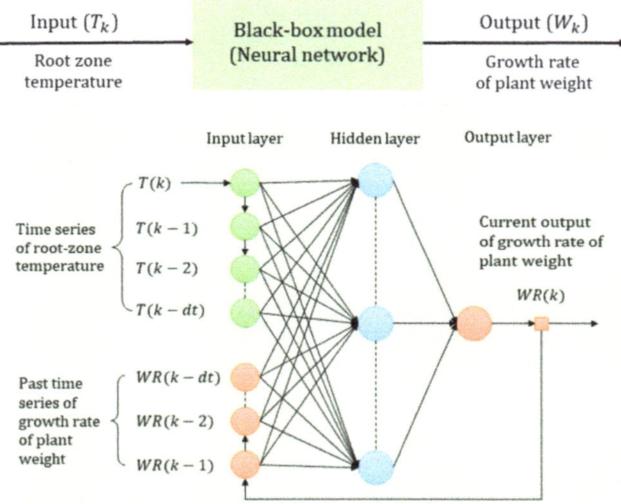

Figure 2. Block diagram of the single input–single output (SISO) system for system identification, and non-linear autoregressive with exogenous input (NARX) network structure with three layers, one input time series of the root zone temperature $T(k)$, one hidden layer (h), one output time series of the growth rate of plant weight $WR(k)$, and a time-delay (dt) neural network for identifying a dynamic model.

The NARX network for identifying the dynamic response of plant growth to RZT and for creating a black box model for simulation is illustrated in Figure 2. In this study, the NARX network structure consists of three layers: input, hidden, and output layers. Moreover, a feedback loop procedure was applied to produce time-series historical input and output data for dynamic identification [23,41].

To identify the dynamic system, the NARX network applies historical input data with time delay (dt) operators, $T(k), T(k-1), \ldots, T(k-dt)$, and historical output data with time delay operators, $WR(k-1), \ldots, GWR(k-dt)$, to the input layer, and applies the current output value $WR(k)$ to the output layer as training signals [23]. Backpropagation with the Levenberg–Marquardt algorithm was then used [46] to train the network. Meanwhile, for prediction, the current output $WR(k)$ was estimated from historical input data $T(k), T(k-1), \ldots, T(k-dt)$ and historical output data $WR(k-1), \ldots, WR(k-dt)$, as shown in Equation (2):

$$WR(k) = f(T(k), T(k-1), \ldots, T(k-dt), \ WR(k-1), \ldots, WR(k-dt)) \tag{2}$$

A program created using the Matlab® Deep Learning Toolbox™ R2019a (MathWorks® Inc., Natick, MA, USA) was used to develop the model [47]. The workflow of model development in this study is illustrated in Figure 3.

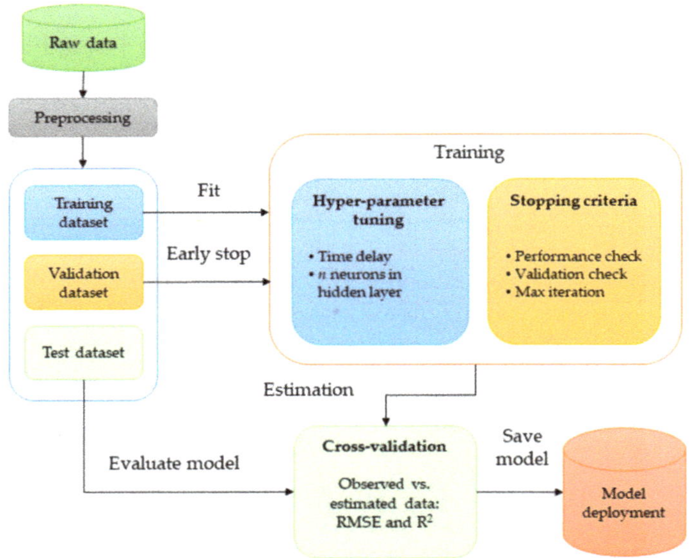

Figure 3. The workflow of model development using NARX neural networks.

2.4.3. Model Validation and Model Structure Selection

To obtain sufficient accuracy of the identified model, data for identification were divided into three independent datasets: training, validation, and test [48]. The training dataset was the sample of data used by the learning algorithm to fit the model. The trained model was then evaluated using the unbiased validation dataset while tuning the model parameters. During the training process, the validation dataset was also used for regularization by the early stopping training process to prevent model overfitting when generalization was not improving [46]. Four patterns of observation datasets were used for the training and validation datasets, with the proportions of 75 and 25%, respectively. This means that in each dataset, as many as 45 and 15 days of data were used for training and validation, respectively.

Because the structure of the neural network model is significant in the performance of the model, the model parameters, which consist of the time-delay operator and the number of neurons in the hidden layer, needed to be determined. Here, the model parameters were determined through trial and error based on cross-validation. Cross-validation provides an unbiased and robust evaluation of the final model by evaluating the performance of the identified model by comparing the test dataset and estimated data obtained from the model simulation [48].

2.4.4. Model Performance

To measure the performance of the proposed model, two performance criteria were used: the root-mean-squared error (RMSE) and the coefficient of determination (R^2) [48]. The RMSE calculates the error variance of the predicted and observed values (the closer to zero, the better the model performance for RMSE), as shown in Equation (3):

$$RMSE = \sqrt{\frac{1}{n}\sum_{i=1}^{n}(y_i - \hat{y}_i)^2}. \tag{3}$$

Meanwhile, the coefficient of determination is used to evaluate a model's ability to explain and predict future outcomes. The value of R^2 ranges from 0 to 1 (the closer to 1, the better the model fits), and is calculated using Equation (4):

$$R^2 = \frac{\left(n \sum_{i=1}^n \hat{y}_i y_i - \sum_{i=1}^n \hat{y}_i \sum_{i=1}^n y_i\right)^2}{\left(n \sum_{i=1}^n \hat{y}_i{}^2 - (\sum_{i=1}^n \hat{y}_i)^2\right)\left(n \sum_{i=1}^n y_i{}^2 - (\sum_{i=1}^n y_i)^2\right)} \tag{4}$$

where n is the number of data points, y_i is the actual value or network output, and $\hat{y}_i$ is the predicted value or network target.

3. Results

3.1. The Response of Plant Growth to Root Zone Temperature (RZT) for Identification

The time series data of plant weight development were successfully collected using the plant weight measurement system based on a micro load cell. However, much unreliable data were also recorded during observation. Figure 4 shows an example of the raw data from sensor number 14, consisting of numerous suspicious records or outliers. These outliers appeared because of a system error and probably because of unintentional actions, such as unintentional touching of the plants or sensor during observation. The application of data preprocessing effectively removed the unreliable records and reduced the noise from the observed data.

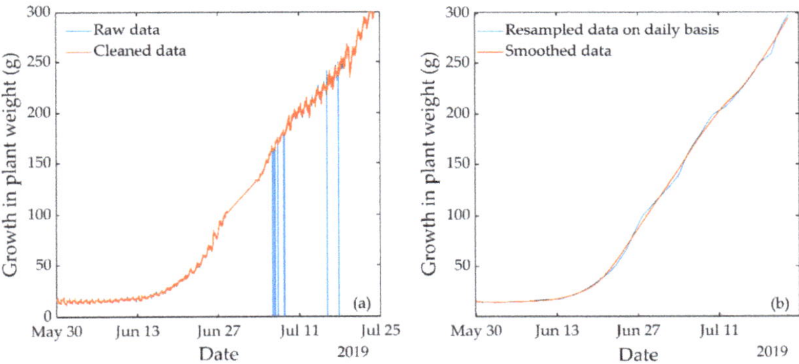

Figure 4. Sample preprocessing of observation data from sensor 14: (**a**) the original data were cleaned of unreliable records using the Hampel Identifier and reconstructed using the moving average method; (**b**) the cleaned data were then resampled on a daily basis and smoothed using the Savitzky–Golay filter.

Figure 5 shows daily changes in plant weight, the growth rate of plant weight, and the RZT of hydroponic cultivation for 60 days of observations, which started at 35 days after seeding. These 60 days of observation correspond to the initial stages of pepper plant growth. The data on plant weight are the average values of three pepper plants. Typical patterns of the growth in plant weight under five different RZT treatments are shown. For Pattern 5, however, data were obtained up to only 55 days because of a faulty sensor that produced unreliable records.

In this study, a model was developed based on the measured data for the input and output variables of the system using neural networks. From the observations, the data on growth in plant weight represent the accumulation of plant weight over time. Meanwhile, the growth rate of plant weight measured the sensitivity of the change in plant weight to the change in time, which means that the dynamic effect of RZT treatment on the growth in plant weight can be identified more clearly by using a growth rate variable. Therefore, to achieve a control strategy, the growth rate of plant weight was chosen as the controlled output for neural network model identification.

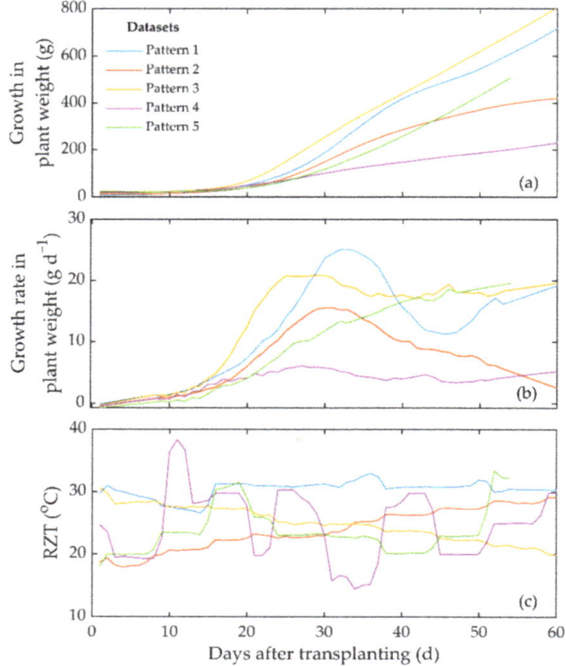

Figure 5. Typical daily change in plant weight (**a**), the growth rate in plant weight (**b**), and the RZT (**c**) in hydroponic pepper cultivation.

3.2. Determination of the Model Structure

The growth rate of plant weight, as determined by the RZT, was then identified by using neural networks, and a black box model for predicting the growth rate of plant weight was developed. Because the final model structure is determined by the model parameters (*dt*), which are a combination of time-delay order and the number of neurons in the hidden layer (h), these parameters were determined to find the best performance of the identified model. The effect of the time-delay order and the number of neurons in the hidden layer on model performance was investigated. Figure 6 shows the effect of the model parameters on the estimation error (RMSE). It was found that the RMSE reached its minimal value with the combination of 2 and 10 for the time-delay order and the number of neurons in the hidden layer, respectively. For the second model performance, identical results were obtained, with R^2 reaching the best value when the time-delay order and number of neurons in the hidden layer were 2 and 10, respectively. Therefore, the foregoing suggests that the neural network structure with time-delay order $dt = 2$ and number of neurons in the hidden layer h = 10 is useful for identification.

3.3. Identification Results

The performance of the identified neural network model was evaluated by comparing the estimated values and the independent test dataset with the observed values. Figure 7 shows a comparison of the estimated dynamic response calculated by the neural network model and the observed response for the growth rate of plant weight. With RMSE and R^2 values of 0.49 g and 0.99, respectively, it was found that the estimated response closely correlated with the observed response.

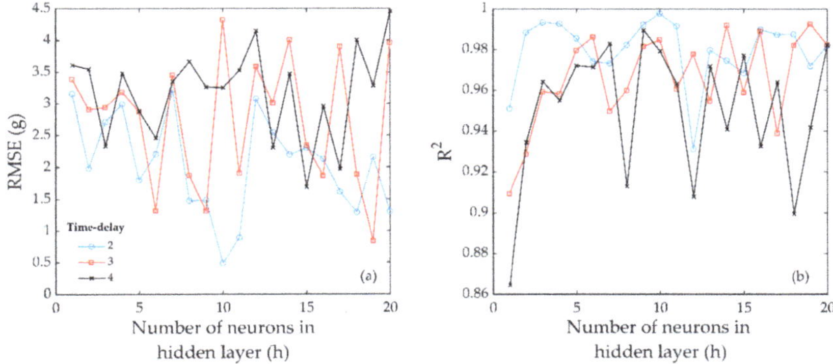

Figure 6. The relationship between the time-delay order (*dt*), the number of neurons in the hidden layer (h), and model performances: (**a**) root-mean-squared error (RMSE) and (**b**) R^2.

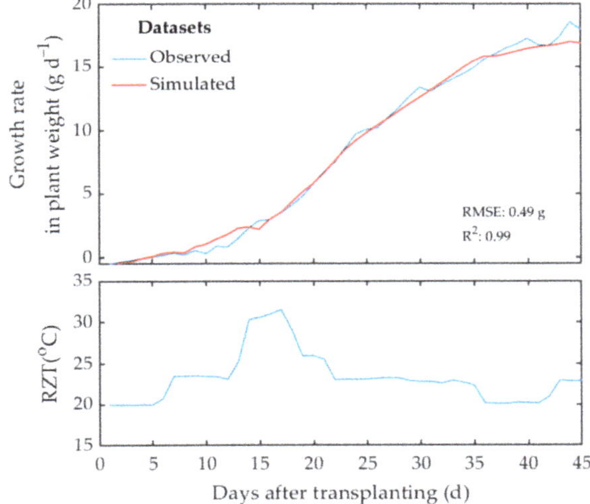

Figure 7. Comparison of the estimated response calculated by the developed neural network model and the observed growth rate of plant weight.

3.4. Estimation of the Characteristics of Plant Response

Using the identified neural network model, the relationship between RZT and the response of plant growth was then estimated to provide an insight into the basic properties of the dynamic system model. Figure 8 shows the estimated step response of the growth rate of plant growth to stepped input from 20 to 21, 23, 25, 27, and 29 °C. In general, the step responses dramatically increased as the step input of RZT increased. However, through the given range of five step inputs, the dynamic system model generated five different characteristics of the step response. The gain of step response from 20 to 21 °C was slightly faster, but then immediately steady. Meanwhile, the gains of step response from 20 to 23 °C and 20 to 25 °C were largest. In general, with change in the input variable of RZT, the gain in the step response of growth rate in plant weight gradually increased with time. These results show how the model simulates the dynamic characteristics of step response for different input conditions, which illustrates the ability of RZT in various ranges to control plant growth in a dynamic system.

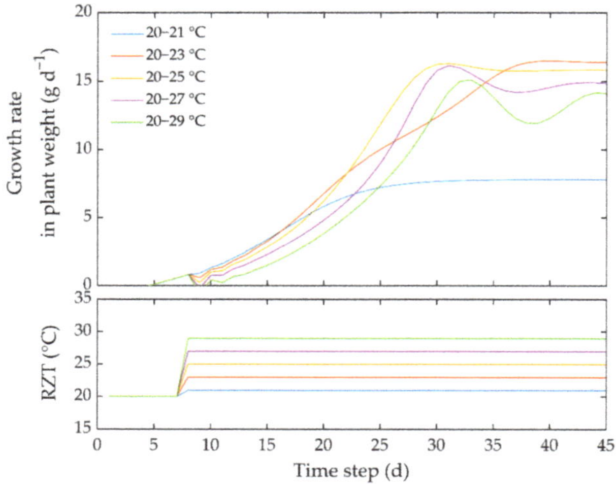

Figure 8. Estimated step response of the growth rate of plant weight to stepped RZT input, obtained via simulation.

3.5. Estimation of the Relationship between RZT and the Growth Rate of Plant Weight

In order to confirm the relevance of the model in this study, the estimated relationship between RZT and the average growth rate in plant weight, which also represents plant growth, was examined via model simulation. Figure 9 shows the estimated static relationship between RZT and the growth rate of plant weight. The estimated value of the growth rate of plant weight was obtained from the average of the step response calculated using the identified model, where the step input value was set as stationary from various values. From the simulation, it can be seen that the growth rate of plant weight increased with RZT. However, above the range 25 to 26 °C, the growth rate tended to decrease with RZT. In general, it was found that the relationship between RZT and the growth rate of the plant showed strong non-linearity and peaked in the range of 24 to 26 °C. This result is consistent with a previous study which reported that the optimal RZT range for bell pepper (*Capsicum annum* L.) is in the range 25 to 27 °C or less in terms of yield [49] and 24 °C in terms of shoot dry weight [50].

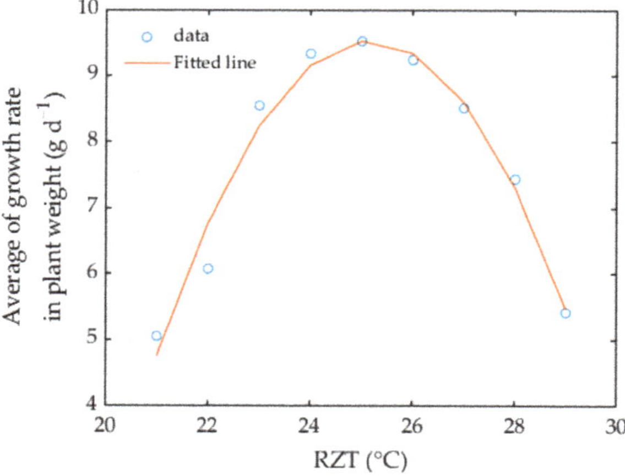

Figure 9. The estimated static relationship between the growth rate of plant weight and RZT, obtained via simulation.

4. Discussion and Conclusions

In this study, the dynamic responses of plant growth as affected by changes in the RZT were examined using an automatic plant weight measurement system based on a load cell. Since plant tissue is composed mostly of water, the fresh weight of a plant is very sensitive to changes in environmental conditions. For this reason, in order to examine the fundamental eco-physiological behavior of the response of plant growth to RZT, experiments were conducted in a strictly controlled environment using a growth chamber, where light intensity, temperature, and relative humidity were controlled precisely. As the environmental conditions inside the growth chamber were maintained constant, it can be assumed that the responses of plant growth were affected by the changes in RZT.

Then, an artificial neural network was used to construct a dynamic model of the responses of plant growth as affected by changes in the RZT in hydroponic cultivation. Based on the examination, it was found that the NARX networks presented in this study were useful in identifying the complex process of the dynamic response of plant growth as affected by changes in the RZT over 45 days of cultivation. The NARX networks were shown to be capable of performing long-term dependency prediction, as also shown by [40,41]. The model also successfully performed step response simulation, which is typically conducted in control studies to estimate the basic properties of a dynamic system [19]. This suggests that dynamic optimization can potentially be applied to the system in order to maximize plant growth. Moreover, from the simulation, estimation of the static relationship between the response of plant growth and RZT can also be generated. The static relationship was determined as having strong non-linearity with an upside-down parabolic curve peaking in the range of 24 to 26 °C, which is consistent with previous studies on the relationship of plant growth and RZT [49,50]. This suggests that a reliable computational model can be useful to predict the dynamic behavior of plant growth as affected by RZT.

In previous studies, the examination of dynamic plant growth control using root zone environmental factors in hydroponic cultivation was limited to nutrient solution concentration [16]. Meanwhile, optimal temperature control of the root zone environment, which is one of the essential determining factors in hydroponic cultivation, had not been thoroughly investigated [18]. Identification of the dynamic responses of plant growth as affected by RZT, which was conducted in this study, may be useful for a better understanding of plant growth control in plant production systems.

However, the artificial neural network model presented in this study cannot be applied in cultivation systems as it is. This is because the model is limited to the specific range of environmental conditions in the setup used during the experiment. More data on plant response in broader environmental conditions are required to construct a more robust model. For an artificial neural network model, more training data are better for identification. Given that one of the significant advantages of an artificial neural network model is the ability to re-train the current model using new data, this model has great potential for further development.

Author Contributions: Conceptualization, G.K.A., K.H. and T.M.; methodology, G.K.A., K.H. and T.M.; software, G.K.A.; formal analysis, G.K.A.; investigation, G.K.A.; resources, K.H. and T.M.; data curation, G.K.A.; writing—original draft preparation G.K.A.; writing—review and editing, G.K.A.; visualization, G.K.A.; supervision, K.H. and T.M. All authors have read and agreed to the published version of the manuscript.

Acknowledgments: The first author would like to thank the Indonesia Endowment Fund for Education (LPDP), Ministry of Finance and Ministry of Research, Technology and Higher Education of the Republic of Indonesia for their support of his study.

Conflicts of Interest: The authors declare no conflict of interest.

References

1. Sambo, P.; Nicoletto, C.; Giro, A.; Pii, Y.; Valentinuzzi, F.; Mimmo, T.; Lugli, P.; Orzes, G.; Mazzetto, F.; Astolfi, S.; et al. Hydroponic Solutions for Soilless Production Systems: Issues and Opportunities in a Smart Agriculture Perspective. *Front. Plant Sci.* **2019**, *10*, 923. [CrossRef] [PubMed]
2. He, J. Root Growth, Morphological and Physiological Characteristics of Subtropical and Temperate Vegetable Crops Grown in the Tropics Under Different Root-Zone Temperature. In *Plant Growth*; Rigobelo, E.C., Ed.; IntechOpen: London, UK, 2016; pp. 131–148.
3. Mortensen, L.M. Growth Responses of Some Greenhouse Plants to Environment. II. The Effect of Soil Temperature on Chrysanthemum Morifolium Ramat. *Sci. Hortic. (Amsterdam)* **1982**, *16*, 47–55. [CrossRef]
4. Gosselin, A.; Trudel, M.J. Interactions between Air and Root Temperatures on Greenhouse Tomato: II. Mineral Composition of Plants. *J. Amer. Soc. Hort. Sci.* **1983**, *108*, 905–909.
5. Ingram, D.L.; Ruter, J.M.; Martin, C.A. Review: Characterization and Impact of Supraoptimal Root-Zone Temperatures in Container-Grown Plants. *HortScience* **2015**, *50*, 530–539. [CrossRef]
6. Kawasaki, Y.; Matsuo, S.; Suzuki, K.; Kanayama, Y.; Kanahama, K. Root-Zone Cooling at High Air Temperatures Enhances Physiological Activities and Internal Structures of Roots in Young Tomato Plants. *J. Jpn. Soc. Hortic. Sci.* **2013**, *82*, 322–327. [CrossRef]
7. Kawasaki, Y.; Matsuo, S.; Kanayama, Y.; Kanahama, K. Effect of Root-Zone Heating on Root Growth and Activity, Nutrient Uptake, and Fruit Yield of Tomato at Low Air Temperatures. *J. Jpn. Soc. Hortic. Sci.* **2014**, *83*, 295–301. [CrossRef]
8. Morimoto, T.; Hashimoto, Y.; Fukuyama, T. Identification and Control of Hydroponic System in Greenhouses. *IFAC Proc. Vol.* **1985**, *18*, 1689–1693. [CrossRef]
9. Morimoto, T.; Hashimoto, Y. Speaking Plant/Fruit Approach for Greenhouses and Plant Factories. *Environ. Control. Biol.* **2009**, *47*, 55–72. [CrossRef]
10. Hashimoto, Y. Recent Strategies of Optimal Growth Regulation By the Speaking Plant Concept. *Acta Hortic.* **1989**, *260*, 115–122. [CrossRef]
11. Morimoto, T.; Hatou, K.; Hashimoto, Y. Intelligent Control for a Plant Production System. *Control. Eng. Pract.* **1996**, *4*, 773–784. [CrossRef]
12. Nishina, H. Development of Speaking Plant Approach Technique for Intelligent Greenhouse. *Agric. Agric. Sci. Procedia* **2015**, *3*, 9–13. [CrossRef]
13. Morimoto, T.; Hashimoto, Y. An Intelligent Control for Greenhouse Automation, Oriented by the Concepts of SPA and SFA—An Application to a Post-Harvest Process. *Comput. Electron. Agric.* **2000**, *29*, 3–20. [CrossRef]
14. Morimoto, T.; Hashimoto, Y. AI Approaches to Identification and Control of Total Plant Production Systems. *Control. Eng. Pract.* **2000**, *8*, 555–567. [CrossRef]
15. Fasol, K.H.; Jörgl, H.P. Principles of Model Building and Identification. *Automatica* **1980**, *16*, 505–518. [CrossRef]
16. Yumeina, D.; Morimoto, T. Dynamic Optimization of Solution Nutrient Concentration to Promote the Initial Growth of Tomato Plants in Hydroponics. *Environ. Control. Biol.* **2014**, *52*, 87–94. [CrossRef]
17. Morimoto, T.; Takeuchi, T.; Hashimoto, Y. Growth Optimization of Plant by Means of the Hybrid System of Genetic Algorithm and Neural Network. In Proceedings of the 1993 International Conference on Neural Networks, Nagoya, Japan, 25–29 October 1993; IEEE: Piscataway, NJ, USA, 1993; Volume 3, pp. 2979–2982.
18. Yumeina, D.; Aji, G.K.; Morimoto, T. Dynamic Optimization of Water Temperature for Maximizing Leaf Water Content of Tomato in Hydroponics Using an Intelligent Control Technique. *Acta Hortic.* **2017**, *1154*, 55–64. [CrossRef]
19. Isermann, R.; Münchhof, M. *Identification of Dynamic Systems: An Introduction with Applications*, 1st ed.; Springer-Verlag: Berlin/Heidelberg, Germany, 2011.
20. Whittaker, A.D.; Thieme, R.H. Editorial: Integration of Knowledge Systems into Agricultural Problem Solving. *Comput. Electron. Agric.* **1990**, *4*, 271–273. [CrossRef]
21. Yin, X.; Laar, H.H. Van. *Crop. Systems Dynamics: An. Ecophysiological Simulation Model. for Genotype-by-Environment Interactions*, 1st ed.; Wageningen Academic Publishers: Wageningen, The Netherlands, 2005.
22. Carson, E.; Feng, D.D.; Pons, M.N.; Soncini-Sessa, R.; van Straten, G. Dealing with Bio- and Ecological Complexity: Challenges and Opportunities. *Annu. Rev. Control.* **2006**, *30*, 91–101. [CrossRef]

23. Isermann, R.; Ernst, S.; Nelles, O. Identification with Dynamic Neural Networks—Architectures, Comparisons, Applications. *IFAC Proc. Vol.* **1997**, *30*, 947–972. [CrossRef]

24. Hinton, G.E. How Neural Networks Learn from Experience. *Sci. Am.* **1992**, *267*, 144–151. [CrossRef]

25. Rumelhart, D.E.; Hinton, G.E.; Williams, R.J. Learning Internal Representations by Error Propagation. In *Readings in Cognitive Science: A Perspective from Psychology and Artificial Intelligence*; Morgan Kaufmann Publishers, Inc.: San Mateo, CA, USA, 1988; pp. 399–421.

26. Sablani, S.S.; Datta, A.K.; Rahman, M.S.; Mujumdar, A.S. *Handbook of Food and Bioprocess. Modeling Techniques*; Sablani, S., Datta, A., Shafiur Rehman, M., Mujumdar, A., Eds.; CRC Press: Boca Raton, FL, USA, 2006; Volume 20065751.

27. Liakos, K.G.; Busato, P.; Moshou, D.; Pearson, S.; Bochtis, D. Machine Learning in Agriculture: A Review. *Sensors* **2018**, *18*, 2674. [CrossRef] [PubMed]

28. Islam, M.P.; Morimoto, T.; Hatou, K. Dynamic Optimization of inside Temperature of Zero Energy Cool Chamber for Storing Fruits and Vegetables Using Neural Networks and Genetic Algorithms. *Comput. Electron. Agric.* **2013**, *95*, 98–107. [CrossRef]

29. FAO. *Good Agricultural Practices for Greenhouse Vegetable Production in the South. East. European Countries*; FAO: Rome, Italy, 2017.

30. Erickson, A.N.; Markhart, A.H. Flower Developmental Stage and Organ Sensitivity of Bell Pepper (*Capsicum annuum* L.) to Elevated Temperature. *Plant. Cell Environ.* **2002**, *25*, 123–130. [CrossRef]

31. Pressman, E.; Shaked, R.; Firon, N. Exposing Pepper Plants to High Day Temperatures Prevents the Adverse Low Night Temperature Symptoms. *Physiol. Plant.* **2006**, *126*, 618–626. [CrossRef]

32. Bosland, P.W.; Votava, E.J. Introduction. In *Peppers: Vegetable and spice capsicums*; CABI: Wallingford, UK, 2012; pp. 1–12.

33. Aloni, B.; Karni, L.; Daie, J. Effect of Heat Stress on the Growth, Root Sugars, Acid Invertase and Protein Profile of Pepper Seedlings Following Transplanting. *J. Hortic. Sci.* **1992**, *67*, 717–725. [CrossRef]

34. Oda, M.; Tsuji, K. Monitoring Fresh Weight of Leaf Lettuce. *Jpn. Agric. Res. Q.* **1992**, *25*, 19–25.

35. Chen, W.-T.; Yeh, Y.-H.F.; Liu, T.-Y.; Lin, T.-T. An Automated and Continuous Plant Weight Measurement System for Plant Factory. *Front. Plant. Sci.* **2016**, *7*, 392. [CrossRef]

36. Helmer, T.; Ehret, D.L.; Bittman, S. CropAssist, an Automated System for Direct Measurement of Greenhouse Tomato Growth and Water Use. *Comput. Electron. Agric.* **2005**, *48*, 198–215. [CrossRef]

37. Wong, W.K.; Guo, Z.X. Intelligent Sales Forecasting for Fashion Retailing Using Harmony Search Algorithms and Extreme Learning Machines. In *Optimizing Decision Making in the Apparel Supply Chain Using Artificial Intelligence (AI)*; Woodhead Publishing Limited: Sawston, UK, 2013; pp. 170–195. [CrossRef]

38. Davies, L.; Gather, U. The Identification of Multiple Outliers. *J. Am. Stat. Assoc.* **1993**, *88*, 782–792. [CrossRef]

39. Dai, W.; Selesnick, I.; Rizzo, J.-R.; Rucker, J.; Hudson, T. A Nonlinear Generalization of the Savitzky-Golay Filter and the Quantitative Analysis of Saccades. *J. Vis.* **2017**, *17*, 10. [CrossRef]

40. Siegelmann, H.T.; Horne, B.G.; Giles, C.L. Computational Capabilities of Recurrent NARX Neural Networks. *IEEE Trans. Syst. Man Cybern. Part. B* **1997**, *27*, 208–215. [CrossRef] [PubMed]

41. Menezes, J.M.P.; Barreto, G.A. Long-Term Time Series Prediction with the NARX Network: An Empirical Evaluation. *Neurocomputing* **2008**, *71*, 3335–3343. [CrossRef]

42. Mohd, N.; Aziz, N. Performance and Robustness Evaluation of Nonlinear Autoregressive with Exogenous Input Model Predictive Control in Controlling Industrial Fermentation Process. *J. Clean. Prod.* **2016**, *136*, 42–50. [CrossRef]

43. Ng, B.C.; Darus, I.Z.M.; Jamaluddin, H.; Kamar, H.M. Dynamic Modelling of an Automotive Variable Speed Air Conditioning System Using Nonlinear Autoregressive Exogenous Neural Networks. *Appl. Therm. Eng.* **2014**, *73*, 1253–1267. [CrossRef]

44. Chan, R.W.K.; Yuen, J.K.K.; Lee, E.W.M.; Arashpour, M. Application of Nonlinear-Autoregressive-Exogenous Model to Predict the Hysteretic Behaviour of Passive Control Systems. *Eng. Struct.* **2015**, *85*, 1–10. [CrossRef]

45. Mai, C.V.; Spiridonakos, M.D.; Chatzi, E.N.; Sudret, B. Surrogate Modeling for Stochastic Dynamical Systems by Combining Nonlinear Autoregressive with Exogenous Input Models and Polynomial Chaos Expansions. *Int. J. Uncertain. Quantif.* **2016**, *6*, 313–339. [CrossRef]

46. Beale, M.H.; Hagan, M.T.; Demuth, H.B. *Neural Network Toolbox (TM) User's Guide*; The MathWorks, Inc.: Natick, MA, USA, 2016.

47. Aji, G.K.; Hatou, K.; Morimoto, T. Matlab Program Script for Modeling the Dynamic Response of Plant Growth to Root Zone Temperature in Hydroponic Chili Pepper Plant using Neural Network. Available online: https://github.com/mradjie/narx-plant-growth. (accessed on 25 May 2020).
48. Kuhn, M.; Johnson, K. *Applied Predictive Modeling*; Springer: New York, NY, USA, 2013.
49. Díaz-Pérez, J.C. Bell Pepper (*Capsicum annum* L.) Grown on Plastic Film Mulches: Effects on Crop Microenvironment, Physiological Attributes, and Fruit Yield. *HortScience* **2010**, *45*, 1196–1204. [CrossRef]
50. Gosselin, A.; Trudel, M.J. Root-Zone Temperature Effects on Pepper. *J. Am. Soc. Hort. Sci.* **1986**, *111*, 220–224.

Article

ANN-Based Continual Classification in Agriculture

Yang Li [1,2,*] and Xuewei Chao [1]

[1] College of Mechanical and Electrical Engineering, Shihezi University, Xinjiang 832003, China; psicnhpu@163.com
[2] School of Electrical and Information Engineering, Tianjin University, Tianjin 300072, China
* Correspondence: liyang328@shzu.edu.cn

Received: 13 April 2020; Accepted: 15 May 2020; Published: 18 May 2020

Abstract: In the area of plant protection and precision farming, timely detection and classification of plant diseases and crop pests play crucial roles in the management and decision-making. Recently, there have been many artificial neural network (ANN) methods used in agricultural classification tasks, which are task specific and require big datasets. These two characteristics are quite different from how humans learn intelligently. Undoubtedly, it would be exciting if the models can accumulate knowledge to handle continual tasks. Towards this goal, we propose an ANN-based continual classification method via memory storage and retrieval, with two clear advantages: Few data and high flexibility. This proposed ANN-based model combines a convolutional neural network (CNN) and generative adversarial network (GAN). Through learning of the similarity between input paired data, the CNN part only requires few raw data to achieve a good performance, suitable for a classification task. The GAN part is used to extract important information from old tasks and generate abstracted images as memory for the future task. Experimental results show that the regular CNN model performs poorly on the continual tasks (pest and plant classification), due to the forgetting problem. However, our proposed method can distinguish all the categories from new and old tasks with good performance, owing to its ability of accumulating knowledge and alleviating forgetting. There are so many possible applications of this proposed approach in the agricultural field, for instance, the intelligent fruit picking robots, which can recognize and pick different kinds of fruits; the plant protection is achieved by automatic identification of diseases and pests, which can continuously improve the detection range. Thus, this work also provides a reference for other studies towards more intelligent and flexible applications in agriculture.

Keywords: similarity; metric; memory; deep learning

1. Introduction

In the field of intelligent agriculture, for instance, plant protection and precision farming, there are incremental progresses in agricultural image processing, e.g., classification of crop pests, and harvest yield forecast. Step advances are catalyzed by the developed various computerized models, which have covered a wide range of technologies, such as machine learning, deep learning, transfer learning, few-shot learning, and so on. For instance, several machine learning methods were adopted in crop pest classification [1,2]. The convolutional neural networks were used to diagnose and identify the plant diseases from leaf images [3,4]. The deep learning neural networks showed a powerful and excellent performance on several agricultural applications, such as plant identification [5], crop classification [6,7], fruit classification [8], weed classification [9], animal classification [10], quality evaluation [11], and field pest classification [12,13]. The transfer learning technology helped fine-tune the pre-trained models to reduce the difficulty of model training [14,15]. The few-shot learning method reduced the requirements for the scale of the training dataset [16]. There were also some related agricultural research surveys [17–19], providing more comprehensive views.

Although the abovementioned methods achieved good performance on some special tasks, they are still far away from the true intelligence in this area. Specifically, one deep neural network is designed for a special task with a static evaluation protocol. The entire dataset will be split in two parts: A training set used for learning and a testing set used for accuracy evaluation. Once the training period completes, the structure and parameters of this model are fixed, and any new knowledge cannot be learned again. This is quite different from how humans learn.

Biological learning is to continually learn new skills (tasks) and accumulate knowledge throughout the lifetime [20]. We can also incorporate new information to expand our cognitive abilities without seriously breaking past memories, which results from the good balance between synaptic plasticity and stability [21,22]. As known, the basic principle of deep learning networks is the back-propagation error and gradient descent. However, from the perspective of biological cognition, our learning process is more likely through similarity matching, rather than back-propagation error or gradient descent in the brain [23]. So, the bio-inspired work in this article will be around metric learning and continual learning. The metric learning aims to learn the internal similarity between input paired data [24], which is suitable for classification and pattern recognition in agriculture. Continual learning requires the designed model to continuously learn new tasks without forgetting old ones, that is, keeping good performance on both new and old tasks.

Continual learning is an approach inspired from the biological factors of the mammalian brain. In this topic, the most important thing is the stability–plasticity dilemma. In detail, plasticity means to integrate novel information to incrementally refine and transfer knowledge, and stability aims not to catastrophically interfere with consolidated knowledge. For a stable continual learning process, two types of plasticity are required: Hebbian plasticity for positive feedback instability and compensatory homeostatic plasticity, which stabilizes neural activity [25]. So far, the main methods to realize continual learning can be divided into three categories: (1) Store and replay, including previous data or memory. The limitation is that the storage of old information will lead to large working memory requirements; (2) regularization approaches, such as learning without forgetting (LWF), elastic weight consolidation (EWC), etc. They alleviate catastrophic forgetting by imposing constraints on the update of the neural weights. However, the additional loss terms for protecting consolidated knowledge will lead to a trade-off on the performance of old and novel tasks; and (3) dynamic architectures, which change the structure of networks in response to new information, e.g., re-training with an increased number of neurons or network layers. Obviously, the limitation is the continuously growing complexity of the architecture with the increasing number of learnt tasks.

In this study, in order to imitate human learning and memory patterns to maintain a good performance on both new and old tasks, we propose an artificial neural network (ANN)-based continual classification method via memory storage and retrieval, including the convolutional neural network (CNN) and generative adversarial network (GAN). Looking at ourselves, how do we remember past events? We only keep the most important information in our brain, throwing out the details and abstracting the inner relationships. These life experiences inspire us to find a way to abstract and preserve prior knowledge in memory. The memory only records the most important information from prior events, automatically ignoring the details. Inspired by this, we used the GAN to extract central information from old tasks and generate abstracted images as memory. For the similarity matching tasks in agriculture, it has a good effect on both new and old tasks, alleviating the forgetting problem. The main contributions of this work are as follows:

(1) The CNN model only requires few raw data, which is helpful for practical applications, such as classification and pattern recognition.

(2) The proposed method based on memory storage and retrieval can deal with the sequential tasks and maintain a good performance on both new and old tasks, without forgetting.

(3) This work has two important advantages: Few data and high flexibility. It provides a foundation for other relevant studies toward more flexible applications in agriculture.

Clearly, there are so many possible applications of this proposed approach in the field of agriculture, for instance, intelligent fruit picking robots, which can recognize and pick different kinds of fruits, and plant protection through automatic identification of diseases and pests, which can continuously improve the detection range to show the ability to upgrade the developed model.

2. Materials and Methods

2.1. Crop Pest and Plant Leaf Datasets

The typical deep neural networks require amounts of data to train the model, while the metric learning method only needs few raw data. In this research work, we collected two common cross-domain agricultural datasets: Crop pests and plant leaves. The crop pest dataset was collected from the open dataset [26], which provides images of important insects in agriculture with natural scenes and complex backgrounds, close to the real world. The plant leaf dataset was collected from the famous open dataset (PlantVillage). Generally, the image preprocessing for deep neural networks includes the target cropping, background operation, gray transform, etc. Here, we used the raw images to make this study closer to the practical application.

In the crop pest dataset, there are 10 categories and the number of samples in each class is 20. The total number of samples is 200, which is a small dataset compared to that required for the traditional deep learning models based on back-propagation error. Some samples of the crop pest dataset are shown in Figure 1.

Figure 1. Samples of the crop pest dataset (from [26]).

The plant leaf dataset also includes 10 classes, and the number of samples in each class is 20. The parameter sizes of these two databases are the same. Some samples of the plant leaf dataset are shown in Figure 2.

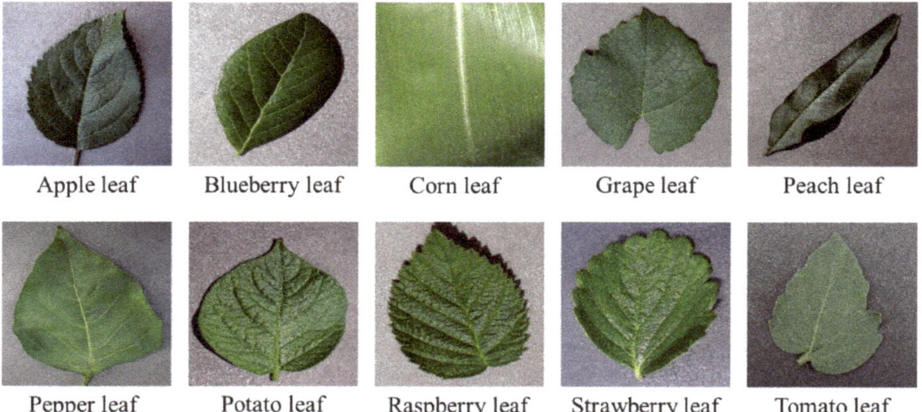

| Apple leaf | Blueberry leaf | Corn leaf | Grape leaf | Peach leaf |
| Pepper leaf | Potato leaf | Raspberry leaf | Strawberry leaf | Tomato leaf |

Figure 2. Samples of the plant leaf dataset (from PlantVillage).

2.2. Classification with Metric Learning Based on CNN

Metric learning learns the inner similarity between input paired data using a distance metric, which is aimed at distinguishing and classifying. The typical metric learning model is the Siamese network [27]. The Siamese network basically consists of two symmetrical neural networks sharing the same weights and architecture, which are joined together at the end using some energy function. During the training period of the Siamese network, the inputs are a pair of images, and the objective is to distinguish whether the input paired images are similar or dissimilar. The workflow of the Siamese network is shown as Figure 3.

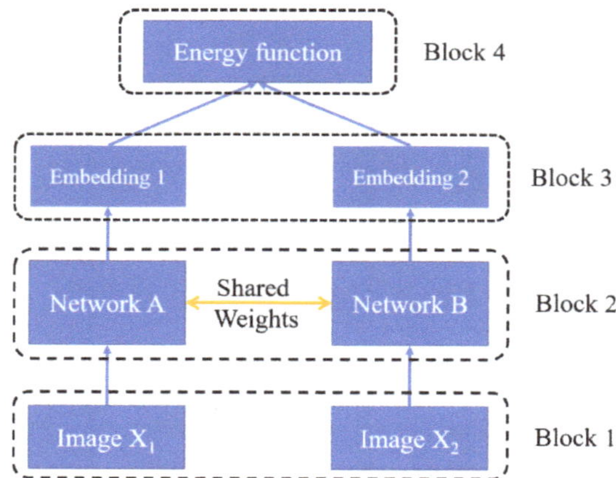

Figure 3. The workflow of the Siamese network.

As shown in Figure 3, there are four blocks. Now, we considered them one by one. For block 1, it means the input paired images, including the images X1 and X2, fed to network A and network B, respectively. They may come from the same category or not.

For block 2, there are two convolutional neural networks (CNNs), named network A and network B. The role of network A and network B is to generate the embeddings (feature vectors) for the input paired images. Since the inputs of the model are images, we used a CNN to generate the embeddings. Remember that the role of the CNNs here is only to extract features but not to classify. This differs with the traditional deep learning classification models. It is required that the two CNNs in the Siamese network have shared weights and structure, which means the two CNNs, in fact, have the same topology, as shown in Figure 4.

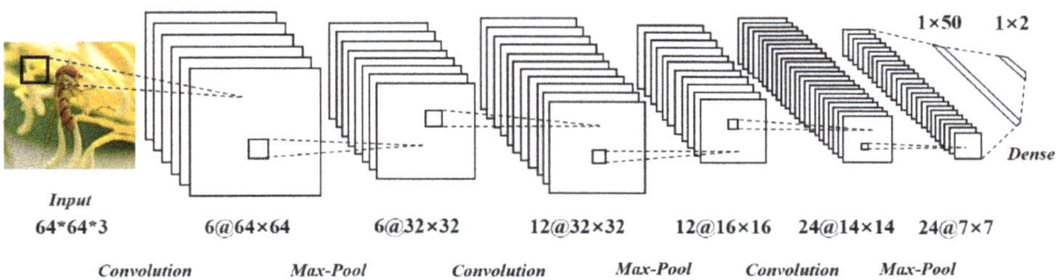

Figure 4. The topology of used convolutional neural network (CNN).

Here, the shared structure and parameters of CNN are shown in Table 1. Specifically, the output shape of the layers in CNN, and the size and number of kernels used in the convolutional layers and max-pooling layers are included. The programming tool used was 'Jupyter Notebook', which is a popular web-based interactive computing environment. We realized the functions with Python language and the environmental backend was TensorFlow. Our programming files and used image dataset were uploaded to the ZENODO.org, which is free and open for other researchers [28].

Table 1. The CNN structure and parameters in the Siamese network.

Layers	Output Shape	Kernel Size	Kernel Num	Stride	Parameters
Input	(64, 64, 3)				0
Conv2D	(64, 64, 6)	(3, 3)	6	1	168
MaxPooling2D	(32, 32, 6)	(2, 2)	0	2	0
Conv2D	(32, 32, 12)	(3, 3)	12	1	660
MaxPooling2D	(16, 16, 12)	(2, 2)	0	2	0
Conv2D	(14, 14, 24)	(3, 3)	24	1	2616
MaxPooling2D	(7, 7, 24)	(2, 2)	0	2	0
Dense	(50)				58,850
Dense	(2)				102

Then, for block 3, the embedding is referred to the output of the last dense layer of CNN, as shown in Figure 4. Network A and network B generate the embeddings for the input images X1 and X2, respectively. These embeddings are fed to block 4, the energy function, which gives the similarity between the paired inputs. The Euclidean distance is adopted as the energy function, which is the most common way to measure the distance between the two embeddings in the high-dimensional space. The expression of block 4, the energy function, can be written as Equation (1):

$$E(X_1, X_2) = \|f_{N_A}(X_1) - f_{N_B}(X_2)\|_2 \tag{1}$$

The value of E represents the similarity between the outputs of the two networks: If X_1 and X_2 are similar (from the same category), the value of E will be less. Otherwise, the value of E will be large if the inputs are dissimilar (from different categories).

To train the Siamese network well, the loss function is very important. The loss function guides the iteration of parameters of CNNs in the Siamese network. Since the goal of the Siamese network is to understand the similarity between the paired input images, we used the contrastive loss function, expressed as Equation (2):

$$Contrastive\ Loss = Y \times E^2 + (1 - Y) \times [max(margin - E, 0)]^2 \tag{2}$$

where E is the energy function and Y is the true label, which is 0 if the two input images are from the same category and 1 if the two input images are from different categories. Some examples of the input pairs are shown in Figure 5.

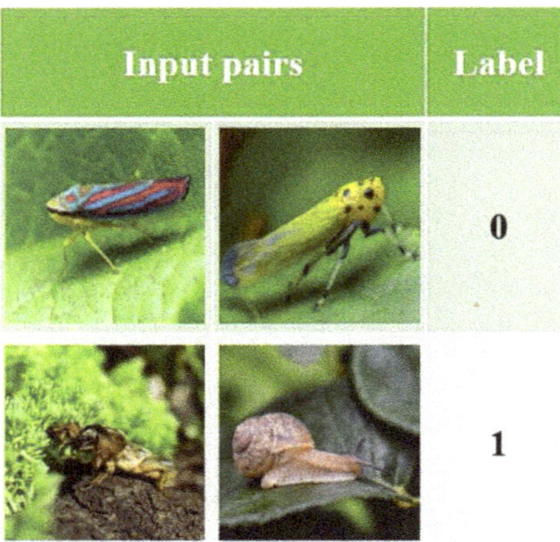

Figure 5. Examples of the input pairs.

In Equation (2), the term margin is used to set the threshold, that is, when input pairs are dissimilar, the Siamese network needs to hold their distance greater than the margin; otherwise, there will be a loss during the training period. Here, the margin was set as 1. When the training period is done, the distribution of embeddings will have a group effect, where different groups represent different categories.

2.3. Continual Classification with Metric Learning Based on CNN and GAN

From the bio-inspired perspective, we aimed for the model to be more flexible and able to handle continual tasks. Continual learning, also called lifelong learning, differs from transfer learning or other traditional networks. As known, a typical deep neural network is designed for some specific task, e.g., crop pest classification. After the training period, the weights and structure of the designed model are fixed, with an excellent performance on the specific task. However, if we want the model to perform another new task directly, e.g., plant leaf classification, it will have a very bad performance unless it is trained again from scratch or uses transfer learning. However, if we train the model by the new dataset, the distribution of weights will change to ensure a good performance on a new task. Since the weights of the network are modified, the network loses the ability to recognize the old task; in other words, it forgets the old knowledge. For transfer learning, the forgetting problem of old knowledges still exists. Obviously, traditional learning way has very poor flexibility.

If we want a model that can continually learn new tasks without forgetting old knowledge, it should have some bio-inspired ability, such as memory. In this study, we proposed a continual classification method based on memory storage and retrieval to maintain a good performance on both new and old tasks. Look at ourselves, how do we remember past events? We only keep the most important information in our brain, throwing out the details and abstracting the inner relationships. These life experiences inspire us to find a way to abstract and preserve prior knowledge in memory.

Here, we used the GAN to perform information abstracting and memory storage, which is a technique to learn to generate new data with the same statistics as the raw dataset, which consisted of two parts: Generator and discriminator. The basic workflow of GAN is shown as Figure 6.

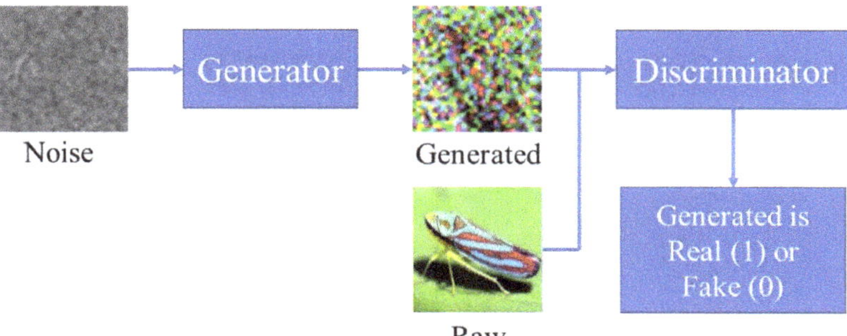

Figure 6. The workflow of generative adversarial network (GAN).

The generator and discriminator are both deep convolutional neural networks, and their structures are shown in Table 2.

Table 2. The generator and discriminator in GAN.

Generator		Discriminator	
Layers	**Output Shape**	**Layers**	**Output Shape**
Input	(64)	Input	(64, 64, 3)
Dense	(32, 32, 128)	Conv2D	(62, 62, 128)
Conv2D	(32, 32, 256)	Conv2D	(30, 30, 128)
Conv2DTranspose	(64, 64, 256)	Conv2D	(14, 14, 128)
Conv2D	(64, 64, 256)	Conv2D	(6, 6, 128)
Conv2D	(64, 64, 3)	Dense	(1)

The GAN chains the generator and discriminator together, expressed as Equation (3):

$$GAN(X) = discriminator(generator(X)) \tag{3}$$

The generator and discriminator contest with each other in a game. We trained the discriminator using samples of raw and generated images with the corresponding labels, such as any regular image classification model. To train the generator, we started with the random noise and used the gradients of the generator's weights, which means, at every step, moving the weights of the generator in a direction that will make the discriminator more likely to classify the images decoded by the generator as "real". In other words, we trained the generator to fool the discriminator.

Since the GAN can carry out the memory storage for old tasks, the workflow of our proposed continual metric learning method can be shown as Figure 7, which is mainly based on memory storage and retrieval.

When the first task comes, the task data will be organized as pairs and fed to the metric learning model (Siamese network). The output result is the similarity between input pairs, that is to say whether the input images are from the same category or not. Besides, the task data will also be fed to the GAN after data augmentation, due to the small scale of the raw database. Then, the GAN generates the abstracted images that represent the most important information of the old tasks, after the amount of iterations. We call this process memory storage. When the second task comes, the new task data and the data from memory will be mixed together, and fed to the metric learning model. We call this process memory retrieval.

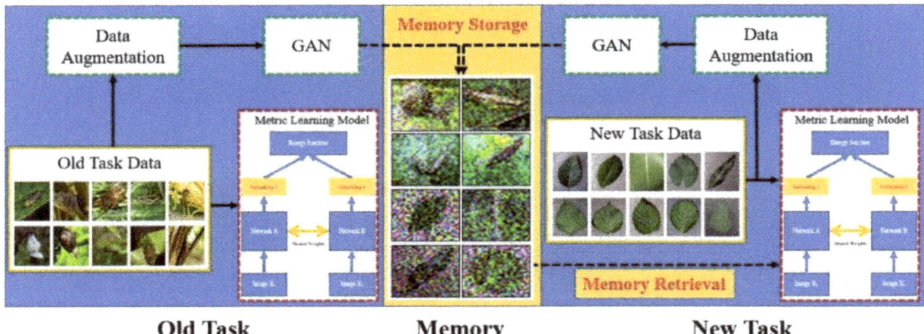

Figure 7. The workflow of continual metric learning.

3. Results

3.1. Single Task Experiment with the Basic CNN Model

In order to testify the performance of the metric learning model on similarity matching for a single task, we carried out experiments on a crop pest dataset and plant leaf dataset, respectively. For these two datasets, we prepared the input data as paired images. In detail, the total number of input pairs was 10,000, which may have contained a small number of duplicates because of the random combinations. We spilt the training set and testing set by the ratio of 8:2, that is, 2000 input pairs were used to test the accuracy. During training, 25% of the training data were taken out for the validation set. In summary, there were 6000 pairs for training, 2000 pairs for validation, and 2000 pairs for testing.

For the crop pest dataset, the loss and accuracy of the CNN model is shown in Figure 8.

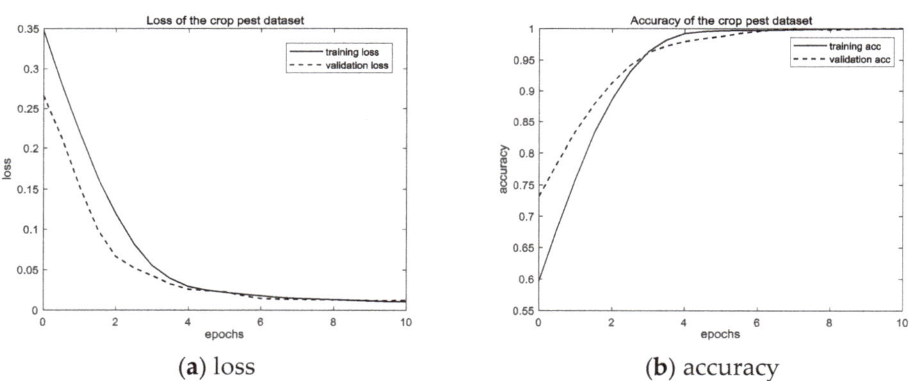

(**a**) loss   (**b**) accuracy

Figure 8. The loss and accuracy on the crop pest dataset.

It is shown that the variation trend of the training loss is consistent with that of the validation loss. The variation trend of the training accuracy is also consistent with that of the validation accuracy. This indicates that there is no overfitting problem in the training. The testing accuracy is 100%, which means the model can distinguish the input paired images well. The distribution of embeddings from the crop pest dataset is shown in Figure 9.

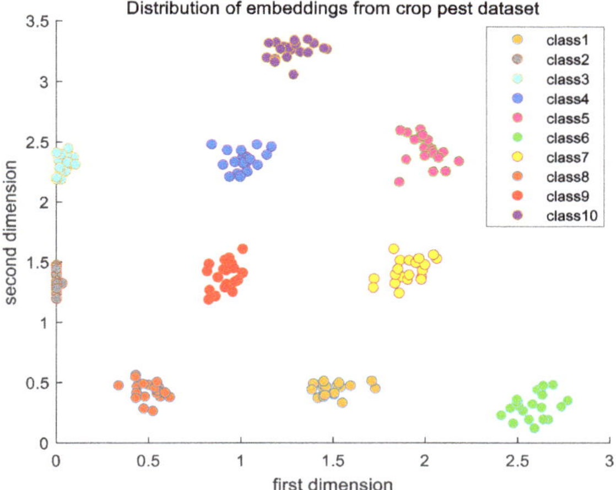

Figure 9. The distribution of embeddings from the crop pest dataset.

Through the distribution of the model's output embeddings, it can be seen that the metric learning model has good ability for similarity matching on the single task, that is, the images from the same category gather while those from different categories are far away from each other.

Similar experiments on the other dataset were also carried out. The loss and accuracy of the CNN model on the plant leaf dataset is shown in Figure 10. The variation trends of the training loss and training accuracy are consistent with those of the validation loss and validation accuracy, which indicates that there is also no overfitting problem in the training period. The distribution of the model's output embeddings of images from the plant leaf dataset is shown as Figure 11, which also shows the good ability of the similarity matching on a single task to distinguish the input paired images well.

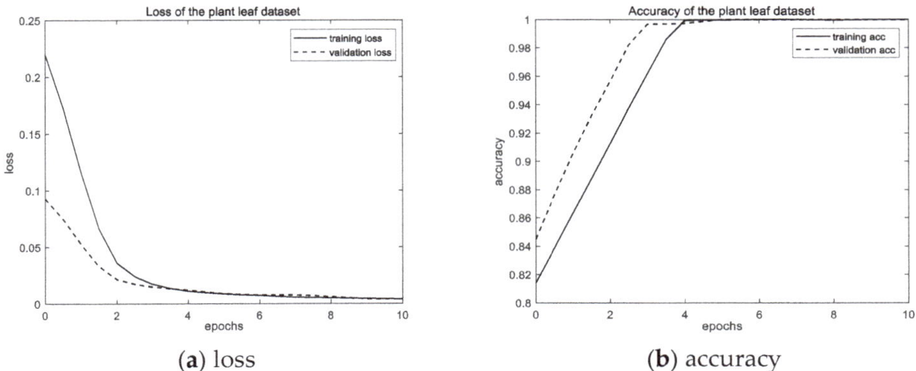

(**a**) loss (**b**) accuracy

Figure 10. The loss and accuracy on the plant leaf dataset.

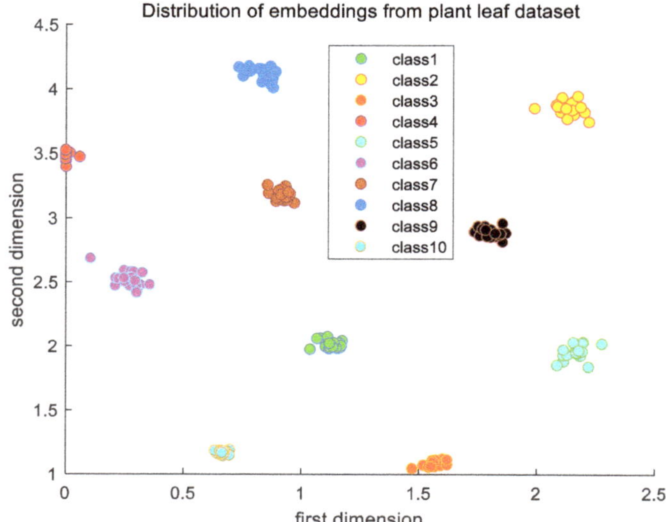

Figure 11. The distribution of embeddings from the plant leaf dataset.

3.2. Continual Tasks Experiment with the Basic CNN Model

As mentioned earlier, we hope that the model can be more flexible and able to handle continuous tasks, accumulating knowledge like humans to perform well on both old and new tasks. So, we carried out the experiments on sequential tasks to testify the continual performance of the CNN model, namely, the basic metric learning model.

For these two datasets, two occurring orders exist, that is, from the crop pest task to the plant leaf task, and the opposite one. For the first case, the testing accuracy of the two tasks is shown in Figure 12.

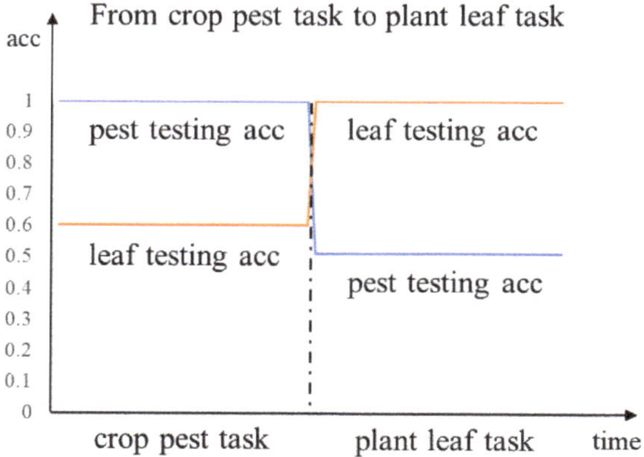

Figure 12. The testing accuracy of the first case.

At the first stage, the model has a good performance on the crop pest dataset, which was verified in Section 3.1. However, it has a very bad performance on the other dataset. The reason is that the other dataset is an unknown task and has never been seen before; this result is understandable and acceptable.

At the second stage, the model begins to learn the plant leaf task. Note that the model also learnt the crop pest task in the past. After the training period, the testing accuracy on the plant leaf task increases to 100% while that of the crop pest dataset decreases to nearly 50%, which is almost a blind guess. So, the extent of catastrophic forgetting for the crop pest task is nearly 50%. The new distribution of output embeddings from the old crop pest task is shown in Figure 13, which indicates that the basic metric learning model has lost the ability to distinguish the similarity between input paired images. The extracted features (embedding) of samples from different categories are mixed, and cannot be separated. This is an undesired forgetting problem!

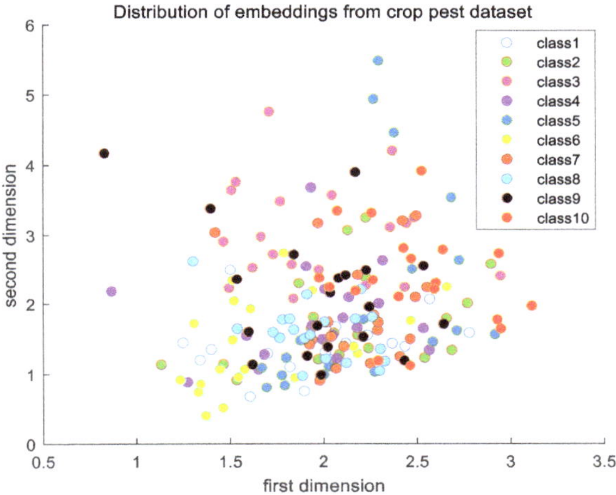

Figure 13. The distribution of embeddings from the old pest task.

For the second case, from the plant leaf task to the crop pest task, the experimental result of the testing accuracy is shown in Figure 14. The testing accuracy on the plant leaf task decreases from 100% to 60%, which means the extent of catastrophic forgetting for the plant leaf task is 40%. We found that, regardless of the occurring order of sequential tasks, the basic metric learning model does have a serious forgetting problem, as shown in Figures 12 and 14. In other words, after new learning, the basic metric learning model can no longer do the previous task well, due to the forgetting.

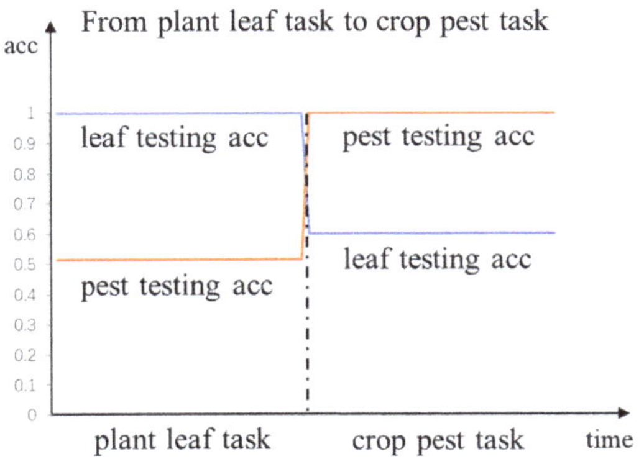

Figure 14. The testing accuracy of the second case.

The distribution of embeddings from the old plant leaf task is shown in Figure 15, which is very mixed and chaotic, losing the ability to distinguish and classify the similarity between input paired images.

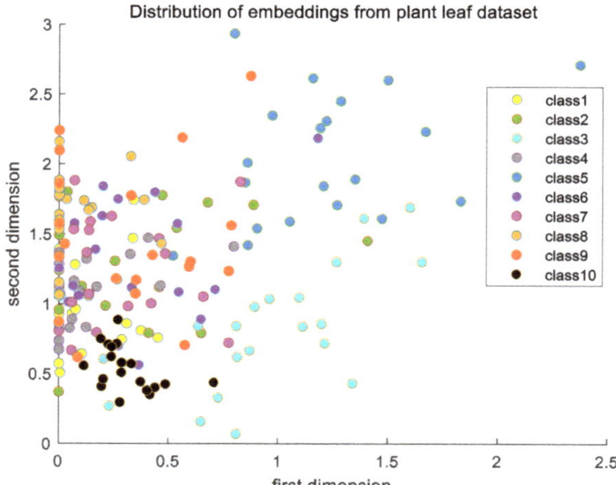

Figure 15. The distribution of embeddings from the old leaf task.

3.3. Continual Tasks Experiment with Our Proposed Method

As known, due to the forgetting problem, the basic CNN model cannot balance new and old tasks. Taking tthe sequential tasks from the crop pest dataset to the plant leaf dataset as an example, we used the designed GAN model to abstract the most important information of the old task (crop pest) and generated the abstracted images as memory for the future task, automatically ignoring the trivial details. When a new task comes, the abstracted images in memory will be retrieved and mixed with the new dataset, and then fed to the metric learning model.

Owing to this mechanism, the metric learning model can accumulate knowledge and better understand what it has learnt. The stored memory can be expanded, as does the increased ability to handle more continual tasks. The distribution of the model's output embeddings, corresponding to the testing images from both new and old tasks, is shown in Figure 16.

The results show the ability of our method to continually distinguish the similarity between input paired images and classify the testing images. All the categories from new and old tasks are separated clearly, which means that the metric learning model has a good performance on both new and old tasks, alleviating the forgetting problem. Compared with Section 3.2, the alleviated extent of catastrophic forgetting for the crop pest task and plant leaf task is 50% and 40%, respectively.

In addition, the results presented above are clear, and easily assessed. However, this is not always the case if we want to go further, e.g., an evaluation for the grouping results. In our opinion, the sum of the nearest distances between centers of groups will be a good choice. In detail, firstly, the center point of each group is calculated by the mean value; then, for every group center, the nearest distance with others is calculated; and finally, the sum of the nearest distances between the centers of groups is calculated, which is called the score. The evaluation metric should be proportional to the score, which means the larger the score is, the better the model's performance is.

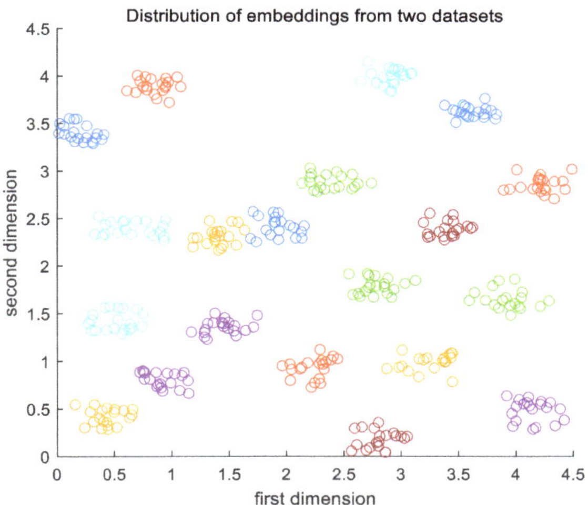

Figure 16. The distribution of embeddings from new and old tasks.

4. Discussion

We conduct the discussion about this work from the following three aspects.

4.1. Idea and Contents

The existing traditional models cannot accumulate the knowledge from old tasks, which means that they are all task specific, only focusing on the current task while forgetting the prior ones. This is a lack of flexibility and is quite different from humans' learning style. Besides, at present, there are mainly two basic types of neural network learning principles: Probability based on back-propagation error and similarity-based metric comparison. The former is more mature, but metric-based similarity learning is closer to biological learning.

So, from the bio-inspired perspective, we imitated the way biology learns and remembers, and proposed a continual metric learning method based on memory storage and retrieval to balance old and new tasks. Through several comparative experiments, it was found that the basic metric learning model can perform a single task excellently, distinguishing different categories well. However, when it is faced with continual tasks, the obvious forgetting problem occurs, and its poor flexibility loses the ability of dealing with old tasks. However, the addition of memory storage and retrieval in our method helps alleviate the forgetting problem, as all the categories from old and new tasks can be separated clearly, with good performance on both old and new tasks.

4.2. Contributions to Existing Research

We proposed an ANN-based continual classification method via memory storage and retrieval, combining the CNN and GAN technology, on the common agricultural datasets, such as the crop pest dataset and plant leaf dataset. The key contributions are two points: Few data and high flexibility.

As known, the big scale of the dataset is the basic requirement for the existing typical deep neural networks. However, the collection and labelling of big datasets are laborious and time consuming. So, research based on few data is a promising way. The metric learning used in this work only requires few raw data, because what it cares about is the paired inputs. Although the size of the raw dataset is small, the number of combinations of pairs from the same category and different categories can be expanded hundreds of times. Besides, the proposed continual learning method based on memory storage and retrieval increases the flexibility of the classification model, allowing it to balance old and new tasks, by accumulating knowledge and alleviating forgetting. This can be regarded as another small step towards more intelligent and flexible studies in agriculture.

4.3. Limitations and Future Works

Although the numerical solution described in this study achieved a good performance on both new and old tasks with its continual learning ability, it still has some limitations. The key structure consisted of CNN and GAN. The CNN is relatively easy to implement with a stable performance, while the GAN is usually not stable and hard to train, and indeed requires experience. Besides, the number of sequential tasks is two, namely the crop pest dataset and plant leaf dataset, hence the work at this stage is only primary continual learning in agriculture. In future, we would like to analyze more tasks to develop the robust continual learning model, considering the complex combination of popular technologies in neural networks and information extraction.

5. Conclusions

In this study, we proposed an ANN-based continual classification method via memory storage and retrieval, combining the CNN and GAN, with two clear advantages. One is few data, as the metric learning model based on CNN works well from few data, which significantly reduces the difficulty of image collection and annotation; the other is flexibility, as continual classification based on memory storage and retrieval can balance old and new tasks through the accumulation of knowledge and alleviation of forgetting. The results show that the regular CNN can deal with a single task well and classify the categories clearly. However, when it comes to continuous tasks, there is a serious forgetting problem. With the addition of memory storage and the retrieval mechanism, the modified continual model can distinguish all the categories from both old and new tasks, without the forgetting problem. There are so many possible applications of this proposed approach in the field of agriculture, for instance, intelligent fruit picking robots, which can recognize and pick different kinds of fruits; and plant protection by the identification of diseases and pests, which can continuously improve the detection range. This work lays a foundation and provides a reference for other relevant studies towards more intelligent and flexible applications in the agricultural area.

Author Contributions: Conceptualization, Y.L.; Methodology, Y.L.; Writing—original draft, Y.L.; Software, X.C.; Validation, X.C. All authors have read and agreed to the published version of the manuscript.

Funding: This research was funded by Natural Science Program of Shihezi University, grant number KX01230101. The APC was funded by Shihezi University.

Conflicts of Interest: The authors declare no conflict of interest.

References

1. Martineau, M.; Conte, D.; Raveaux, R.; Arnault, I.; Munier, D.; Venturini, G. A survey on image-based insect classification. *Pattern Recogn.* **2017**, *65*, 273–284. [CrossRef]
2. Wang, J.; Lin, C.; Ji, L.; Liang, A. A new automatic identification system of insect images at the order level. *Knowl. Based Syst.* **2012**, *33*, 102–110. [CrossRef]
3. Ferentinos, K. Deep learning models for plant disease detection and diagnosis. *Comput. Electron. Agric.* **2018**, *145*, 311–318. [CrossRef]
4. Lu, Y.; Yi, S.; Zeng, N.; Liu, Y.; Zhang, Y. Identification of rice diseases using deep convolutional neural networks. *Neurocomputing* **2017**, *267*, 378–384. [CrossRef]
5. Ghazi, M.; Yanikoglu, B.; Aptoula, E. Plant identification using deep neural networks via optimization of transfer learning parameters. *Neurocomputing* **2017**, *235*, 228–235. [CrossRef]
6. Wu, Y.; Xu, L. Crop Organ Segmentation and Disease Identification Based on Weakly Supervised Deep Neural Network. *Agronomy* **2019**, *9*, 737. [CrossRef]
7. Pourdarbani, R.; Sabzi, S.; García-Amicis, V.M.; García-Mateos, G.; Molina-Martínez, J.M.; Ruiz-Canales, A. Automatic Classification of Chickpea Varieties Using Computer Vision Techniques. *Agronomy* **2019**, *9*, 672. [CrossRef]

8. Koirala, A.; Walsh, K.B.; Wang, Z.; Anderson, N. Deep Learning for Mango (*Mangifera indica*) Panicle Stage Classification. *Agronomy* **2020**, *10*, 143. [CrossRef]

9. Knoll, F.J.; Czymmek, V.; Harders, L.O.; Hussmann, S. Real-time classification of weeds in organic carrot production using deep learning algorithms. *Comput. Electron. Agric.* **2019**, *167*, 105097. [CrossRef]

10. Jwade, S.A.; Guzzomi, A.; Mian, A. On farm automatic sheep breed classification using deep learning. *Comput. Electron. Agric.* **2019**, *167*, 105055. [CrossRef]

11. Przybylak, A.; Kozłowski, R.; Osuch, E.; Osuch, A.; Rybacki, P.; Przygodziński, P. Quality Evaluation of Potato Tubers Using Neural Image Analysis Method. *Agriculture* **2020**, *10*, 112. [CrossRef]

12. Liu, Z.; Gao, J.; Yang, G.; Zhang, H.; He, Y. Localization and classification of paddy field pests using a saliency map and deep convolutional neural network. *Sci. Rep.* **2016**, *6*, 20410. [CrossRef] [PubMed]

13. Xie, C.; Wang, R.; Zhang, J.; Chen, P.; Dong, W.; Li, R.; Chen, H. Multi-level learning features for automatic classification of field crop pests. *Comput. Electron. Agric.* **2018**, *152*, 233–241. [CrossRef]

14. Abdalla, A.; Cen, H.; Wan, L.; Rashid, R.; Weng, H.; Zhou, W.; He, Y. Fine-tuning convolutional neural network with transfer learning for semantic segmentation of ground-level oilseed rape images in a field with high weed pressure. *Comput. Electron. Agric.* **2019**, *167*, 105091. [CrossRef]

15. Thenmozhi, K.; Reddy, U.S. Crop pest classification based on deep convolutional neural network and transfer learning. *Comput. Electron. Agric.* **2019**, *164*, 104906. [CrossRef]

16. Li, Y.; Yang, J. Few-shot cotton pest recognition and terminal realization. *Comput. Electron. Agric.* **2020**, *169*, 105240. [CrossRef]

17. Too, E.C.; Yujian, L.; Njuki, S.; Yingchun, L. A comparative study of fine-tuning deep learning models for plant disease identification. *Comput. Electron. Agric.* **2019**, *161*, 272–279. [CrossRef]

18. Kamilaris, A.; Prenafeta-Boldú, F.X. Deep learning in agriculture: A survey. *Comput. Electron. Agric.* **2018**, *147*, 70–90. [CrossRef]

19. Lima, M.C.F.; de Almeida Leandro, M.E.D.; Valero, C.; Coronel, L.C.P.; Bazzo, C.O.G. Automatic Detection and Monitoring of Insect Pests—A Review. *Agriculture* **2020**, *10*, 161. [CrossRef]

20. Fagot, J.; Cook, R.G. Evidence for large long-term memory capacities in baboons and pigeons and its implications for learning and the evolution of cognition. *Proc. Natl. Acad. Sci. USA* **2006**, *103*, 17564–17567. [CrossRef]

21. Grutzendler, J.; Kasthuri, N.; Gan, W.B. Long-term dendritic spine stability in the adult cortex. *Nature* **2002**, *420*, 812–816. [CrossRef] [PubMed]

22. Abraham, W.C.; Robins, A. Memory retention—The synaptic stability versus plasticity dilemma. *Trends Neurosci.* **2005**, *28*, 73–78. [CrossRef] [PubMed]

23. Carpenter, G.A. Looking to the future: Learning from experience, averting catastrophe. *Neural Netw.* **2019**, *120*, 5–8. [CrossRef]

24. Kaya, M.; Bilge, H.Ş. Deep metric learning: A survey. *Symmetry* **2019**, *11*, 1066. [CrossRef]

25. Parisi, G.I.; Kemker, R.; Part, J.L.; Kanan, C.; Wermter, S. Continual lifelong learning with neural networks: A review. *Neural Netw.* **2019**, *113*, 54–71. [CrossRef]

26. Li, Y.; Wang, H.; Dang, L.M.; Sadeghi-Niaraki, A.; Moon, H. Crop pest recognition in natural scenes using convolutional neural networks. *Comput. Electron. Agric.* **2020**, *169*, 105174. [CrossRef]

27. Berlemont, S.; Lefebvre, G.; Duffner, S.; Garcia, C. Class-balanced siamese neural networks. *Neurocomputing* **2018**, *273*, 47–56. [CrossRef]

28. Li, Y. Code and dataset for the ANN-based Continual Classification in Agriculture. *Zenodo* **2020**. [CrossRef]

Article

Application of Artificial Neural Networks to Analyze the Concentration of Ferulic Acid, Deoxynivalenol, and Nivalenol in Winter Wheat Grain

Gniewko Niedbała [1,*], Danuta Kurasiak-Popowska [2], Kinga Stuper-Szablewska [3] and Jerzy Nawracała [2]

[1] Institute of Biosystems Engineering, Faculty of Agronomy and Bioengineering, Poznań University of Life Sciences, Wojska Polskiego 50, 60-627 Poznań, Poland

[2] Department of Genetics and Plant Breeding, Faculty of Agronomy and Bioengineering, Poznan University of Life Sciences, Dojazd 11, 60-632 Poznań, Poland; danuta.kurasiak-popowska@up.poznan.pl (D.K.-P.); jnawrac@up.poznan.pl (J.N.)

[3] Department of Chemistry, Faculty of Wood Technology, Poznan University of Life Sciences, Wojska Polskiego 75, 60-625 Poznań, Poland; kinga.stuper@up.poznan.pl

* Correspondence: gniewko@up.poznan.pl

Received: 24 February 2020; Accepted: 11 April 2020; Published: 14 April 2020

Abstract: Biotic stress, which includes infection by pathogenic fungi, causes losses of wheat yield in terms of quantity and quality. Ear Fusarium is caused by strains of *F. graminearum* and *F. culmorum*, which can produce mycotoxins—deoxynivalenol (DON) and nivalenol (NIV). One of the wheat's defense mechanisms against stressors is the activation of biosynthesis pathways of antioxidant compounds, including ferulic acid. The aim of the study was to conduct pilot studies on the basis of which neural models were created that would examine the impact of the variety and weather conditions on the concentration of ferulic acid, and link its content with the concentration of deoxynivalenol and nivalenol. The plant material was 23 winter wheat genotypes with different Fusarium resistance. The field experiment was conducted in 2011–2013 in Poland in three experimental combinations, namely: with full chemical protection; without chemical protection, but infested with natural disease (control); and in the absence of fungicidal protection, with artificial inoculation by genus *Fusarium* fungi. As a result of the pilot studies, three neural models—FERUANN analytical models (ferulic acid content), DONANN (deoxynivalenol content) and NIVANN (nivalenol content)—were produced. Each model was based on 14 independent features, 12 of which were in the form of quantitative data, and the other two were presented as qualitative data. The structure of the created models was based on an artificial neural network (ANN) of the multilayer perceptron (MLP) with two hidden layers. The sensitivity analysis of the neural network showed the two most important features determining the concentration of ferulic acid, deoxynivalenol, and nivalenol in winter wheat seeds. These are the experiment variant (VAR) and winter wheat variety (VOW).

Keywords: winter wheat; grain; artificial neural network; ferulic acid; deoxynivalenol; nivalenol; MLP network; sensitivity analysis; precision agriculture; machine learning

---

1. Introduction

During the growing season, wheat is exposed to numerous biotic and abiotic stresses. The factors causing abiotic stress are intense solar radiation, low or high temperature, excess or shortage of water, strong winds, etc. Biotic stress includes pests or diseases [1,2]. The plant's response to stress depends on many factors, including the applied variety, age, and developmental stage of plant. Numerous developmental, morphological, and physiological adaptations enable the passive avoidance of stress. The active interaction of the plant and stressor cause defense responses that prevent or tolerate changes.

One of the most important biogenic stressors is the infestation of crops by pathogenic fungi. This results in crop losses, the level of which depends on the variety, meteorological conditions, and cultivation technology. Potential losses can be minimized using appropriate agrotechnical measures and fungicide protection [3]. The most effective method for limiting the effects of disease infestation is the use of varieties containing resistance genes for individual diseases. In such cases, during plant cultivation, the accumulation of various combinations of many genes are applied [4].

A completely different strategy should be adopted in the case of Fusarium, because of the specificity of both the pathogen and the mechanism of pathogenesis. Resistance to Fusarium is a quantitative trait, associated with the presence of many quantitative trait loci (QTL) [5–9]. Almost every wheat chromosome has been identified with this type of QTL [10]. One of the most effective R genes that fight Fusarium is Fhb1 from the Sumai 3 wheat variety. Unfortunately, this variety has many unfavorable agronomic traits strongly linked to the Fhb1 gene. Growers are therefore looking for other sources to fight this disease.

Pathogenic fungi cause not only quantitative, but also qualitative losses in the wheat yield as well as a reduction in grain quality parameters (thousand grain weight, falling number, sedimentation number, and total protein content) [11–16]. Quality losses are not only the result of the development of pathogen mycelium, but also the effect of secondary metabolite production by fungi called mycotoxins. *Fusarium* fungi, mainly *F. culmorum* and *F. graminearum*, produce deoxynivalenol (DON) and zearalenone, which are highly toxic to humans and animals [17]. In 2005, the European Union introduced a standard where the maximum allowable DON concentration was set at 1250 µg/kg of unprocessed wheat grain (1.25 ppm (parts per million)).

In addition to genetic studies on wheat resistance to Fusarium, research has also been conducted on the biochemical aspects of plant responses to a massive pathogen attack. Based on the literature sources [18,19] and our own [20,21] research, it was found that antioxidative processes based on the significantly increased biosynthesis of the low-molecular antioxidants of the plant have a significant impact on pathogenesis, and constitute the first line of defense against pathogens. Based on the concentration of the selected phenolic acids, including ferulic acid, it is possible to assess the degree of disease risk at an early stage of pathogenesis [20,22]. Taking into account all of the current information on the mechanism of wheat resistance to Fusarium [20,23,24], as well as the results of chemical analyzes from three years of field experiments on wheat, it was decided to use modern methods of data analysis to determine the relationship between the concentration of ferulic acid; the concentration of deoxynivalenol and nivalenol (NIV), depending on weather conditions; and the infestation degree by fungal diseases.

Recently, increasing interest in microbiological forecasting is being observed, which has been mainly used for the examination of bacterial pathogens in the context of food. Modern computer technologies are used in other interdisciplinary research areas. Many new methods, such as artificial neural network (ANN), fuzzy logic, and genetic algorithms, which are part of artificial intelligence methods, are being used for multidimensional data analysis. Recent years show that there has been a significant increase in the use of nonlinear data modeling methods in agriculture. Particularly important analysis results are observed when using artificial neural networks, the results of which are often compared with classical statistical methods, e.g., multiple regression. As a result of the possibility of learning and generalizing data, the use of ANN gives better results than statistical methods. Neural modeling methods are used in classification, identification, and prediction, therefore, their potential is significant for practical application in broadly understood agriculture [25–39].

One of the most interesting uses of ANN is crop yield prediction. Forecasting of winter rapeseed and winter wheat yield has been applied in many works [40–45]. Because of the fact that plant yield is affected by many factors, such as meteorological conditions, fertilization level, and soil cultivation method, the use of modern data analysis techniques brings even more accurate results. In addition, because of the unique properties of data processing, ANN can simultaneously analyze empirical data in a quantitative and qualitative form. Such data are not possible to achieve using classical statistical methods, as they are usually limited only to the quantitative interpretation of the analyzed data. Therefore, the purpose of the work is to conduct pilot studies on the basis of which neural models will be produced that examine the impact of the variety and weather conditions on the concentration of ferulic acid, and to link its content with the concentration of deoxynivalenol and nivalenol.

2. Materials and Methods

2.1. Site Location

The field experiment was conducted during the 2011–2013 growing seasons in Poland at the Mikulice Breeding and Production Facility belonging to the Małopolska Plant Breeding Station (50°00′26.7″ N 22°26′25.2″ E; Figure 1).

Figure 1. Location of the study site.

The plant material was 23 winter wheat genotypes (Table S1) with different Fusarium wilt resistance, as follows: 13 wheat lines obtained from Polish plant breeding companies; seven genotypes from six European countries; and three from the Department of Genetics and Plant Breeding, the Poznań University of Life Sciences [20]. The experiment had the following two factors: the first was the variant of protection and the second was the wheat genotype. All of the wheat genotypes were sown onto 3 m^2 plots in three replications in three different variants:

- full chemical crop protection (CH)—fungicides: Duet Ultra 497 SC (epoxyconazole, thiophanate methyl) at a concentration of 0.6 l·ha^{-1}, and Capalo 337.5 SE (fenpropimorph, epoxyconazole, and metrafenone) at 2 l·ha^{-1};
- no chemical crop protection, natural infestation (K);
- no chemical crop protection, artificial inoculation with fungi from the genus *Fusarium* (I).

No fungicides were applied to the experimental plots of K and I.

The genotypes were sown on 29 September 2010, 27 September 2011, and 28 September 2012, and harvested on 5 August 2011, 2 August 2012, and 7 August 2013.

The inoculum used in the inoculation was produced from three isolates obtained from *F. graminearum* and three isolates obtained from *F. culmorum*.

After the appearance of first symptoms, the severity of the Fusarium wilt was assessed. The disease index (DI) was calculated from the following formula [46,47]:

$$DI = \frac{\text{number of infected plants}}{\text{total number of plants}} \times 100 \tag{1}$$

The average monthly temperatures and monthly precipitation, measured according to the (World Meteorological Organization) WMO guidelines for the years 2011–2013, were obtained from a Vantage

Vue 6357 UE 9 meteorological station (Davis Instruments) located approximately 400 m from the experimental field (Table S2).

After harvest, the plant height was measured and chemical analyzes were carried out (Table 1).

Table 1. Data structure in the neural models.

Symbol	Unit of Measure	Variable Name	The Scope of Data
		Quantitative data	
P1-3	mm	Sum of precipitation from 1 January to 31 March	90–102
T1-3	°C	Average air temperature from 1 January to 31 March	−1.6--−1.1
P4	mm	Sum of precipitation from 1 April to 30 April	25–29
T4	°C	Average air temperature from January 1 April to 30 April	9.9–10
P5	mm	Sum of precipitation from 1 May to 31 May	4–70
T5	°C	Average air temperature from 1 May to 31 May	15.6–16
P6	mm	Sum of precipitation from 1 June to 31 June	102–103
T6	°C	Average air temperature from 1 June to 31 June	18.3–18.7
P7	mm	Sum of precipitation from 1 July to 31 July	33–58
T7	°C	Average air temperature from 1 July to 31 July	19.5–21.4
WH	cm	Wheat height	67–122
DI	%	Disease index	0–95
		Qualitative data	
VAR	word	Experimental variant	Inoculation Protection Control
VOW	word	Variety of wheat	MUSZELKA SMH 8489 KBP 08.17 ARKADIA STH 9011 NAD 08104 STH 9035 AND 394/07 BAMBERKA SMH 8540 KBP 08.8 SVPC 87185 CHD 7143/04 82/2011 TARKUS 91/2011 PRAAG 8 20816 83/2011 FREGATA ERTUS 20818 UNG 136.6.1.1.

VAR: experiment variant; VOW: winter wheat variety. FERUANN analytical models (ferulic acid content), DONANN (deoxynivalenol content) and NIVANN (nivalenol content).

2.2. Determination of Ferulic Acid

The ferulic acid in the samples was analyzed after alkaline and acidic hydrolysis [20]. Analysis was performed using an Aquity H class (Ultra Performance Liquid Chromatography) UPLC system equipped with a Waters Acquity PDA detector (Waters Corporation, Parsippany, NJ, USA). Chromatographic separation was performed on an Acquity UPLC® BEH C18 column (100 mm × 2.1 mm, particle size 1.7 μm; Waters CorporationWaters, Dublin, Ireland). The elution was done with a gradient using the following mobile phase composition: (A) acetonitryl with 0.1% formic acid and (B) 1% aqueous formic acid mixture (pH = 2). The concentrations of phenolic compounds were determined using an internal standard at wavelengths of λ = 320 nm and 280 nm. Compounds were identified based on a comparison of the retention time of the analyzed peak with the retention time of the standard, and by adding a specific amount of the standard to the analyzed samples. The detection level was 1 µg/g. The retention time for ferulic acid was 17.50 min.

2.3. Determination of Trichothecenes

The grain samples were analyzed for the presence of trichothecenes, according to Perkowski et al. [48]. The trichothecenes of group B (DON and NIV) were analyzed as TMS (trimethylsilylsilyl ethers) derivatives. The analyses were run on a gas chromatograph (Hewlett Packard GC 6890) hyphenated to a mass spectrometer (Hewlett Packard 5972 A, Waldbronn, Germany), using an HP-5MS, 0.25 mm × 30 m capillary column. Quantitative analysis was performed in the single ion monitored mode (SIM) using the following ions for the detection of DON, 103 and 512, and NIV, 191. Qualitative analysis was performed in the SCAN mode (100–700 amu).

2.4. The Method of Constructing Neural Models

For the construction of three FERUANN analytical models (ferulic acid content), DONANN (deoxynivalenol content), and NIVANN (nivalenol content), artificial neural networks were applied using an Automatic Network Designer (AND) from Statistica v7.1 (StatSoft Inc., Tulsa, OK, USA). Each model was the result of learning 10,000 networks, one of which was selected for further analysis. For creation of the models, 138 data were used, and were divided into three sets, namely: learning, validation, and test. The structure of sets was divided into appropriately 70%:15%:15%. (96:21:21 cases for each set). Based on previous research [40–44], MLP (multilayer perceptron) topology networks with two hidden layers were selected for the analysis. This type of network is mainly used for regression or classification data analysis. Because of the use of AND, the assessment parameters for each neural model were adopted based on the following indicators: standard deviation, mean error, error deviation, mean absolute error, quotient deviations, and correlation. The best neural model was selected based on the highest correlation value and the lowest mean absolute error value. After selecting one neural model for each variant, a sensitivity analysis of the neural network was performed. This analysis determines the value of each independent variable (network input) in the FERUANN, DONANN, and NIVANN models produced. To determine the extent of the independent variable, the "error quotient" indicator was used. This indicator describes the ratio of error to error obtained using all of the independent features. This means that the greater the value, the greater the importance of the feature. If the value of the error quotient was less than 1, a given variable could be removed from the model to improve its quality. However, this is not mandatory. The indicator of the error ratio is "rank", which indicates the order of variables through a decreasing error value—rank 1 for a specific independent variable is of greatest importance for the network.

3. Results

As a result of the analyzes, three independent neural models, FERUANN, DONANN, and NIVANN, were created. Each model was based on 14 independent variables, 12 of which were in the form of quantitative data, and the other 2 were presented as qualitative data (Table 1). The structure of the generated models was based on the multi-layer perceptron (MLP) ANN type with two hidden layers. The FERUANN model had nine neurons in the first hidden layer and six neurons in the second hidden layer. Accordingly, the DONANN model had 13 and 7 neurons, and the NIVANN model 13 and 4 neurons (Figure 2).

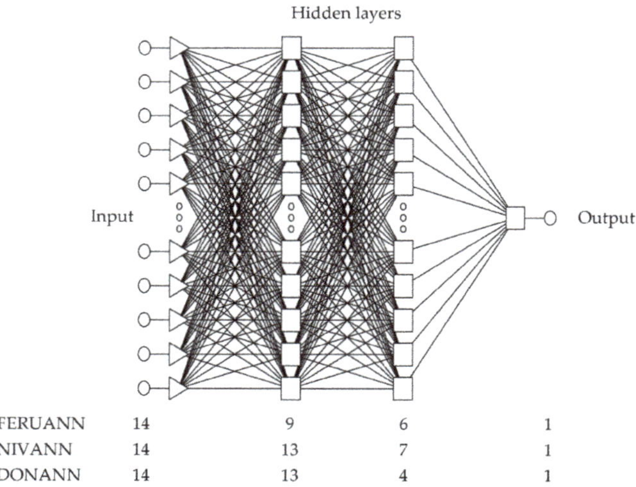

Figure 2. General diagram of the network, divided into hidden and input layers, and one output. FERUANN: ferulic acid content; NIVANN: nivalenol content; DONANN: deoxynivalenol content.

The best network fit for each model was selected from 10,000 networks, based on the best qualitative indicators. In the FERUANN model, the correlation coefficient was 0.9887, while in the DONANN model, the correlation coefficient was 0.9919. The last NIVANN model had a slightly lower correlation coefficient of 0.8106, however, it was still at an acceptable level (Table 2).

Table 2. The quality and structure of the neural models produced.

	FERUANN	**DONANN**	**NIVANN**
Neural Network Structure	**MLP** 14:38-9-6-1:1	**MLP** 14:38-13-7-1:1	**MLP** 14:38-13-4-1:1
Learning error	0.0210	0.0175	0.0244
Validation error	0.0349	0.0308	0.0301
Test error	0.0492	0.0356	0.2288
Mean	1646.79	2.9708	0.0705
Standard deviation	1034.54	4.2031	0.1146
Average error	12.84	0.0541	0.0076
Deviation error	156.89	0.5336	0.0672
Mean Absolute error	114.34	0.3705	0.0220
Quotient deviations	0.1516	0.1269	0.5861
Correlation	0.9887	0.9919	0.8106

FERUANN: ferulic acid content; DONANN: deoxynivalenol content; NIVANN: nivalenol content; MLP: multilayer perceptron.

In the next step, sensitivity analysis of the generated FERUANN, DONANN, and NIVANN neural models was performed (Table 3). As a result, the independent variables that had the greatest impact on the concentration of ferulic acid, deoxynivalenol, and nivalenol were identified. In the FERUANN model, the largest error quotient was achieved by experiment variant (VAR), which amounted to 7.0823. Another feature in this model was winter wheat variety (VOW), which reached a much lower level of just 3.1471. Other independent variables in this model have reached values slightly higher than 1, which indicates z low impact on the ferulic acid content, but they should be left in the model. The situation is slightly different with the DONANN model. As before, the independent variables VAR and WOV achieved the highest values 1.3778 and 1.1069, respectively, but the other five independent features (T1-3, P6, T7, WH, and DI) achieved an error quotient smaller than 1. In view of the above, these features can be removed from the model, as their impact on the final result of the analysis is scant. The NIVANN model pointed out the two most important independent qualitative variables—VOW and VAR—whose error quotients were 1.6315 and 1.4793, respectively. Other independent variables, as in the FERUANN model, have reached an error quotient above 1. This means that they should be left in the model. All of the sensitivity analysis results of neural networks are presented in Table 3.

Table 3. Sensitivity analysis of neural networks.

Variable	Model					
	FERUANN		**DONANN**		**NIVANN**	
	Quotient	Rank	Quotient	Rank	Quotient	Rank
P1-3	1.5973	4	1.0917	3	1.1436	7
T1-3	1.1977	9	0.9860	13	1.0629	11
P4	1.3120	7	1.0347	7	1.2597	4
T4	1.0780	11	1.0477	6	1.1246	8
P5	1.0027	14	1.0754	4	1.3281	3
T5	1.2976	8	1.0731	5	1.0381	12
P6	1.3718	5	0.9784	14	1.0977	9
T6	1.6332	3	1.0335	8	1.1894	6
P7	1.1242	10	1.0122	9	1.2163	5
T7	1.5276	5	0.9930	12	1.0853	10
WH	1.0355	12	0.9963	11	1.0025	14
DI	1.0163	13	0.9979	10	1.0292	13
VAR	7.0823	1	1.3778	1	1.4793	2
VOW	3.1471	2	1.1069	2	1.6315	1

FERUANN: ferulic acid content; DONANN: deoxynivalenol content; NIVANN: nivalenol content; WH: DI: VAR: experiment variant; VOW: winter wheat variety.

4. Discussion

One may find many relationships in the literature between genotype resistance and the resulting *Fusarium* wilt infestation, and the level of mycotoxin contamination [49]. Most often, ear *Fusarium* wilt is caused by *F. graminearum* and *F. culmorum*, which can produce mycotoxins from the group of trichothecenes, including DON and NIV. The production of toxins by individual strains of fungi depends on many factors, which include, first of all, the pathogen–plant interaction or the prevailing climatic conditions. The degree of ear infestation observed in the field is the result of the resistance of types I and II of the genotype. Type I is the resistance of the plant to the infection itself, and type II is resistance against spreading the pathogen in the ear [50].

Miedaner et al. [49] observed a significant effect of genotypic variability on DON accumulation in wheat. The most resistant varieties significantly affected DON production in the Mesterházy [51] experiment. In the most resistant varieties, a very low toxin contamination of grain was observed, despite the high toxin production by pathogenic fungi. Similarly, in the experiments by Miedaner et al. [49] and Paul et al. [3], the most resistant varieties accumulated less DON. In our own research, 23 wheat genotypes were used. The lines obtained from Polish breeding companies, according to the breeders, were characterized by varied resistance to Fusarium and other fungal diseases. Three winter wheat lines brought from the Department of Genetics and Plant Breeding of the Poznań University of Life Sciences were characterized by a high resistance to both mildew and Fusarium. These lines come from crossing English and French half-dwarf forms with leading Polish varieties. The remaining seven genotypes were varieties and lines from six European countries (Germany, the Czech Republic, Austria, the Netherlands, Hungary, and Sweden), which were used as sources of resistance to Fusarium. The current independent observations indicate a significant role of the variety in shaping the quantitative profile of the analyzed toxins. In the built neural models of DONANN and NIVANN, the error quotient for the variety reached a value of 1.1 in DON production, and 1.63 in the production of NIV. In turn, high correlations between field infestation and DON content in the grain were observed by Miedaner et al. [52] and Mesterházy [51]. The content of DON and NIV has been presented in detail in the publication Stuper-Szablewska et al. [21]. The content of DON was not dependent on the index disease calculated on the basis of the field observations. Index disease also had a very low significance in cases of NIV accumulation (13th position in the constructed NIVANN neural model). Potential differences are probably as a result of the large race diversity of pathogens, genotypes used, and environmental conditions [49].

The results of the studies on the effect of fungicides as a research factor on the extent of Fusarium infestation and the content of DON are not unambiguous, this may result from the use of fungicides containing various active substances, different doses and terms of application, diversity of pathogens attacking the plant, differences between varieties, and various weather conditions [53]. In the studies of Homdork et al. [54], the use of triazole fungicide lowered the concentration of DON in grain after inoculation compared with the control group. In turn, the use of fungicide in conditions of natural infestation did not have a significant impact on the level of DON and NIV reduction in the grain [55]. In studies conducted at various locations in the USA regarding the effect of tebuconazole on the content of DON, Paul et al. [3] found that the use of fungicide was more effective in reducing Fusarium wilt infestation in comparison with the level of DON reduction. Independent research found a significant impact of the experimental combination on the concentration of DON and NIV. The presence of a fungicide during a pathogen attack is an additional stressor for both plants and the pathogen itself. The use of chemical plant protection limits the development of the pathogen, which in response increases mycotoxin biosynthesis.

Genetically determined plant resistance to Fusarium can be modified by humidity and air temperature. In the conducted neural network sensitivity analysis, weather conditions from early January to the end of the vegetation period had an impact on the level of mycotoxins and ferulic acid. Only winter temperature (T1-3), end of vegetation period in July (T7), and rainfall from June (P6) affected the content of DON in the conducted experiment.

Field tests aimed at demonstrating the relationship between the level of pathogens, mycotoxins, and weather conditions are very difficult, because of the inability to prepare an experiment in which only one factor will change. In view of the above, literature data on the subject are very divergent, e.g., Miedaner et al. [49] could not relate weather conditions with the content of DON and NIV.

However, some authors suggest that some morphological features are important during Fusarium infection [56]. Their occurrence protects plants against necroses to which the pathogen may lead. Snijders [57] believed that on the one hand, tall plants tend to have a lower level of natural infection by Fusarium; on the other, many semi-dwarf plants with increased resistance to Fusarium can be selected. Probably, which is confirmed by our own research, the genotype of the plant itself is more important than its height. In the sensitivity analysis of neural models, the content of DON was not dependent on the height of the plants, and its impact on the NIV content was very small.

One of the most important defense mechanisms of wheat against stressors of various etiologies is the activation of the biosynthesis pathways of antioxidant compounds. One may find verification in literature sources, which has also been confirmed by our own research, that ferulic acid is one of the main compounds actively participating in the plant's immune mechanisms [58].

In the study [58], the profiles of bound phenolic acids were similar in all of the wheat genotypes analyzed, and ferulic acid accounted for 72% to 85% of all acids. The authors observed a strong genotypic diversification in the ferulic acid content in the analyzed winter wheat genotypes, ranging from 181 to 742 mg/kg. In the research by Hernández et al. [59], the ferulic acid content ranged from 439 mg/kg (Raposo variety) to 1450 mg/kg (Colorado variety). The influence of the variety on the ferulic acid content is confirmed by numerous studies; however, differences in the examined content of this acid are observed [60–62]. Stuper-Szablewska et al. [21] showed that the average content of ferulic acid under control conditions was 844 mg/kg, 780 mg/kg, and 771 mg/kg in the years of the study. After inoculation, these values increased to 2248 mg/kg, 2574 mg/kg, and 3145 mg/kg, respectively [21].

The large range of variability of ferulic acid content in wheat varieties is also caused by the impact of environmental conditions and the interaction of the genotype with the environment [63]. An additional factor affecting the content of this acid is the occurrence of infection [21,64].

5. Conclusions

The harmful effect of mycotoxins imposes actions aimed at their minimization in raw materials and products intended for humans and animals. The most effective preventive action is limiting their formation in the field, by the cultivation of resistant genotypes. In the constructed FERUANN, DONANN, and NIVANN neural models, variety was a factor that significantly affected the content of ferulic acid, deoxynivalenol, and nivalenol in winter wheat seeds. This confirms the significant role of growing new varieties. Resistance breeding is carried out in many countries, and despite the difficulties, numerous resistant varieties are available, characterized by slower and the subsequent development of symptoms of Fusarium compared with sensitive varieties. As part of the pilot study discussed in this work, inoculation and fungicide spraying were treated as stressors. Both of these factors activate nonenzymatic resistance mechanisms, including phenolic acid biosynthesis. As a result, it was observed in our independent research in the FERUANN neural model that the applied experimental variant determined the ferulic acid content to the greatest extent. The variety feature also had a significant impact on the presence of ferulic acid.

The results of the conducted research using artificial neural networks indicate the possibility of the practical application of neural modeling methods to analyze the concentration of ferulic acid, deoxynivalenol, and nivalenol in winter wheat grain.

Supplementary Materials: The following are available online at http://www.mdpi.com/2077-0472/10/4/127/s1, Table S1: Pedigree of analysed winter wheat genotypes. Table S2: Rainfall and average air temperatures between January 2011 and December 2013.

Author Contributions: Conceptualization, G.N., D.K.-P., K.S.-S., and J.N.; data curation, G.N., D.K.-P., K.S.-S., and J.N.; formal analysis, G.N., D.K.-P., and K.S.-S.; funding acquisition, G.N.; investigation, D.K.-P. and J.N.; methodology, G.N., D.K.-P., and K.S.-S.; project administration, G.N.; resources, D.K.-P., K.S.-S., and J.N.; software, G.N.; supervision, G.N.; validation, G.N., D.K.-P., K.S.-S., and J.N.; visualization, G.N.; writing (original draft), G.N., D.K.-P., and K.S.-S.; writing (review and editing), K.S.-S. and J.N. All authors have read and agreed to the published version of the manuscript.

Funding: This research was funded by The Ministry of Agriculture and Rural Development R, grant number 801-13/12 (Basic research for biological progress in crop production).

Acknowledgments: The publication was co-financed within the framework of the Ministry of Science and Higher Education program titled "Regional Initiative Excellence" in 2019–2022, project no. 005/RID/2018/19.

Conflicts of Interest: The authors declare no conflict of interest. The funders had no role in the design of the study; in the collection, analyses, or interpretation of data; in the writing of the manuscript; or in the decision to publish the results.

References

1. Fernandez, O.; Béthencourt, L.; Quero, A.; Sangwan, R.S.; Clément, C. Trehalose and plant stress responses: Friend or foe? *Trends Plant Sci.* **2010**, *15*, 409–417. [CrossRef] [PubMed]
2. Mahajan, S.; Tuteja, N. Cold, salinity and drought stresses: An overview. *Arch. Biochem. Biophys.* **2005**, *444*, 139–158. [CrossRef] [PubMed]
3. Paul, P.A.; Lipps, P.E.; Hershman, D.E.; McMullen, M.P.; Draper, M.A.; Madden, L.V. A Quantitative Review of Tebuconazole Effect on Fusarium Head Blight and Deoxynivalenol Content in Wheat. *Phytopathology* **2007**, *97*, 211–220. [CrossRef] [PubMed]
4. McDonald, B.A.; Linde, C. Pathogen population genetics, evolutionary potential, and durable resistance. *Annu. Rev. Phytopathol.* **2002**, *40*, 349–379. [CrossRef] [PubMed]
5. Xue, S.; Xu, F.; Tang, M.; Zhou, Y.; Li, G.; An, X.; Lin, F.; Xu, H.; Jia, H.; Zhang, L.; et al. Precise mapping Fhb5, a major QTL conditioning resistance to Fusarium infection in bread wheat (*Triticum aestivum* L.). *Theor. Appl. Genet.* **2011**, *123*, 1055–1063. [CrossRef]

6. Xue, S.; Li, G.; Jia, H.; Xu, F.; Lin, F.; Tang, M.; Wang, Y.; An, X.; Xu, H.; Zhang, L.; et al. Fine mapping Fhb4, a major QTL conditioning resistance to Fusarium infection in bread wheat (*Triticum aestivum* L.). *Theor. Appl. Genet.* **2010**, *121*, 147–156. [CrossRef]

7. Dhokane, D.; Karre, S.; Kushalappa, A.C.; McCartney, C. Integrated Metabolo-Transcriptomics Reveals Fusarium Head Blight Candidate Resistance Genes in Wheat QTL-Fhb2. *PLoS ONE* **2016**, *11*, e0155851.

8. Giancaspro, A.; Giove, S.L.; Zito, D.; Blanco, A.; Gadaleta, A. Mapping QTLs for Fusarium Head Blight Resistance in an Interspecific Wheat Population. *Front. Plant Sci.* **2016**, *7*, 1381. [CrossRef]

9. Arruda, M.P.; Brown, P.; Brown-Guedira, G.; Krill, A.M.; Thurber, C.; Merrill, K.R.; Foresman, B.J.; Kolb, F.L. Genome-Wide Association Mapping of Fusarium Head Blight Resistance in Wheat using Genotyping-by-Sequencing. *Plant Genome* **2016**, *9*. [CrossRef]

10. Buerstmayr, H.; Ban, T.; Anderson, J.A. QTL mapping and marker-assisted selection for Fusarium head blight resistance in wheat: A review. *Plant Breed.* **2009**, *128*, 1–26.

11. Riley, R.T.; An, N.H.; Showker, J.L.; Yoo, H.S.; Norred, W.P.; Chamberlain, W.J.; Wang, E.; Merrill, A.H.; Motelin, G.; Beasley, V.R.; et al. Alteration of Tissue and Serum Sphinganine to Sphingosine Ratio: An Early Biomarker of Exposure to Fumonisin-Containing Feeds in Pigs. *Toxicol. Appl. Pharmacol.* **1993**, *118*, 105–112. [CrossRef]

12. Bal, G. Scab of Wheat: Prospects For Control. *Plant Dis.* **1994**, *78*, 760.

13. McMullen, M.; Jones, R.; Gallenberg, D. Scab of Wheat and Barley: A Re-emerging Disease of Devastating Impact. *Plant Dis.* **1997**, *81*, 1340–1348. [PubMed]

14. Gang, G.; Miedaner, T.; Schuhmacher, U.; Schollenberger, M.; Geiger, H.H. Deoxynivalenol and Nivalenol Production by Fusarium culmorum Isolates Differing in Aggressiveness Toward Winter Rye. *Phytopathology* **1998**, *88*, 879–884.

15. D'Mello, J.P.F.; Placinta, C.M.; Macdonald, A.M.C. Fusarium mycotoxins: A review of global implications for animal health, welfare and productivity. *Anim. Feed Sci. Technol.* **1999**, *80*, 183–205.

16. Muthomi, J.W.; Schütze, A.; Dehne, H.-W.; Mutitu, E.W.; Oerke, E.-C. Characterization of Fusarium culmorum isolates by mycotoxin production and aggressiveness to winter wheat/Charakterisierung von Isolaten von Fusarium culmorum anhand der Mykotoxinbildung und Aggressivität an Winterweizen. *Z. Pflanzenkrankh. Pflanzenschutz/J. Plant Dis. Prot.* **2000**, *107*, 113–123.

17. Peng, W.-X.; Marchal, J.L.M.; van der Poel, A.F.B. Strategies to prevent and reduce mycotoxins for compound feed manufacturing. *Anim. Feed Sci. Technol.* **2018**, *237*, 129–153.

18. Martínez, G.; Regente, M.; Jacobi, S.; Del Rio, M.; Pinedo, M.; de la Canal, L. Chlorogenic acid is a fungicide active against phytopathogenic fungi. *Pestic. Biochem. Physiol.* **2017**, *140*, 30–35.

19. Siranidou, E.; Kang, Z.; Buchenauer, H. Studies on Symptom Development, Phenolic Compounds and Morphological Defence Responses in Wheat Cultivars Differing in Resistance to Fusarium Head Blight. *J. Phytopathol.* **2002**, *150*, 200–208.

20. Stuper-Szablewska, K.; Kurasiak-Popowska, D.; Nawracała, J.; Perkowski, J. Response of non-enzymatic antioxidative mechanisms to stress caused by infection with Fusarium fungi and chemical protection in different wheat genotypes. *Chem. Ecol.* **2017**, *33*, 949–962.

21. Stuper-Szablewska, K.; Kurasiak-Popowska, D.; Nawracała, J.; Perkowski, J. Quantitative profile of phenolic acids and antioxidant activity of wheat grain exposed to stress. *Eur. Food Res. Technol.* **2019**, *245*, 1595–1603. [CrossRef]

22. Stuper-Szablewska, K.; Kurasiak-Popowska, D.; Nawracała, J.; Perkowski, J. Study of metabolite profiles in winter wheat cultivars induced by Fusarium infection. *Cereal Res. Commun.* **2016**, *44*, 572–584. [CrossRef]

23. Goral, T.; Stuper-Szablewska, K.; Busko, M.; Boczkowska, M.; Walentyn-Goral, D.; Wisniewska, H.; Perkowski, J. Relationships between Genetic Diversity and Fusarium Toxin Profiles of Winter Wheat Cultivars. *Plant Pathol. J.* **2015**, *31*, 226–244. [CrossRef] [PubMed]

24. Stuper-Szablewska, K.; Perkowski, J. Phenolic acids in cereal grain: Occurrence, biosynthesis, metabolism and role in living organisms. *Crit. Rev. Food Sci. Nutr.* **2019**, *59*, 664–675. [CrossRef] [PubMed]

25. Abdipour, M.; Ramazani, S.H.R.; Younessi-Hmazekhanlu, M.; Niazian, M. Modeling Oil Content of Sesame (*Sesamum indicum* L.) Using Artificial Neural Network and Multiple Linear Regression Approaches. *J. Am. Oil Chem. Soc.* **2018**, *95*, 283–297. [CrossRef]

26. Mohammadi Torkashvand, A.; Ahmadi, A.; Gómez, P.A.; Maghoumi, M. Using artificial neural network in determining postharvest LIFE of kiwifruit. *J. Sci. Food Agric.* **2019**, *99*, 5918–5925. [CrossRef] [PubMed]

27. Taner, A.; Öztekin, Y.; Tekgüler, A.; Sauk, H.; Duran, H. Classification of Varieties of Grain Species by Artificial Neural Networks. *Agronomy* **2018**, *8*, 123. [CrossRef]

28. Torkashvand, A.M.; Ahmadi, A.; Nikravesh, N.L. Prediction of kiwifruit firmness using fruit mineral nutrient concentration by artificial neural network (ANN) and multiple linear regressions (MLR). *J. Integr. Agric.* **2017**, *16*, 1634–1644. [CrossRef]

29. Kujawa, S.; Dach, J.; Kozłowski, R.J.; Przybył, K.; Niedbała, G.; Mueller, W.; Tomczak, R.J.; Zaborowicz, M.; Koszela, K. Maturity classification for sewage sludge composted with rapeseed straw using neural image analysis. In Proceedings of the SPIE—The International Society for Optical Engineering, Chengu, China, 29 August 2016; Volume 10033, p. 100332H.

30. Kozłowski, R.J.; Kozłowski, J.; Przybył, K.; Niedbała, G.; Mueller, W.; Okoł, P.; Wojcieszak, D.; Koszela, K.; Kujawa, S. Image analysis techniques in the study of slug behaviour. In Proceedings of the SPIE—The International Society for Optical Engineering, Chengu, China, 29 August 2016; Volume 10033, p. 100332I.

31. Niedbała, G.; Mioduszewska, N.; Mueller, W.; Boniecki, P.; Wojcieszak, D.; Koszela, K.; Kujawa, S.; Kozłowski, R.J.; Przybył, K. Use of computer image analysis methods to evaluate the quality topping sugar beets with using artificial neural networks. In Proceedings of the SPIE—The International Society for Optical Engineering, Chengu, China, 29 August 2016; Volume 10033, p. 100332M.

32. Boniecki, P.; Koszela, K.; Świerczyński, K.; Skwarcz, J.; Zaborowicz, M.; Przybył, J. Neural Visual Detection of Grain Weevil (*Sitophilus granarius* L.). *Agriculture* **2020**, *10*, 25. [CrossRef]

33. Ryu, G.-A.; Nasridinov, A.; Rah, H.; Yoo, K.-H. Forecasts of the Amount Purchase Pork Meat by Using Structured and Unstructured Big Data. *Agriculture* **2020**, *10*, 21. [CrossRef]

34. Apolo-Apolo, O.E.; Pérez-Ruiz, M.; Martínez-Guanter, J.; Egea, G. A Mixed Data-Based Deep Neural Network to Estimate Leaf Area Index in Wheat Breeding Trials. *Agronomy* **2020**, *10*, 175. [CrossRef]

35. Abbaspour-Gilandeh, Y.; Molaee, A.; Sabzi, S.; Nabipur, N.; Shamshirband, S.; Mosavi, A. A Combined Method of Image Processing and Artificial Neural Network for the Identification of 13 Iranian Rice Cultivars. *Agronomy* **2020**, *10*, 117. [CrossRef]

36. Abdipour, M.; Younessi-Hmazekhanlu, M.; Ramazani, S.H.R.; omidi, A. hassan Artificial neural networks and multiple linear regression as potential methods for modeling seed yield of safflower (*Carthamus tinctorius* L.). *Ind. Crops Prod.* **2019**, *127*, 185–194. [CrossRef]

37. Ray, A.; Halder, T.; Jena, S.; Sahoo, A.; Ghosh, B.; Mohanty, S.; Mahapatra, N.; Nayak, S. Application of artificial neural network (ANN) model for prediction and optimization of coronarin D content in Hedychium coronarium. *Ind. Crops Prod.* **2020**, *146*, 112186. [CrossRef]

38. Niazian, M.; Sadat-Noori, S.A.; Abdipour, M. Modeling the seed yield of Ajowan (*Trachyspermum ammi* L.) using artificial neural network and multiple linear regression models. *Ind. Crops Prod.* **2018**, *117*, 224–234. [CrossRef]

39. Niazian, M.; Sadat-Noori, S.A.; Abdipour, M. Artificial neural network and multiple regression analysis models to predict essential oil content of ajowan (*Carum copticum* L.). *J. Appl. Res. Med. Aromat. Plants* **2018**, *9*, 124–131. [CrossRef]

40. Niedbała, G. Simple model based on artificial neural network for early prediction and simulation winter rapeseed yield. *J. Integr. Agric.* **2019**, *18*, 54–61. [CrossRef]

41. Niedbała, G. Application of artificial neural networks for multi-criteria yield prediction of winter rapeseed. *Sustainability* **2019**, *11*, 533. [CrossRef]

42. Niedbala, G.; Kozlowski, R.J. Application of Artificial Neural Networks for Multi-Criteria Yield Prediction of Winter Wheat. *J. Agric. Sci. Technol.* **2019**, *21*, 51–61.

43. Niedbała, G.; Nowakowski, K.; Rudowicz-Nawrocka, J.; Piekutowska, M.; Weres, J.; Tomczak, R.J.; Tyksiński, T.; Pinto, A.Á. Multicriteria prediction and simulation of winter wheat yield using extended qualitative and quantitative data based on artificial neural networks. *Appl. Sci.* **2019**, *9*, 2773. [CrossRef]

44. Niedbała, G.; Piekutowska, M.; Weres, J.; Korzeniewicz, R.; Witaszek, K.; Adamski, M.; Pilarski, K.; Czechowska-Kosacka, A.; Krysztofiak-Kaniewska, A. Application of Artificial Neural Networks for Yield Modeling of Winter Rapeseed Based on Combined Quantitative and Qualitative Data. *Agronomy* **2019**, *9*, 781. [CrossRef]

45. Wojciechowski, T.; Niedbala, G.; Czechlowski, M.; Nawrocka, J.R.; Piechnik, L.; Niemann, J. Rapeseed seeds quality classification with usage of VIS-NIR fiber optic probe and artificial neural networks. In Proceedings of the 2016 International Conference on Optoelectronics and Image Processing, ICOIP 2016, Warsaw, Poland, 10–12 June 2016.

46. Kaushik, P.; Dhaliwal, M. Diallel Analysis for Morphological and Biochemical Traits in Tomato Cultivated under the Influence of Tomato Leaf Curl Virus. *Agronomy* **2018**, *8*, 153. [CrossRef]

47. Viriyasuthee, W.; Saksirirat, W.; Saepaisan, S.; Gleason, M.L.; Jogloy, S. Jogloy Variability of Alternaria Leaf Spot Resistance in Jerusalem Artichoke (*Helianthus Tuberosus* L.) Accessions Grown in a Humid Tropical Region. *Agronomy* **2019**, *9*, 268.

48. Perkowski, J.; Wiwart, M.; Buśko, M.; Laskowska, M.; Berthiller, F.; Kandler, W.; Krska, R. Fusarium toxins and total fungal biomass indicators in naturally contaminated wheat samples from north-eastern Poland in 2003. *Food Addit. Contam.* **2007**, *24*, 1292–1298. [CrossRef]

49. Miedaner, T.; Reinbrecht, C.; Lauber, U.; Schollenberger, M.; Geiger, H.H. Effects of genotype and genotype-environment interaction on deoxynivalenol accumulation and resistance to Fusarium head blight in rye, triticale, and wheat. *Plant Breed.* **2001**, *120*, 97–105. [CrossRef]

50. Foroud, N.; Eudes, F. Trichothecenes in Cereal Grains. *Int. J. Mol. Sci.* **2009**, *10*, 147–173. [CrossRef]

51. Mesterházy, Á. Role of Deoxynivalenol in Aggressiveness of *Fusarium graminearum* and *F. culmorum* and in Resistance to Fusarium Head Blight. *Eur. J. Plant Pathol.* **2002**, *108*, 675–684. [CrossRef]

52. Miedaner, T.; Schneider, B.; Geiger, H.H. Deoxynivalenol (DON) content and Fusarium head blight resistance in segregating populations of winter rye and winter wheat. *Crop Sci.* **2003**, *43*, 519–526. [CrossRef]

53. Champeil, A.; Doré, T.; Fourbet, J. Fusarium head blight: Epidemiological origin of the effects of cultural practices on head blight attacks and the production of mycotoxins by Fusarium in wheat grains. *Plant Sci.* **2004**, *166*, 1389–1415. [CrossRef]

54. Homdork, S.; Fehrmann, H.; Beck, R. Effects of Field Application of Tebuconazole on Yield, Yield Components and the Mycotoxin Content of Fusarium-infected Wheat Grain. *J. Phytopathol.* **2000**, *148*, 1–6. [CrossRef]

55. Champeil, A.; Fourbet, J..; Doré, T.; Rossignol, L. Influence of cropping system on Fusarium head blight and mycotoxin levels in winter wheat. *Crop Prot.* **2004**, *23*, 531–537. [CrossRef]

56. Rudd, J.C.; Horsley, R.D.; McKendry, A.L.; Elias, E.M. Host Plant Resistance Genes for Fusarium Head Blight. *Crop Sci.* **2001**, *41*, 620. [CrossRef]

57. Snijders, C.H.A. Resistance in wheat to Fusarium infection and trichothecene formation. *Toxicol. Lett.* **2004**, *153*, 37–46. [CrossRef] [PubMed]

58. Li, L.; Shewry, P.R.; Ward, J.L. Phenolic Acids in Wheat Varieties in the HEALTHGRAIN Diversity Screen. *J. Agric. Food Chem.* **2008**, *56*, 9732–9739. [CrossRef] [PubMed]

59. Hernández, L.; Afonso, D.; Rodríguez, E.M.; Díaz, C. Phenolic Compounds in Wheat Grain Cultivars. *Plant Foods Hum. Nutr.* **2011**, *66*, 408–415. [CrossRef]

60. Gallardo, C.; Jiménez, L.; García-Conesa, M.-T. Hydroxycinnamic acid composition and in vitro antioxidant activity of selected grain fractions. *Food Chem.* **2006**, *99*, 455–463. [CrossRef]

61. Van Hung, P.; Maeda, T.; Miyatake, K.; Morita, N. Total phenolic compounds and antioxidant capacity of wheat graded flours by polishing method. *Food Res. Int.* **2009**, *42*, 185–190. [CrossRef]

62. Zhou, K.; Laux, J.J.; Yu, L. Comparison of Swiss Red Wheat Grain and Fractions for Their Antioxidant Properties. *J. Agric. Food Chem.* **2004**, *52*, 1118–1123. [CrossRef] [PubMed]

63. Martini, D.; Taddei, F.; Nicoletti, I.; Ciccoritti, R.; Corradini, D.; D'Egidio, M.G. Effects of Genotype and Environment on Phenolic Acids Content and Total Antioxidant Capacity in Durum Wheat. *Cereal Chem. J.* **2014**, *91*, 310–317. [CrossRef]

64. Kasote, D.M.; Katyare, S.S.; Hegde, M.V.; Bae, H. Significance of Antioxidant Potential of Plants and its Relevance to Therapeutic Applications. *Int. J. Biol. Sci.* **2015**, *11*, 982–991. [CrossRef]

 *agriculture*

Article

Neural Visual Detection of Grain Weevil (*Sitophilus granarius* L.)

Piotr Boniecki [1], Krzysztof Koszela [1,*], Krzysztof Świerczyński [1], Jacek Skwarcz [2], Maciej Zaborowicz [1] and Jacek Przybył [1]

[1] Institute of Biosystems Engineering, Poznan University of Life Sciences, Wojska Polskiego 50, 60-625 Poznan, Poland; bonie@up.poznan.pl (P.B.); krzys.swierczynski@gmail.com (K.Ś.); maciej.zaborowicz@up.poznan.pl (M.Z.); jacek.przybyl@up.poznan.pl (J.P.)

[2] Faculty of Production Engineering, Lublin University of Life Sciences, 20-950 Lublin, Poland; jacek.skwarcz@up.lublin.pl

* Correspondence: koszela@up.poznan.pl; Tel.: +48-502-288-097

Received: 23 November 2019; Accepted: 14 January 2020; Published: 20 January 2020

Abstract: A significant part of cereal production is intended for agri-food processing, which implies a necessity to search for and implement modern storage systems for this product. Stored grain is exposed to many unfavorable factors, particularly caryopsis macro-damage caused mainly by grain weevil (*Sitophilus granarius* L.). This triggers a substantial decrease in the value of the stored material, thus resulting in serious economic losses. Due to this fact, it is necessary to take steps to effectively detect this pest's presence when grain is delivered to storage facilities. The purpose of this work was to identify the representative physical characteristics of wheat caryopsis affected by grain weevil. An automated visual system was developed to ease the detection of damaged kernels and adult weevils. In order to obtain the empirical data, a decision was made to take advance of SKCS 4100 (the Perten Single Kernel Characterization System). The measurements obtained were used to build the training sets necessary in the process of ANN (artificial neural network) learning with digital neural classifiers. Next, a set of identifying neural models was created and verified, and then the optimal topology was selected. The utilitarian goal of the research was to support the decision-making process taking place during grain storage.

Keywords: artificial neural network (ANN); *Grain weevil* identification; neural modelling classification

---

1. Introduction

Cereal grain storage is mainly conditioned by ambient temperature and humidity, which constitute crucial parameters affecting the quality of stored grain. The presence of living organisms that cause direct losses resulting from their infestation is also important. There are also notable indirect losses in the stored material resulting from contamination with excretions and secretions, as well as from moistening and heating. Stored grain may be infested by many different organisms, such as bacteria, fungi, mites, and insects. One of the most damaging grain pest species in Europe is the grain weevil (*Sitophilus granarius* L.) [1]. It can cause up to 5% of losses in stored crops. This pest is often a cleverly hidden grain destroyer. Although its beetles can be easily detected during sieving, unfortunately, the identification of eggs, larvae, and pupae is difficult and requires appropriate laboratory tests. Weevil feeding on grains significantly reduces the germination capacity of grain from 93% to 7%. Its presence and development cause an increase in humidity and temperature of the stored mass. These conditions foster the development of fungi, as a result of which grain mold appears, which in turn causes a drastic decrease in grain quality and consequently its value.

An important task for managers of any grain storage facility is to prevent the establishment and development of grain weevil in stored crops. The identification and recognition of pests should therefore take place before the grain is taken to silos. This procedure must be quick, efficient, and precise, which implies a necessity to automate this identification process [2–6].

An effective identification of the effects of grain weevil feeding on stored grain entailed the designing, manufacturing, and verification of the new, original classification model [7–11]. The commonly recognized separation properties, such as those represented by artificial neural networks, imply the legitimacy of using them in the process of identifying grain damage caused by grain weevil [12]. It ought to be emphasized that the source literature provides no examples of any effective use of neural modeling in the process of identifying internal macro-damage to wheat caryopsis caused by pests feeding on them. Moreover, there is no scientific knowledge concerning the determination of the type and level of significance of the representative properties of the infected wheat grain, which would enable the identification of the negative effects of this pest [13–15].

2. Materials and Methods

2.1. Materials

Grain weevil (*Sitophilus granarius* L.) (Figure 1) is a winged cosmopolitan beetle from the Curculionidae family. Although it belongs to the subclass of the winged insects, it is deprived of volatile wings. It belongs to the group of the most damaging storage pests. Depending on the age of the individual, the grain weevil's color ranges from light-brown to black.

Figure 1. Grain weevil (*Sitophilus granarius* L.) in the context of the grain storage process [16].

Young individuals are light-brown, whereas adults are dark. The body length of grain weevils ranges from 2 to 5 mm. Most often grain weevils hide in small cracks and avoid sunlight. Grain weevils are characterized by their high resistance to hunger. At a temperature of about 12 °C, a grain weevil can survive 115 days without food. A female grain weevil makes a hole in grain and deposits her egg inside, and then covers it with an adhesive/a viscous substance mixed with starch coming from grain. The sealed hole made by the pest is not visible with the naked eye. Grain weevils are characterized by their high fertility. During one day, a female can lay from one to nine eggs, and during its life, can lay about 150 eggs. Usually a typical female lays one egg in one grain, thus providing the right amount of food for its offspring. A grain weevil egg has an oval shape, with a pointed edge, with a size of 0.65–0.30 mm. The intensity of grain weevil procreation depends on the following factors:

- the air temperature,
- the air humidity,
- the amount of available food.

Becoming a member state of the European Union, Poland was obliged to adopt the mandatory standards related to food products as well as the requirements set out by institutions accepting and purchasing cereals. These standards also define the methods of assessing the quality of the material provided. All these standards can be found in the Commission Regulation No. 687/2008 [RK (EC) 687/2008], and they clearly state that the goods delivered to storage facilities should be free of any storage pests. In the regulation (Official Journal No. 29 item 189) issued in 2007, there is a statement that the presence of a storage pest disqualifies a given product as being seed that is sellable to markets.

It is due to the fact (among others) that one of the criteria for assessing the seed quality is the minimum germination capacity, which is 85% for wheat. In case of grain weevil infestation, this capacity decreases to as low as 7%.

The research conducted in this study involved four varieties of spring wheat showing signs of damage caused by grain weevil (*Sitophilus granarius* L.), namely the following varieties: Torka, Narwa, Banti, and Symfonia. The empirical material was obtained from two selected plant breeding stations: (Plant Breeding Strzelce Ltd. 99-307 Strzelce, Poland) and (Agricultural Plant Breeding–Kobierzyce Seeds Ltd. 55-040 Kobierzyce, Poland) The samples were selected randomly. The characteristics of the above varieties were prepared on the basis of the data published on the producer's website (www.hr-strzelce.pl):

Torka—This kind of wheat belongs to the elite group of wheat that possesses very good flour baking value. It is characterized by a large 1000 grain mass (45–52 g), good resistance to lodging, average protein content, good to very good flour yield, very good wholesomeness, and average to good prolificacy.

Narwa—This variety is characterized by very good baking value, very high protein and gluten content, 1000 grain mass over 50 g, good shattering and fouling resistance, good wholesomeness, and early maturation.

Banti—This variety is characterized by high protein content, good baking value, tendency to ear-sprouting, and average 1000 grain mass.

Symfonia—This variety has a 1000 grain mass of approx. 45 g, as well as high fouling and shatter resistance (8 in the 9-point scale). Additionally, it is resistant to powdery mildew, blight, leaf and chaff septaria, and stem base diseases. It is recommended for cultivation all over the country and in the areas where it is necessary to grow varieties with very good winter hardiness (6.5 in the 9-point scale).

The empirical studies were conducted in the specialist laboratory of the Department of Entomology of the Plant Protection Institute—the National Research Institute in Poznan (IOR–PIB Poznan). For the experiments, an appropriate number of beetles were prepared via special breeding. Then, using a stereoscopic microscope, the males were separated from the females on the basis of the morphological differences in the structure of the snout and abdomen. The study used 50 polypropylene containers (height 65 mm, diameter 38 mm, capacity 60 mL), which had a suitable construction enabling air inlet but prevented the tested pests from getting out of the container. Four hundred randomly selected caryopses were put into each container, and then 20 pairs of grain weevil beetles were placed inside (20 males and 20 females). The period of time when the beetles were kept in particular containers was 5, 10, 15, and 20 days. During that time, all the grain weevil individuals were feeding, copulating, and laying eggs, which in turn developed into larvae. The containers were placed in an incubator in order to make it possible to take measurements and obtain empirical data at the same time. After the incubation period of the grain weevils feeding on the caryopses, the samples were cleared of the beetles and other pollutants. Four hundred caryopses were randomly selected from each sample and placed in a SKCS 4100 (Single-Kernel Characterization System) Perten Instruments, Sweden appliance. The scheme of obtaining empirical data is presented in Figure 2:

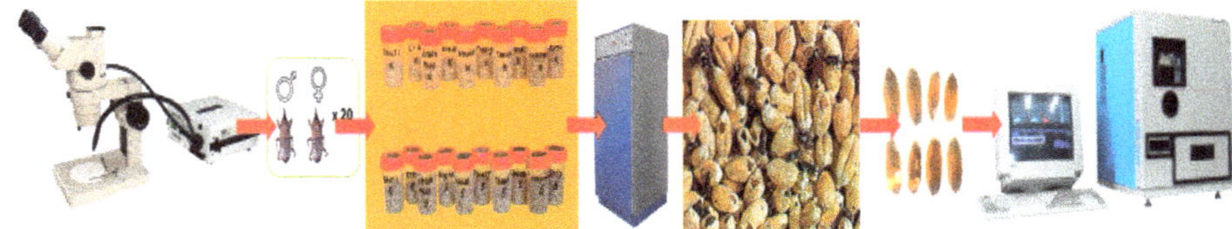

Figure 2. Scheme for obtaining empirical data [17].

The experimental data were saved on the hard drive of the SKCS 4100 appliance, and then collected and organized/segregated into Microsoft Corporation Excel spreadsheets (in the form of .csv files).

The empirical data obtained was converted to the training set relevant to the ANN (artificial neural network) simulator implemented in the commercial package from Statistica v. 10- StatSoft Polska. The generated training set, essential in the neural modeling process, was then used to generate the ANN file.

2.2. Method

In order to create the optimal neural model, the following scheme of procedure was put in place (Figure 3).

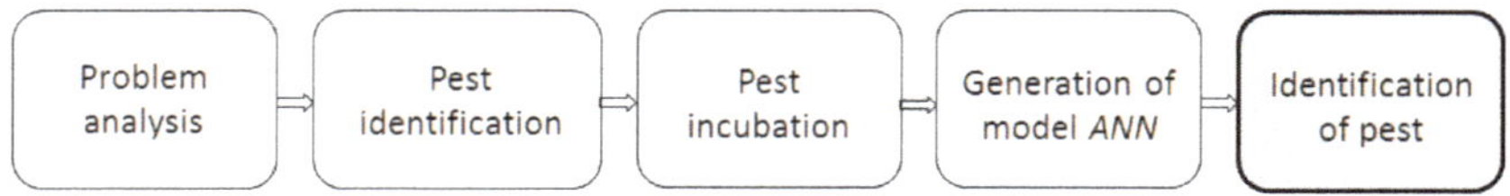

Figure 3. Scheme of procedure. ANN: artificial neural network.

In order to create the classification neural models, a neural network simulator implemented in Statistica v. 10 package was used [1,18–20]. The most important stage of the ANN generation was the preparation of the training files containing encoded selected representative properties, constituting the empirical basis for the classification [20–23]. For this purpose, four numerical input variables and one nominal output variable were determined, which resulted from the nature of the formulated scientific problem and represented the characteristic parameters of the process examined. As the input variables, the following four descriptors were adopted and used, which constituted the basic physical characteristics of the stored caryopses:

- mass (mg),
- equivalent diameter (mm),
- humidity (%),
- hardness (according to the selected hardness scale).

The values of all the aforementioned physical properties were obtained from 400 caryopses randomly selected from each sample. The tests were repeated in the following time intervals: 0 day (control—a combination without pests), and 5, 10, 15, and 20 days from the moment the pest came into contact with the grain. The data collected and classified in this way was converted to a form of training set, which was necessary to generate the ANN. The set of the neural models was created in the Statistica v. 10 program with a simulator implemented in the "Neural Networks" module [20,24–27].

The output variable was encoded in the form of a dichotomous binary nominal variable. The training set consisted of 1800 randomly selected data, and it was divided proportionally 2:1:1 into training, validation, and test sets. These tests included, respectively, 900, 450, and 450 training files. According to the procedure adopted, the test set was not used in the network training process, therefore it was important in the final assessment of the optimal neural model. The structure of the training file is presented in Table 1.

Table 1. Part of a training file.

No.	Mass (mg)	Equivalent Diameter (mm)	Humidity (%)	Hardness	Grain Weevil (Yes, No)
1	34.7	3.0	14.4	118.6	yes
2	34.6	3.1	12.8	69.9	yes
3	34.7	2.9	14.3	81.3	no
...	...	...	...	...	...
1799	34.5	2.8	14.3	60.1	no
1800	34.5	2.7	15.3	86.6	yes

3. Results and Discussion

A set of 100 different neural topologies was generated. The selection of the neural network typology at the initial phase was conducted with an automatic designer, which did the experiments on its own with different network architectures with the use of different learning processes for a given network. Next, two different types of neural networks were tested for each kind of dataset. It was the parameters of the given neural network, such as correlation, coefficient of total determination, and quotient of standard deviations, that determined the selection of the best neural network. Following the selection of the given network, the process of learning the network was implemented. During this process, based on the selected algorithm, special attention was put on its ability of approximation and generalization, based on quality measurements with the lowest root mean square (RMS) error. Also, during this same process of learning/training, an error curve of both the training set and the valuation set was observed. In the event that there was an increase in those errors, the training/learning process was stopped and all the necessary modifications of the network architecture were made by adding or removing neurons or hidden layers. A change of learning algorithms was also applied. Those actions are aimed at eliminating the phenomenon of network "overlearning". Otherwise, the network will not come up to the expected results. RBF (radial basis function) topology with structure 4:10:1 turned out to be the best ANN topology generated. Nowadays, RBF networks belong to the category of basic types of neural networks. These types of networks are most commonly applied in the non-linear approximation of numerical variables. Additionally, they are applied in cases concerning classification (Bishop [1], Nabney [4]), where they reconstruct the density function of the distribution of informing variables. The redial neuron is defined by its core and the parameter called "radius". The point in n-dimensional space is defined with N numbers, which precisely corresponds to the number of weighs in the linear neuron; on this account, the core of the radial neuron is stored in the set of parameters determined in the software Statistica, also as "weighs" (though when determining common "weigh" activation, only the distance between the weigh vector and the input signal vector is determined). Radius (or, in other words, deviation) is stored in the neuron as the so-called threshold [28]. The input layer was composed of four neurons with a PSP (postsynaptic function), a linear function, and a linear activation function. The hidden layer was composed of four radial neurons with a radial PSP function and an exponential activation function. The network output consisted of one neuron with a linear PSP function and a linear (saturated) activation function, representing a two-state nominal variable [29,30]. The generated neural model was trained using optimization algorithms implemented in Statistica v. 10 package. The centers were determined with the k-means method, whereas the deviations were determined with the k-nearest neighbors method. The output layer was optimized with the pseudo-inverse technique. The structure of the generated RBF network is presented in Figure 4. Where:

A—input layer: PSP function: linear, activation function: linear saturated function;
B—hidden layer: PSP function: radial activation function–Gauss function;
C—output layer: PSP function: linear saturated function, activation function: linear saturated function;
PSP—postsynaptic potential.

The basic properties of RBF networks are as follows:

- RBFs are one-direction networks;
- RBFs are trained with a "no teacher" technique;
- they have a three-layer architecture: there is an input, hidden (radial) layer, and output layer,
- the declared connections in the network allow only communication between the neurons from the neighboring layers;
- the input neuron activation functions are of linear nature, the hidden neuron functions are of nonlinear (radial) nature, and the output neurons are fully linear.

RBF networks are characterized by one hidden layer, a short training process, and a small size (Figure 4).

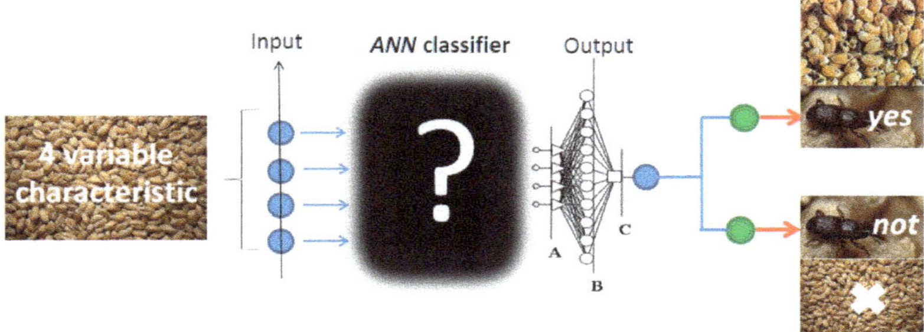

Figure 4. The structure of the generated ANN (Artificial Neural Network) type radial basis function (RBF).

A hidden neuron in these networks performs a function that is changing radially around the selected center and the assuming non-zero values only in the vicinity of this center. The mathematical basis for RBF network functioning is T. Cover's theorem on separability of patterns, which posits that a complex classification problem cast non-linearly into the high-dimensional space is more likely to be linearly separable than in the projection into a low-dimensional space.

A standard measure of the quality of the generated neural model is the RMS (root mean square) error, which is defined in the following way [28,31,32]:

$$\text{RMSE} = \sqrt{\frac{\sum_{i=1}^{n}(y_i - z_i)^2}{n}}, \tag{1}$$

where

- n—number of cases;
- y_i—real values;
- z_i—values set with network.

The RMS error values for the generated neural model were as follows:

- 0.2322 for the training file;
- 0.2462 for the validation file;
- 0.2502 for the test file.

The estimation of the level of significance of the ANN input parameters is usually identified by means of the procedure of the analyzing the neural model's sensitivity to the input variables. This procedure is used to assess to what extent a selected input signal affects the identification process of the input variable. In this way, the information on the rank of the input signal is obtained in the form of error increase quotient [33,34]. If the error increase quotient is below 1, it means that a given property has no influence on the grain weevil identification process. In the described case, all input signals had an error quotient above one. Therefore, it can be concluded that each of the properties did take part in the identification process of the output signal of the optimal ANN, albeit, each of them with a different rank. Humidity was in the first place in the ranking of the input signals of the neural model, obtaining the highest level of the error increase quotient. Such a high rank may prove that the occurrence of pests causes observable changes in the level of water that a caryopsis contains. The following ranks (from highest to lowest) are, respectively, mass, equivalent diameter, and hardness (Table 2).

Table 2. Sensitivity analysis of the model RBF 4:10:1.

	Mass	**Equivalent Diameter**	**Humidity**	**Hardness**
Ratio	1.075	1.022	1.118	1.002
Rank	2	3	1	4

4. Conclusions

The generated RBF-type ANN topology of the 4:10:1 structure was verified in terms of the quality of the identification process and the possibility of its application to solve the problem of identifying grain weevil in stored wheat grain. By applying RBF (instead of Multi-Layer Perceptron MLP), it was observed that the neural network finds approximation that is better suited to the local features of the dataset but with worse extrapolation. Neural networks with radial base functions were applied to deal with classification issues, approximation of functions with numerous variables, and in issues concerning predictions. In those areas of application where sigmoid functions have an established position, the implementation of the generated ANN model will enable automation of the process involving recognition of grain weevil occurrence, which will allow to undertake appropriate measures to protect the stored grain from further losses caused by this pest.

The conclusions resulting from the analysis conducted are as follows:

1. The results of the study demonstrated that artificial neural networks can be used as effective tools supporting the process of identifying pests feeding on stored wheat grain.
2. The generated RBF-type neural model with the structure 4:10:1 proved to be optimal in the process of solving problems of grain weevil identification, based on four selected physical properties of the caryopses.
3. The analysis of the sensitivity of the generated neural model to the input variables demonstrated the existence of various ranks for individual representative signals. The fact that humidity ranked the highest indicates a high degree of significance of this variable. The subsequent parameters are mass, equivalent diameter, and hardness (in this order).
4. The results of the study indicate the possibility of supporting the decision-making processes taking place during agricultural products storage. In particular, the generated RBF-type (4:10:1) neural model may constitute the kernel of the IT system supporting the process of identification of the degree the stored grain is infected by grain weevil.

Author Contributions: Conceptualization: K.Ś. and P.B.; methodology: P.B. and K.K.; software: K.Ś.; validation: M.Z.; formal analysis: M.Z. and P.B.; investigation: J.P.; resources: J.P.; data curation: K.K.; writing—original draft preparation: K.K. and K.Ś.; writing—review and editing: K.K. and J.S.; visualization: K.K.; supervision: P.B.; project administration: J.S. All authors have read and agreed to the published version of the manuscript.

Funding: The authors are grateful for financial support provided by the Poznań University of Life Sciences, Poland, within the framework of fund no. 506.752.03.00.

Acknowledgments: The authors are grateful for the financial support provided by the Poznań University of Life Sciences, Poland, within the framework of fund no. 508.752.00.0.

Conflicts of Interest: The authors declare no conflict of interest.

References

1. Boniecki, P.; Piekarska-Boniecka, H.; Świerczyński, K.; Koszela, K.; Zaborowicz, M.; Przybył, J. Detection of the granary weevil based on X-ray images of damaged wheat kernels. *J. Stored Prod. Res.* **2014**, *56*, 38–42. [CrossRef]
2. Sun, J.; He, X.; Ge, X.; Wu, X.; Shen, J.; Song, Y. Detection of key organs in tomato based on deep migration learning in a complex background. *Agriculture* **2018**, *8*, 196. [CrossRef]
3. Kang, S.H.; Cho, J.H.; Lee, S.H. Identification of butterfly based on their shapes when viewed from different angles using an artificial neural network. *J. Asia Pac. Entomol.* **2014**, *17*, 143–149. [CrossRef]

4. Kodogiannis, V.S.; Kontogianni, E.; Lygouras, J.N. Neural network based identification of meat spoilage using Fourier-transform infrared spectra. *J. Food Eng.* **2014**, *142*, 118–131. [CrossRef]
5. Zhang, Y.; Wang, S.; Ji, G.; Phillips, P. Fruit classification using computer vision and feedforward neural network. *J. Food Eng.* **2014**, *143*, 167–177. [CrossRef]
6. Zaborowicz, M.; Boniecki, P.; Koszela, K.; Przybylak, A.; Przybył, J. Application of neural image analysis in evaluating the quality of greenhouse tomatoes. *Sci. Hortic.* **2017**, *218*, 222–229. [CrossRef]
7. Kashaninejad, M.; Dehghani, A.A.; Kashiri, M. Modeling of wheat soaking using two artificial neural networks (MLP and RBF). *J. Food Eng.* **2009**, *91*, 602–607. [CrossRef]
8. Przybylak, A.; Ślósarz, P.; Boniecki, P.; Koszela, K.; Zaborowicz, M.; Przybył, K.; Wojcieszak, D.; Szulc, R.; Ludwiczak, A.; Górna, K. Marbling classification of lambs carcasses with the artificial neural image analysis. In Proceedings of the Seventh International Conference on Digital Image Processing (ICDIP 2015), Los Angeles, CA, USA, 6 July 2015.
9. Boniecki, P.; Przybył, J.; Zaborowicz, M.; Górna, K.; Dach, J.; Okoń, P.; Przybył, K.; Mioduszewska, N.; Idziaszek, P. SOFM-type artificial neural network for the non-parametric quality-based classification of potatoes. In Proceedings of the SPIE—The International Society for Optical Engineering, Chengu, China, 29 August 2016; Volume 10033.
10. Przybył, K.; Zaborowicz, M.; Koszela, K.; Boniecki, P.; Mueller, W.; Raba, B.; Lewicki, A. Organoleptic damage classification of potatoes with the use of image analysis in production process. In Proceedings of the SPIE—The International Society for Optical Engineering, Athens, Greece, 16 April 2014; Volume 9159.
11. Boniecki, P.; Koszela, K.; Piekarska-Boniecka, H.; Weres, J.; Zaborowicz, M.; Kujawa, S.; Majewski, A.; Raba, B. Neural identification of selected apple pests. *Comput. Electron. Agric.* **2015**, *110*, 9–16. [CrossRef]
12. Piekarska-Boniecka, H.; Kadłubowski, W.; Siatkowski, I. A study of bionomy of the privet sawfly (*Macrophya punctumalbum* (L.)) (Hymenoptera, Tenthredinidae)—A pest of park plants. *Acta Sci. Pol. Hortorum Cult.* **2008**, *7*, 3–11.
13. Riverol, C.; Cooney, J. Estimation of the ester formation during beer fermentation using neural networks. *J. Food Eng.* **2007**, *82*, 585–588. [CrossRef]
14. Cheng, H.D.; Jiang, X.H.; Sun, Y.; Wang, J. Color image segmentation: Advances and prospects. *Pattern Recognit.* **2001**, *34*, 2259–2281. [CrossRef]
15. Niewiada, A.; Nawrot, J.; Szafranek, J.; Szafranek, B.; Synak, E.; Jeleń, H.; Wąsowicz, E. Some factors affecting egg-laying of the granary weevil (*Sitophilus granarius* L.). *J. Stored Prod. Res.* **2005**, *41*, 544–555. [CrossRef]
16. Olejarski, P.; Horoszkiewicz-Janka, J.; Bocianowski, J. Influence of fungi on feeding and development of granary weevil (*Sitophilus granarius* L.)/Wpyw grzybów zasiedlajacych ziarno zbóz na zerowanie i rozwój woka zbozowego (*Sitophilus granarius* L.). *Prog. Plant Prot.* **2010**, *50*, 1711–1718.
17. Olejarski, P. Detection of pests in grain stores. *Ochr. Rosl.* **2005**, *50*, 12–14.
18. Liu, D.; Ning, X.; Li, Z.; Yang, D.; Li, H.; Gao, L. Discriminating and elimination of damaged soybean seeds based on image characteristics. *J. Stored Prod. Res.* **2015**, *60*, 67–74. [CrossRef]
19. Mahesh, S.; Jayas, D.S.; Paliwal, J.; White, N.D.G. Hyperspectral imaging to classify and monitor quality of agricultural materials. *J. Stored Prod. Res.* **2015**, *61*, 17–26. [CrossRef]
20. Koszela, K.; Raba, B.; Zaborowicz, M.; Przybył, K.; Wojcieszak, D.; Czekała, W.; Ludwiczak, A.; Przybylak, A.; Boniecki, P.; Przybył, J. Computer image analysis in caryopses quality evaluation as exemplified by malting barley. In Proceedings of the SPIE—The International Society for Optical Engineering, Los Angeles, CA, USA, 6 July 2015; Volume 9631.
21. Przybył, K.; Górna, K.; Wojcieszak, D.; Czekała, W.; Ludwiczak, A.; Przybylak, A.; Boniecki, P.; Koszela, K.; Zaborowicz, M.; Janczak, D.; et al. The recognition of potato varieties using of neural image analysis method. In Proceedings of the SPIE—The International Society for Optical Engineering, Los Angeles, CA, USA, 6 July 2015; Volume 9631.
22. Muñoz, I.; Rubio-Celorio, M.; Garcia-Gil, N.; Guàrdia, M.D.; Fulladosa, E. Computer image analysis as a tool for classifying marbling: A case study in dry-cured ham. *J. Food Eng.* **2015**, *166*, 148–155. [CrossRef]
23. Koszela, K.; Łukomski, M.; Mueller, W.; Górna, K.; Okoń, P.; Boniecki, P.; Zaborowicz, M.; Wojcieszak, D. Classification of dried vegetables using computer image analysis and artificial neural networks. In Proceedings of the SPIE—The International Society for Optical Engineering, Hong Kong, China, 21 July 2017; Volume 10420, p. 1042031.

24. Zaborowicz, M.; Boniecki, P.; Koszela, K.; Przybył, J.; Mazur, R.; Kujawa, S.; Pilarski, K. Use of artificial neural networks in the identification and classification of tomatoes. In Proceedings of the SPIE—The International Society for Optical Engineering, Beijing, China, 19 July 2013; Volume 8878.
25. Zaborowicz, M.; Boniecki, P.; Koszela, K.; Przybył, J.; Mazur, R.; Kujawa, S.; Pilarski, K. Computer Image Analysis in Obtaining Characteristics of Images Greenhouse Tomatoes in the Process of Generating Learning Sets of Artificial Neural Networks. In Proceedings of the 6TH International Conference on Digital Image Processing (ICDIP 2014), Athens, Greece, 16 April 2014; Volume 9159.
26. Koszela, K.; Otrząsek, J.; Zaborowicz, M.; Boniecki, P.; Mueller, W.; Raba, B.; Lewicki, A.; Przybył, K. Quality assessment of microwave-vacuum dried material with the use of computer image analysis and neural model. In Proceedings of the SPIE—The International Society for Optical Engineering, Athens, Greece, 16 April 2014; Volume 9159, p. 915913.
27. Raba, B.; Nowakowski, K.; Lewicki, A.; Przybył, K.; Zaborowicz, M.; Koszela, K.; Boniecki, P.; Mueller, W. The non-touching method of the malting barley quality evaluation. In Proceedings of the—The International Society for Optical Engineering, Athens, Greece, 16 April 2014; Volume 9159.
28. Nabney, I.T. Efficient training of RBF networks for classification. *Int. J. Neural Syst.* **2004**, *14*, 201–208. [CrossRef]
29. Przybył, K.; Gawałek, J.; Koszela, K.; Wawrzyniak, J.; Gierz, L. Artificial neural networks and electron microscopy to evaluate the quality of fruit and vegetable spray-dried powders. Case study: Strawberry powder. *Comput. Electron. Agric.* **2018**, *155*, 314–323. [CrossRef]
30. Mueller, W.; Idziaszek, P.; Boniecki, P.; Zaborowicz, M.; Koszela, K.; Kujawa, S.; Kozłowski, R.J.; Przybył, K.; Niedbała, G. An IT system for the simultaneous management of vector and raster images. In Proceedings of the SPIE—The International Society for Optical Engineering, Chengu, China, 29 August 2016; Volume 10033.
31. Czekała, W.; Dach, J.; Ludwiczak, A.; Przybylak, A.; Boniecki, P.; Koszela, K.; Zaborowicz, M.; Przybył, K.; Wojcieszak, D.; Witaszek, K. The use of image analysis to investigate C: N ratio in the mixture of chicken manure and straw. In Proceedings of the SPIE—The International Society for Optical Engineering, Los Angeles, CA, USA, 6 July 2015; Volume 9631.
32. Jordan, M.I.; Bishop, C.M. Neural networks. In *Computer Science Handbook*, 2nd ed.; CRC Press: Basingstoke, Hampshire, UK, 2004.
33. Schölkopf, B.; Sung, K.K.; Burges, C.J.; Girosi, F.; Niyogi, P.; Poggio, T.; Vapnik, V. Comparing support vector machines with gaussian kernels to radial basis function classifiers. *IEEE Trans. Signal Process.* **1997**, *45*, 2758–2765. [CrossRef]
34. Schwenker, F.; Kestler, H.A.; Palm, G. Three learning phases for radial-basis-function networks. *Neural Netw.* **2001**, *14*, 439–458. [CrossRef]

MDPI

St. Alban-Anlage 66

4052 Basel

Switzerland

Tel. +41 61 683 77 34

Fax +41 61 302 89 18

www.mdpi.com

Agriculture Editorial Office

E-mail: agriculture@mdpi.com

www.mdpi.com/journal/agriculture